Modeling and Simulation-Based Systems Engineering Handbook

Engineering Management
Book Series

Series Editors

Timothy George Kotnour – Associate Professor
Waldemar Karwowski – Professor & Chair

Department of Industrial Engineering and Management Systems
University of Central Florida (UCF), Orlando, FL

Published Titles:

Modeling and Simulation-Based Systems Engineering Handbook,
edited by Daniele Gianni, Andrea D'Ambrogio, and Andreas Tolk

Systems Life Cycle Costing: Economic Analysis, Estimation, and Management,
John Vail Farr

Transforming Organizations: Strategies and Methods
Timothy George Kotnour

Modeling and Simulation-Based Systems Engineering Handbook

editors **Daniele Gianni**
Andrea D'Ambrogio
Andreas Tolk

CRC Press
Taylor & Francis Group
Boca Raton London New York

CRC Press is an imprint of the
Taylor & Francis Group, an **informa** business

CRC Press
Taylor & Francis Group
6000 Broken Sound Parkway NW, Suite 300
Boca Raton, FL 33487-2742

First issued in paperback 2017

© 2015 by Taylor & Francis Group, LLC
CRC Press is an imprint of Taylor & Francis Group, an Informa business

No claim to original U.S. Government works

ISBN-13: 978-1-4665-7145-7 (hbk)
ISBN-13: 978-1-138-74894-1 (pbk)

Dedication

To Giorgio, my father, for igniting in me the passion for studies and the thirst for knowledge.

Daniele Gianni

To Raffaella, Adriano, and Arianna, the rocks that I stand on and the sparks of my life.

Andrea D'Ambrogio

To Florian, Christopher, and E. J. A., that they may pick up the baton for the next generation.

Andreas Tolk

Contents

Foreword

What a wonderful treat it is to see a book on modeling and simulation and systems engineering. I would like to share with you my enthusiasm about the significance and importance of this book, which explores an important aspect of the synergy of these two very powerful disciplines.

Model/reality dichotomy has been very remarkable since antiquities. To understand or control reality, we need to build models. Engineering of reality also necessitates building and implementing models. In systemic parlance, these three cases correspond, respectively, to analysis, control, and design problems. The centrality of models in simulation, hence, model-based simulation was established in late 1970s and early 1980s, and the concept of *computer-aided modeling and simulation* was introduced in early 1980s. Because mental inertia is very difficult to overcome—at least for some people—it took some time to have this concept widely accepted. Once the concept of *model-based* was well established, it started to diffuse to other disciplines. Thus model-based systems engineering, engineering, science, software development, fault-detection and diagnosis, process control, design, testing, analysis, and meta-analysis are also well recognized.

Acceptance of the importance and vital roles of system theories, systems engineering, cybernetics, and computational (artificial) intelligence followed similar fate. However, progress is inevitable and now all these obstacles have been overcome. For some time, the importance of the contribution of modeling and simulation to systems engineering has been recognized, albeit for military applications only. This makes the current book a very timely and promising development as a powerful infrastructure for a large number of webs of complex systems we need to intelligently manage for the well-being, sustainability, and survival of our environment for ourselves as well as for future generations.

Some areas that this book paves the way to include are the following:

Complex scientific and engineering system problems would definitely benefit from the contribution of model-based simulation as well as from systems engineering and system-theory-based modeling paradigms to assure their integrity, robustness, and reliability. This approach would also allow *built-in quality assurance* in computer-aided modeling and simulation systems. Furthermore, *failure avoidance* can be vital beyond the limits of traditional validation and verification techniques and would reveal several benefits once fully explored.

Social systems are perfect examples of complex systems where among other features *emergence* is common. *Simulation-based predictive displays* can be used in simulation-based systems engineering to provide fact-based anticipatory power to policy makers. I am confident that some of the authors and readers of this volume are capable of developing such anticipatory scopes for use in social systems. Simulation-based predictive displays for social and financial systems can be used—even in near future—to train future decision makers about complex social systems, to foresee abnormal deviations before they actually happen, and to test several plausible scenarios for corrective actions.

So far as the experimentation aspect of simulation is concerned, "simulation is *goal-directed experimentation* with dynamic models." Sometimes the *goal* aspect of experimentation is overlooked and reuse as well as interoperability of simulation models can be considered successful once execution on a computer is achieved. (This is one of the reasons, execution-related definitions of simulation are detrimental.) If the goal is not appropriate, a model *correct and acceptable* under certain conditions may become completely *irrelevant* under different conditions. However, once systems engineering is in practice, the goal of the study will be scrutinized. Sometimes, completely different solution(s) can be found to satisfy the goal of the study even better compared to the one advocated by a simulation model. Furthermore, proper reuse and interoperability of models can be achieved in systems engineering by using high-level model specification languages and automatic generation of simulation software. In this case, reuse of specification of elements of simulation studies can be realized instead of just reuse of simulation software.

Web-based simulation, which is one of the chapters of this book, coupled with appropriate display devices such as e-glass, may lead to ubiquitous simulation or m-sim (for mobile simulation). This may be the essence of next generation of access to knowledge. Currently, with mobile and sedentary devices, access to *existing* knowledge is commonplace. With the advent of mobile devices (and cloud computing) access to simulation models having the power to generate knowledge under different experimental conditions may become a reality.

The authors have to be congratulated for the high quality of their contributions. Professors Tolk, Gianni, and D'Ambrogio, as editors, deserve our acclamation for having provided such a valuable platform and for having invited such talented colleagues to contribute to the book, as well as for their valuable contributions in several chapters. Professor Tolk, as a seasoned scientist, already contributed several very important volumes in advanced topics in modeling and simulation, and this volume that he edited and contributed with Professors Gianni and D'Ambrogio is another appreciated contribution.

I would highly recommend this book to professionals, researchers, and graduate students involved in complex systems. They may find valuable knowledge about simulation-based systems engineering, which is indispensable for the study of complex systems.

Dr. Tuncer Ören
SCS Modeling and Simulation Hall of Fame
(Lifetime achievement award)
Professor Emeritus of Computer Science
University of Ottawa, Ottawa, Ontario, Canada

Preface

The idea for a handbook on modeling and simulation (M&S)–based systems engineering originated from several experiences in which, as M&S experts, we participated in systems engineering activities in various domains. In our roles, we were in charge of supporting the investigative studies of domain experts, designing and implementing new M&S methodologies and technologies that could make the studies more convenient or simply feasible (considering the temporal and budget scopes). Aside from the personal enjoyment derived from the variety of challenges, these experiences gave us also the opportunity to reach a professional point of view that overlooks several domains and research communities. Already from the first glance, it was evident that the M&S practices are commonly embedded within individual disciplines/communities and that these practices are often developed in isolation with respect to the widest scientific and engineering community. Gradually, with the years, we have noticed that recurring problems were addressed in various communities, with a minimal capitalization on the results and experiences gained by researchers and engineers in other domains. As our society moves further in the information era, knowledge and M&S capabilities become key enablers for the engineering of complex systems and systems of systems. Nevertheless, knowledge and M&S methodologies and technologies become themselves valuable output in an engineering activity, and their cross-domain capitalization is the key to further advance the future practices in systems engineering. As a consequence, the well-known saying "reinventing the wheel" should not be interpreted only in the end-product view of the preinformation era, but it should be contextualized in a continuous improvement systems engineering process as "reinventing the M&S methods for designing the wheel."

The aim of this book is to collect reusable M&S lessons for systems engineering, offering an initial mean for cross-domain capitalization of the knowledge, methodologies, and technologies developed in several communities. This objective is per se challenging: not only to recruit key representatives from several communities, but also to harmonize the different perspectives deriving from individual backgrounds, within the book's vision. In reaching the conclusion of this book, we would like to express our sincere gratitude to all the authors who, believing in this idea, made this book come alive. We are also very thankful to external reviewers (in alphabetical order), Joachim Fuchs, Joseph Giampapa, Cristiano Leorato, Steve McKeever, and Quirien Winjands, for their support in this initiative and for bringing their views and comments into some of the chapters. A note of gratitude goes also to the CRC publishing team, for their availability and punctual assistance during the entire project. Finally, a special and sincere acknowledgment goes to all our colleagues, customers, and advisors who, in the years, have provided us with interesting problems and new stimuli, offering us precious and rare opportunities to stretch our skills and areas of competencies.

Daniele Gianni
Guglielmo Marconi University
Andrea D'Ambrogio
University of Rome Tor Vergata
Andreas Tolk
SimIS

MATLAB® is a registered trademark of The MathWorks, Inc. For product information, please contact:
The MathWorks, Inc.
3 Apple Hill Drive
Natick, MA 01760-2098 USA
Tel: 508 647 7000
Fax: 508-647-7001
E-mail: info@mathworks.com
Web: www.mathworks.com

Editors

Daniele Gianni is a requirements engineering consultant at the European Organization for the Exploitation of Meteorological Data (EUMETSAT) (DE), where he supports the requirements management activities of future space programs with the definition of requirements processes and requirements models. He also lectures part-time in information systems engineering. Previously, Gianni held research appointments at the European Space Agency (NL), the University of Oxford (UK), and the Imperial College (UK), where he introduced new modeling and simulation (M&S) methods for the systems engineering activities in several domains, including space, software architectures, software performance modeling, biomedical engineering, and emergency management. On similar activities, he also held visiting positions at the Auckland Bioengineering Institute (NZ) and the Georgia Institute of Technology (USA). He is the co-organizer of several international workshops on the topics related to M&S for systems engineering—the Mod4Sim workshop in SpringSim, on model-driven simulation engineering, and the CoMetS Workshop in the Institute of Electrical and Electronics Engineers (IEEE) WETICE, on collaborative M&S. Gianni holds a PhD in computer engineering from the University of Rome Tor Vergata (Italy).

Andrea D'Ambrogio is an associate professor at the Department of Enterprise Engineering of the University of Roma Tor Vergata. He is director of the postgraduate master's degree program in systems engineering at the University of Roma Tor Vergata. His research interests are in the areas of model-driven systems and software engineering, systems dependability engineering, business process management, and distributed and web-based simulation. In such areas he has participated in several projects at both the European and overseas levels, and has authored more than 80 journal/conference papers. He has served as a member of the program committee of various international conferences, among which are IEEE WETICE, the Association for Computing Machinery (ACM) WOSP, ACM ICPE, the Society for Modeling and Simulation (SCS) SpringSim, ACM PADS, WinterSim, and SIMUTools. He has been general chair of SCS/ACM/IEEE TMS 2014 and IEEE WETICE 2008. In 2010, he started the

IEEE International Workshop on Collaborative Modeling and Simulation (CoMetS), and in 2011 the SCS/ACM/IEEE International Workshop on Model-Driven Approaches for Simulation Engineering (Mod4Sim). He does scientific advisory work for industries and for national and international organizations, and he is a member of IEEE, IEEE Computer Society, ACM, SCS, and the International Council on Systems Engineering (INCOSE).

Andreas Tolk is the chief scientist for SimIS, Inc. in Portsmouth, Virginia, where he is responsible for implementing long-term visions regarding developments in simulation. He is also an adjunct professor of engineering management and systems engineering at Old Dominion University in Norfolk, Virginia. He holds an MS and PhD in computer science from the University of the Federal Armed Forces in Munich, Germany. He received the Excellence in Research Award from the Frank Batten College of Engineering and Technology in 2008, the Technical Merit Award from the Simulation Interoperability Standards Organization in 2010, the Outstanding Professional Contributions Award from the SCS in 2012, and the Distinguished Professional Achievement Award from SCS in 2014. He is on the board of directors of the SCS as well as of the ACM Special Interest Group Simulation (SIGSIM). He is a senior member of IEEE and SCS.

Contributing Authors

Paolo Bocciarelli is a postdoc researcher at the University of Rome Tor Vergata. He has held positions as software engineer and project manager in both private industries and government agencies. His research interests include software and systems engineering and business process management (BPM), specifically in the areas of modeling and simulation (M&S) and model-driven development.

Sergey Bolshchikov earned his MSc degree in information systems from Technion–Israel Institute of Technology in 2013, where he dedicated most of his time to the research and development of Vivid OPM—the animation toolkit for conceptual models. He also holds a bachelor's degree in computer science from Ural State Technical University, Russia. Currently, he holds the position of lead front-end engineer at New ProImage (Agfa), developing the solution for workflow automation in the publishing sphere.

Dídac Busquets is a research associate in the Intelligent Systems and Networks research group of the Department of Electrical & Electronic Engineering at Imperial College London. He joined Imperial College in 2011 as a Marie Curie Fellow, after having been at Universitat de Girona as a research scientist for five years. His research background is on resource allocation mechanisms in multiagent systems. He has applied this background in diverse areas such as robotics, auction mechanisms, and most recently, in self-organized resource allocation with specific applications to intelligent transportation and energy systems.

Olivier Dalle is Maître de Conférences in the computer science department of the Faculty of Sciences at the University of Nice Sophia Antipolis (UNS). He earned his BSc from the University of Bordeaux 1 and MSc and PhD from UNS. From 1999 to 2000, he was a postdoctoral fellow at the French Space Agency Center in Toulouse (CNES-CST), where he started working on component-based discrete-event simulation of multimedia telecommunication systems. In 2000, Dalle joined the I3S Joint Laboratory of UNS and CNRS, in Sophia Antipolis, where he has been doing research on telecommunications systems, discrete-event simulation and methodology, and component-based software engineering. In 2005, he initiated the Open Simulation Architecture project, in which he investigated new software engineering approaches for building simulations, including aspect-oriented programming, component-based software engineering, and layered approach for the separation of concerns. In 2008, he coinitiated the ICST series of International Conferences on Simulation Tools and Techniques (SIMUTools) and in 2014, he agreed to serve as the editor in chief of a new open access journal called *EAI Transactions on Simulation Tools and Techniques* that features the ability for authors to submit additional material along with the papers to allow reproducibility of the simulation results.

Saikou Diallo is a research assistant professor at the Virginia Modeling Analysis and Simulation Center (VMASC) of Old Dominion University (ODU). He received his MS and PhD in M&S from ODU. His research focuses on the theory and practice of interoperability and the advancement of M&S science. Dr. Diallo has authored or coauthored over 80 publications including a number of awarded papers and articles in conferences, journals, and book chapters. He is a member of SCS, IEEE, and ACM.

Dov Dori is a visiting professor at Engineering Systems Division, Massachusetts Institute of Technology. He is the Harry Lebensfeld Chair in Industrial Engineering and the head of the Enterprise System Modeling Laboratory at the Faculty of Industrial Engineering and Management, Technion–Israel Institute of Technology. His research interests include model-based systems engineering, conceptual modeling of complex systems, systems architecture and design, software and systems engineering, and systems biology. He invented and developed Object-Process Methodology (OPM), the emerging ISO 19450 Standard. He has authored about 300 publications, including journal and conference publications, books, and book chapters. He has mentored 48 graduate students. He was chairperson of nine international conferences or workshops. Among his many editorial duties, he was an associate editor of *IEEE Transaction on Pattern Analysis and Machine Intelligence*, and currently he is an associate editor of *Systems Engineering*. He is a fellow of INCOSE—International Council on Systems Engineering, a fellow of IAPR—International Association for Pattern Recognition, a member of Omega Alpha Association—International Honor Society for Systems Engineering, and a senior member of IEEE and ACM.

Philipp M. Fischer is working in the field of model-based system and software engineering. His research focused on formal verification of space systems as well as visualization within concurrent engineering environments. He received his diploma in electrical and computer engineering from the Leibniz Universität Hannover in Germany, in 2007. In addition, he spent one year of studies in computer science at the Swinburne University of Technology in Melbourne, Australia. Since 2007, he has worked for several companies and institutions in the field of M&S.

Andreas Gerndt is the head of the department of *Software for Space Systems and Interactive Visualization* at the German Aerospace Center (DLR). He received his MS in computer science from Technical University, Darmstadt, Germany, in 1993. In the position of a research scientist, he also worked at the Fraunhofer Institute for Computer Graphics in Germany. Thereafter, he was a software engineer for different companies with a focus on software engineering and computer graphics. In 1999, he continued his studies in virtual reality and scientific visualization at RWTH Aachen University, Germany, where he received his doctoral degree in computer science. After two years of interdisciplinary research activities as a postdoctoral fellow at the University of Louisiana, Lafayette, he returned to Germany in 2008.

Ross Gore holds a PhD and a MS in computer science from the University of Virginia and a BS in computer science from the University of Richmond. Dr. Gore has nine years of research experience in problems that lie at the intersection of software engineering and M&S. His work has yielded authorships on more than 20 conference and journal publications and has been recognized by the ARCS (Achievement Rewards for College Scientists) Society as an impactful and novel research avenue. He is currently working as a visiting assistant professor at Gettysburg College.

Jan Himmelspach received his PhD from the University of Rostock, Germany. He worked for more than a decade on software and efficient algorithms as well as on the credibility of M&S. He started the work on the open-source M&S framework JAMES II.

Taylor K. Hughes is a senior enterprise architect and program manager for SimIS, Inc. in Stafford, Virginia, where he is responsible for managing the execution of a contract for the development of solution-level architectures for the United States Marine Corps. He is a United States Army veteran who holds a BS from Auburn University, Alabama, an MA from Christendom College, Virginia, an MS from Florida Institute of Technology, Melbourne, Florida, and the Master Certificate in Business Analysis/Project Management for IS/IT Professionals from Villanova University, Villanova, Pennsylvania. He received the Certified Professional Logistician designation from the International Society of Logistics Engineers in 1998 and the Certification in Enterprise Architecture from California State University—East Bay, Hayward, California (in association with the FEAC Institute) in 2013. He has served as an officer for the nonprofit Reliability, Maintainability, Supportability (RMS) Partnership as well as serving in leadership positions in other trade societies and charities.

Giuseppe Iazeolla is a full professor of Computer Science, Software Engineering, and System Performance Modeling Chairs. Formerly with the University of Pisa, Faculty of Science (through 1984), and with the University of Roma Tor Vergata, Faculty of Engineering (through 2012), he is presently with the Faculty of Science and Applied Technology,

Guglielmo Marconi University, Roma, Italy. Professor Iazeolla has held computer science research appointments with the National Research Council, Italy, with the MIT, Project MAC, Cambridge, Massachusetts, with the University of West Virginia, Concurrent Engineering Research Center, with the University of Cape Town, Data Network Architecture Center, and industrial positions in computer and digital electronics. His research is in the areas of performance and reliability modeling of computer systems, networks, and software systems, and in the areas of parallel and distributed simulation.

Joost-Pieter Katoen is a distinguished professor at the RWTH Aachen University (since 2004) and is part-time affiliated with the University of Twente, the Netherlands. His research interests are concurrency theory, model checking, timed and probabilistic systems, and semantics. He has coauthored more than 200 journals and conference papers, and coauthored the book *Principles of Model Checking*. He is a member of the Academia Europaea, IFIP WG 1.8 and WG 2.2, a member of the steering committees of the conferences ETAPS (chair), CONCUR, FORMATS, and QEST (chair), and a member of the editorial board of STTT.

Marcello La Rosa is an associate professor and the Information Systems School academic director for corporate engagements at Queensland University of Technology (QUT) in Brisbane, Australia. He is also a researcher at the National ICT Australia. His research interests focus on process consolidation, configuration, mining, and automation. He has published over 60 refereed papers on these topics including papers in top journals like *ACM TOSEM, Formal Aspects of Computing and Information Systems*. He leads the Apromore initiative (http://www.apromore.org)—a strategic collaboration between various universities for the development of an advanced process model repository. He has taught BPM to students and practitioners in Australia for over eight years. He is a coauthor of *Fundamentals of Business Process Management* (Springer, 2013), the first comprehensive textbook on BPM. He was awarded with the best paper award at the 11th International Conference on BPM.

Daniel Lüdtke received the diploma degree, Dipl.-Ing. in computer engineering, from Technische Universität (TU) Berlin (Germany) in 2003. He worked as a research assistant at the Department of Computer Engineering and Microelectronics, TU Berlin. Since 2010, he has been with the DLR in the Department of Software for Space Systems and Interactive Visualization. His current research interests include model-based software and simulation engineering for space systems with an emphasis on reconfigurable embedded systems.

Sam Macbeth is a final year PhD student in the Intelligent Systems and Networks research group of the Department of Electrical & Electronic Engineering at Imperial College London. His focus is on the simulation of sociotechnical systems, particularly those concerned with resource provision, appropriation, and allocation. He has developed the Presage2 simulation platform for this purpose and currently leads its development.

Il-Chul Moon is an assistant professor at the Department of Industrial and Systems Engineering, KAIST, South Korea. His theoretic research focuses on the overlapping area of computer science, management, sociology, and operations research. His practical research includes military command and control analysis, counterterrorism analysis, intelligence analysis, and disaster management.

Viet Yen Nguyen is a scientist at the Fraunhofer Institute for Experimental Software Engineering (Germany) since 2014. His research interests are formal semantics, architectural modeling, and model checking timed and probabilistic systems. He was a postdoc at the RWTH Aachen University, Germany (2012–2014), and obtained his PhD in computer science (2012) there as well. In between, he was a visiting researcher at the European Space Agency (2010–2012). Prior to that, he obtained both his MSc and BSc in computer science from the University of Twente, the Netherlands (2007, 2004), and received several awards for his work on software model checking.

Thomas Noll is a staff member of the Software Modeling and Verification Group at RWTH Aachen University. His area of expertise is formal semantics, program analysis, software model checking, and compositional modeling languages. He received his PhD and habilitation degree in computer science from RWTH Aachen University in 1995 and 2003, respectively. He was a guest professor at Philipps University of Marburg, Germany, and lecturer at KTH Stockholm, Sweden. In 2012, he became an associate professor. He has coauthored more than 60 journals and conference papers, acted on numerous program committees, and organized several workshops.

Jose Padilla is a research assistant professor with the Virginia Modeling, Analysis and Simulation Center (VMASC) at ODU. He holds a PhD in engineering management. His main research interest is on the nature of understanding and its implications in conceptual modeling and human social cultural behavioral modeling.

Alessandro Pellegrini received his BS in computer engineering in 2008, his MS in distributed systems and computer architectures in 2010, and his PhD in computer engineering in 2014, all at Sapienza, University of Rome. Dr. Pellegrini collaborates with the High Performance and Dependable Computing Systems research group, where he is working in the area of distributed systems and parallel simulation. His other research interests include autonomic computing, compilers, and debuggers.

Luiz Felipe Perrone is an associate professor of computer science at Bucknell University, where he has been since 2003, following a postdoctoral research associate appointment at Dartmouth College. He holds PhD and MSc degrees from the College of William and Mary (USA), as well as an MSc in systems engineering and computer science and the degree of electrical engineer from the Universidade Federal do Rio de Janeiro (Brazil). He is an associate editor of the *ACM TOMACS* and has served the simulation community in several organizational roles in conferences, including proceedings editor of the *Winter Simulation Conference 2006* and program and general chair of *SIMUTools* (2009 and 2010, respectively). His interests include M&S, computer networks, and high-performance computing.

Alessandra Pieroni received the master's degree in computer science engineering at the University of Rome Tor Vergata (Italy) in November 2008. In June 2013, she earned the PhD in telecommunication engineering at the University of Rome Tor Vergata (Italy). Currently, she is an assistant researcher at "Guglielmo Marconi" University of Study, Rome, Italy. Her research interests currently include software technologies for distributed simulation, M&S, and performance evaluation of complex systems.

Jeremy Pitt is a reader in intelligent systems in the Department of Electrical & Electronic Engineering at Imperial College, London. He has been at the forefront of pioneering research in computational logic and its interdisciplinary applications for 25 years. His current research interests are in self-organizing multiagent systems and their application to computational sustainability. He has collaborated widely, having worked on over 30 national and international projects, and has been involved in much interdisciplinary research, having worked with lawyers, philosophers, psychologists, physiologists, and fashion designers. He also has a strong interest in the social implications of technology, and has edited two volumes in this concern: *This Pervasive Day* (IC Press: 2012) and *The Computer After Me* (IC Press: 2014).

Francesco Quaglia received the Laurea degree (MS level) in electronic engineering in 1995 and the PhD in computer engineering in 1999 from the University of Rome "La Sapienza." From summer 1999 to summer 2000, Quaglia held an appointment as a researcher at the Italian National Research Council (CNR). Since January 2005, he has worked as an associate professor at the School of Engineering of the University of Rome "La Sapienza," where he previously worked as an assistant professor from September 2000 to December 2004. He has been an elected member of the Academic Senate at the same university (2009–2013). His research interests span from theoretical to practical aspects concerning distributed systems and applications, distributed protocols, middleware platforms, parallel discrete event simulation, federated simulation systems, parallel computing applications, cloud computing, fault-tolerant programming, transactional systems, web-based systems, and performance evaluation. In these areas, he has authored or coauthored more than 130 technical articles. He has been the leader, or coleader, of several national and European research projects targeting topics in the previously mentioned areas. He regularly serves as program committee member for prestigious international conferences. Dr. Quaglia has also served as program co-chair of PADS 2002 and 2010, as program co-chair of NCA 2007, as general chair of PADS 2008, as general co-chair of SIMUTools 2011, as program co-chair of SIMUTools 2012, and as program chair of NCCA 2014. He has also been a member of PADS's Steering Committee (2008–2011). Starting from 2003, he has regularly served as consultant expert of the National Center for Informatics in the Public Administration.

Stefan Rybacki received his diploma in computer science at the University of Rostock. He currently pursues his PhD in the field of M&S focusing on software development as well as credibility. He is involved in the development of the open-source M&S framework JAMES II.

Volker Schaus is working as a research scientist at the DLR Institute for Simulation and Software Technology and is leading the working group on M&S. He is also responsible for the Virtual Satellite software development project. He holds the diploma degree Dipl.-Ing. in aerospace engineering from the University of Stuttgart, Germany. He did his final thesis at the University of Sydney, Australia, working on adaptive neural network control algorithms for unmanned aerial vehicles. Before joining the DLR in 2010, he worked in VIP outfitting projects of wide-body aircrafts as a mechanical engineer. His research interests are digital product design and evaluation, model-based systems engineering, and concurrent engineering.

Arthur H. M. ter Hofstede is a professor in the Information Systems School in the Science and Engineering Faculty, QUT, Brisbane, Australia, and is the head of the BPM Discipline. He is also a professor in the Information Systems Group of the School of Industrial Engineering of Eindhoven University of Technology, Eindhoven, the Netherlands. His research interests are in the areas of business process automation and process mining.

Wil M. P. van der Aalst is a full professor of information systems at the Technische Universiteit Eindhoven (TU/e). He is also the academic supervisor of the International Laboratory of Process-Aware Information Systems of the National Research University, Higher School of Economics in Moscow. Moreover, since 2003, he has a part-time appointment at QUT. At TU/e, he is the scientific director of the Data Science Center Eindhoven. His personal research interests include process mining, Petri nets, and BPM. Many of his papers are highly cited (he has an H-index of more than 107 according to Google Scholar). In 2012, he received the degree of doctor honoris causa from Hasselt University. In 2013, he was appointed as distinguished university professor of TU/e and was awarded an honorary guest professorship at Tsinghua University. He is also a member of the Royal Holland Society of Sciences and Humanities and the Academy of Europe.

Roberto Vitali received his bachelor's degree in computer engineering in 2007 and the master's in computer engineering, with focus on distributed systems and computer architectures in 2009. In 2013, he achieved the PhD in computer science—the topic has been about high performance simulation with special focus on multicore architectures. He is a member of the *High Performance and Dependable Computing Systems*. All of his studies have been carried on in the University of Rome "La Sapienza." His research interests are primarily high performance computing, operating systems, simulation, computer architectures, and computer dependability.

Gabriel Wainer (SMSCS, SMIEEE) is a full professor at the Department of Systems and Computer Engineering at Carleton University. He is the author of three books and over 260 research articles; he edited four other books, and helped organize over 120 conferences, including being one of the founders of SIMUTools and SimAUD. He is the principal investigator of different research projects. He is the vice president of conferences. He is special issues editor of *SIMULATION*. He is the head of the Advanced Real-Time Simulation lab, located at Carleton University's Centre for Advanced Simulation and Visualization (V-Sim). He has been the recipient of various awards, including the IBM Eclipse Innovation Award, SCS Leadership Award, and various Best Paper awards. His current research interests are related with modeling methodologies and tools, parallel/distributed simulation, and real-time systems.

Sixuan Wang is a PhD candidate in electrical and computer engineering with the Department of Systems and Computer Engineering at Carleton University. He received his BEng and MEng from Harbin Institute of Technology, China, and an MSc from the University of Bordeaux 1, France. His academic experience spreads over a wide range of areas such as M&S, software engineering, semantic Web and cloud computing, including six months research in the NSERC Engage project (Integrating Building Information Modeling, DEVS Remote Simulation, and 3D Visualization) with Autodesk Research on Toronto (2012, Canada), eight months R&D in the European REMICS project (Reuse and Migration of legacy applications to Interoperable Cloud Services) within SINTEF institute (2011, Norway). His current research interests are simulation as a service, DEVS/Cell-DEVS theory, semantic mash-ups, service composition, ontology learning, and cloud computing.

Niva Wengrowicz received her PhD from Bar-Ilan University, Israel, in 2012. She also holds a bachelor's degree in computer science and education from Bar-Ilan University, an EMBA degree from Tel Aviv University, and she graduated summa cum laude from Bar-Ilan University with another master's in education. Currently, she is a postdoctoral fellow at the Engineering Systems Division at Massachusetts Institute of Technology, and a postdoctoral fellow at the Department of Education in Science and Technology, at the Technion–Israel Institute of Technology. Her current research focuses on alternative assessment at systems engineering higher education programs in technology-enhanced learning environments, and her current interests include, in addition, transactional distance, pedagogical decision making and change, engineering education, and cognitive info-communications.

Moe T. Wynn is a researcher in the field of BPM within the Information Systems School at QUT, Australia. She received her PhD in the area of business process automation/workflow management in 2007. Her current research interests include cost-aware BPM, risk-aware BPM, process automation, and process mining. She works on interdisciplinary research projects (e.g., in the healthcare domain) and carries out collaborative research with industry partners. She has published over 45 refereed papers in international journals and conferences in the field of BPM in the past 10 years. Her research has appeared in the following journals: *Information Sciences, Data and Knowledge Engineering, Information and Software technology, Formal Aspects of Computing, Journal of Computer and System Sciences, International Journal of Cooperative Information Systems, Transactions on Petri Nets and Other Models of Concurrency, Computers in Industry,* and *Journal of Information Technology Theory and Applications.*

List of Contributors

Paolo Bocciarelli
Department of Enterprise
  Engineering
University of Rome "Tor Vergata"
Rome, Italy

Sergey Bolshchikov
Technion, Israel Institute of
  Technology
Haifa, Israel

Dídac Busquets
Intelligent Systems & Networks
  Group
Department of Electrical &
  Electronic Engineering
Imperial College London
London, England

Andrea D'Ambrogio
Department of Enterprise
  Engineering
University of Rome "Tor Vergata"
Rome, Italy

Olivier Dalle
University of Nice Sophia
  Antipolis
Sophia Antipolis, France

Saikou Diallo
Virginia Modeling Analysis and
  Simulation Center
Old Dominion University
Suffolk, Virginia, USA

Dov Dori
Technion, Israel Institute of
  Technology
Haifa, Israel
and
Massachusetts Institute of
  Technology
Cambridge, Massachusetts, USA

Philipp M. Fischer
Software for Space Systems and
  Interactive Visualization
Simulation and Software
  Technology
German Aerospace Center (DLR)
Braunschweig, Germany

Andreas Gerndt
Software for Space Systems and
  Interactive Visualization
Simulation and Software
  Technology
German Aerospace Center (DLR)
Braunschweig, Germany

Daniele Gianni
Guglielmo Marconi University
Rome, Italy

Ross Gore
Virginia Modeling Analysis and
  Simulation Center
Old Dominion University
Suffolk, Virginia, USA

Jan Himmelspach
Hamburg, Germany

Taylor K. Hughes
SimIS, Inc.
Stafford, Virginia, USA

Giuseppe Iazeolla
Guglielmo Marconi University
Rome, Italy

Joost-Pieter Katoen
Software Modeling and
  Verification Group
RWTH Aachen University
Aachen, Germany

Marcello La Rosa
Queensland University of
  Technology
Brisbane, Australia
and
NICTA Queensland Research Lab
Brisbane, Australia

Daniel Lüdtke
Software for Space Systems and
  Interactive Visualization
Simulation and Software
  Technology
German Aerospace Center (DLR)
Braunschweig, Germany

Sam Macbeth
Intelligent Systems & Networks
  Group
Department of Electrical &
  Electronic Engineering
Imperial College London
London, England

Il-Chul Moon
Department of Industrial and
  Systems Engineering
KAIST
Daejeon, South Korea

Viet Yen Nguyen
Embedded Systems Quality
  Assurance Department
Fraunhofer Institute for
  Experimental Software
  Engineering
Kaiserslautern, Germany

Thomas Noll
Software Modeling and
  Verification Group
RWTH Aachen University
Aachen, Germany

Jose Padilla
Virginia Modeling Analysis and
  Simulation Center
Old Dominion University
Suffolk, Virginia, USA

Alessandro Pellegrini
DIAG
University of Rome "La Sapienza"
Rome, Italy

Luiz Felipe Perrone
Department of Computer Science
Bucknell University
Lewisburg, Pennsylvania, USA

Alessandra Pieroni
Guglielmo Marconi University
Rome, Italy

Jeremy Pitt
Intelligent Systems & Networks
  Group
Department of Electrical &
  Electronic Engineering
Imperial College London
London, England

Francesco Quaglia
DIAG
University of Rome "La Sapienza"
Rome, Italy

Stefan Rybacki
University of Rostock
Rostock, Germany

Volker Schaus
Software for Space Systems and
  Interactive Visualization
Simulation and Software
  Technology
German Aerospace Center (DLR)
Braunschweig, Germany

Arthur H. M. ter Hofstede
Queensland University of
  Technology
Brisbane, Australia
and
Eindhoven University of
  Technology
Eindhoven, the Netherlands

Andreas Tolk
SimIS, Inc.
Portsmouth, Virginia, USA
and
Virginia Modeling Analysis and
  Simulation Center
Old Dominion University
Suffolk, Virginia, USA

Wil M. P. van der Aalst
Eindhoven University of
  Technology
Eindhoven, the Netherlands
and
Queensland University of
  Technology
Brisbane, Australia

Roberto Vitali
DIAG
University of Rome "La Sapienza"
Rome, Italy

Gabriel Wainer
Department of Systems and
  Computer Engineering
Carleton University Centre for
  Visualization and Simulation
  (V-Sim)
Ottawa, Canada

Sixuan Wang
Department of Systems and
  Computer Engineering
Carleton University Centre for
  Visualization and Simulation
  (V-Sim)
Ottawa, Canada

Niva Wengrowicz
Technion, Israel Institute of
  Technology
Haifa, Israel
and
Massachusetts Institute of
  Technology
Cambridge, Massachusetts, USA

Moe T. Wynn
Queensland University of
  Technology
Brisbane, Australia

chapter one

Introduction to the modeling and simulation-based systems engineering handbook

Daniele Gianni, Andrea D'Ambrogio, and Andreas Tolk

Contents

Systems engineering—"… an interdisciplinary collaborative approach to derive, evolve, and verify a life-cycle balanced system solution which satisfies customer expectations and meets public acceptability"(IEEE 1998, p. 11)—has been taught and practiced for several decades, and a body of knowledge (http://www.sebokwiki.org/) has been developed to share and centralize experiences and successful practices. However, despite the fact that they are applying systems engineering processes, many projects still overrun their budgets, fail to deliver in time, and sometimes even do not deliver something really useful or do not even meet the requirements. Many team members continued to work in *silos of domain excellence*, and the systems engineering processes did not connect the team members as intended. New emerging ideas were gradually introduced under the efforts of model-based systems engineering (MBSE) (Estefan 2008; Wymore 1993), bringing a significant improvement with respect to the aforementioned shortcomings. With MBSE, a common model becomes a centerpiece of the engineering process. This common model is used to communicate with various stakeholders, users in different phases of the life cycle, and in particular team members of the systems engineering team in charge of defining, developing, operating, maintaining, upgrading, and finally retiring the system. The model is rooted in system requirements and is modified step-by-step through analysis and real-world constraints to become a blueprint of the real system.

Introducing models to capture the efforts of the various systems engineering processes instead of pure documents is mathematically and organizationally a significant step forward. Using the findings of isomorphism, the equivalency of models can be evaluated precisely, replacing educated guesses of subject-matter experts with precise model-transformation rules and formal evaluations. In the software domain, the model-driven architecture (MDA) approach of the Object Management Group was a milestone when introduced in 2003. The core idea of MDA is to base software engineering on a series of model transformation, from platform-independent models via platform-dependent models to code and executables. The interested reader is referred to Kleppe et al. (2003) for more details. In the domain of modeling and simulation (M&S), the idea to use model transformation as the backbone of simulation development was recently featured in detail by Cetinkaya (2013). The book of Mittal and Martin (2013) is another example that the idea starts to not only be accepted, but also has a significant influence for practitioners in the field, as also witnessed by the various editions of the *International Workshop on Model-Driven Simulation Engineering*, started in 2010 by D'Ambrogio and Petriu (2011) within the SCS SpringSim multi-conference. The theoretic foundations for such efforts can be traced back to the mathematical branch of model theory, as shown by Tolk et al. (2013). Models, in particular formal models and their transformations, are claiming a central role in systems engineering. Theoretically and practically, we made some significant progress.

Looking at these success stories of MBSE and our own experience with this topic, a question arose spontaneously: is it possible to add more value when using M&S methods and computational capabilities to numerically evaluate the properties of complex systems? It is well known that the earlier mistakes are uncovered in system designs, the cheaper it is to correct them. Introducing M&S-based systems engineering, as an extension of MBSE practices to also include simulation methods, offers the opportunity for a more comprehensive coverage of the design space by enabling systems engineers to identify a wider set of early mistakes. Currently, M&S already has its place in testing, in particular in the defense domain (Zeigler et al. 2005), but could we move it even closer to the system design? Is it possible to start thinking about M&S-based systems engineering?

Toward M&S-based systems engineering

The aim of this handbook is to start laying the foundations for such an effort, which may lead to a textbook in the future, when many of the currently open research questions will be answered. To structure the chapter contributions—and also hopefully a research agenda for a future discipline of M&S-based systems engineering—we utilized the taxonomy already utilized by MBSE, as starting point for possible extensions too. Of the identified topics, the handbook covers the following thematic areas:

- *Methods to support system modeling*: How can the process of modeling a system be supported, in particular when M&S shall be applied in this process, or the use of M&S methods shall be prepared to optimize these processes? This topic is addressed in Chapters 2–4.
- *Improving system architecture*: System architectures are well known in the systems engineering community. Frameworks such as the Zachman framework (Inmon et al. 1997), The Open Group Architecture Framework (Harrison 2007), the United States Department of Defense Architecture Framework and related efforts in NATO and other nations (U.S. DoD 2014), and related systems engineering efforts are well established, but can they support the idea of using M&S and systems engineering better? This topic is addressed in Chapter 2.
- *Domain-specific languages*: General languages such as the Unified Modeling Language (UML) or the System Modeling Language (SysML) are well established, but are there new developments that support M&S-based systems engineering better? Will ontologies and metamodeling play a larger role here? This topic is addressed in Chapters 6 and 10.
- *Model-driven simulation engineering*: Systems engineering principles already are applied often to drive simulation solutions. How can the latest research insights drive the ideas forward to use models and simulation to drive the engineering of systems? This topic is addressed in Chapters 6 and 7.
- *Collaborative environments*: A common and well-orchestrated set of tools and methods supporting M&S experts as well as systems engineering is the ultimate vision. However, until we reach this vision, we need better support of integrative collaborative environments that bring current methods and tools of experts of all domains together. This topic is addressed in Chapter 7.
- *Simulation algorithms and performance engineering*: To support systems engineering effectively, simulation not only needs to be reliable and informative, but also needs to be performing. High-performance computing, performance engineering, and better simulation algorithms will support engineers with thousands of simulation runs to cover the decision space with numerical evaluations. This topic is addressed in Chapters 8 and 9.
- *Simulation software architecture*: When simulation systems are used for decision support for systems engineering, new requirements for the simulation software architecture will emerge. Part of these will be derived from the need for the reuse and integration of systems engineering tools; others will be derived from performance

challenges. The integration into the operational environment and the fully calibratable accessibility of measures of metrics for performance and efficiency are part of them. This topic is addressed in Chapters 10 and 11.

- *Processes:* Western philosophy of science has been heavily influenced by the idea that substantials are the main carriers of knowledge. Objects and their attributes, and their relations to other objects dominate the world of knowledge representation. Processes play a subordinate role as they are merely seen as the things that create, change, or destroy objects. Several alternative approaches are using objects and processes as equally important concepts and have successfully been applied to systems engineering (Dori 1995). We assume that such balanced approaches will be advantageous for supporting system engineers, their stakeholders, and other team members. This topic is addressed in Chapters 12 and 13.

- *Verification and validation (V&V)*: Systems engineering as well as M&S use the terms V&V, but the interpretation thereof is slightly different. They agree, however, that verification ensures that the system is built correctly, and that validation ensures that the correct system is built. Nonetheless, for systems engineers, the requirements and their fulfillment build a trusted foundation for these V&V processes. If they can show that a system fulfills the requirements, they are good. Simulation, on the other hand, is the result of modeling, which is the task-oriented purposeful abstraction and simplification of a perception of reality. Ensuring that simulations are correct and useful for the system or representing the system appropriately is an important task with many open research points. This topic is addressed in Chapters 14 and 15.

- *Enterprise architecture*: The era of net-centric systems introduced a new set of challenges. Systems are no longer providing their functionality exclusively in the context of the original requirements, but they can be composed into federations or system of systems in which the individual functionalities of systems are combined to satisfy new customer needs. Systems have to be resilient and adaptive, and enterprise architectures provide one of the means to support this. The topic is closely related to systems architectures, but adds a new level of complexity to it.

- *Model repositories*: We assumed that models can be reused before, so it makes sense to provide the reusable models in common repositories. Although this topic can be seen as part of the collaborative environment, we decided to give it its own bullet. The challenges on how to describe a model that can be identified as a candidate, the selection of a good set of candidates based on problem descriptions, composing the models to provide a solution, and many more challenges belong to this topic. This topic is addressed in Chapters 7 and 16.

• *Advanced concepts*: There are likely many more emerging categories that have to be captured in the body of knowledge that describes what to do to provide better M&S support for systems engineering, and we captured contributions that are not easily mappable to the earlier bullets here. Chapter 17 falls under this topic.

In summary, we tried to identify a way forward to use the potentials of M&S for better systems engineering in the domains of architecture alignment and improvement, collaboration tools and repositories, and theoretical foundations. It may be too early to coin the term M&S-based systems engineering as the necessary concepts are still in the state of infancy. However, by bringing the various experts together to contribute to a common book to become the foundation of future research and collaboration, we hope to have contributed to make this happen in the foreseeable future.

Chapter contributions

The book consists of 16 chapters (Chapters 2 through 17) that deal with the aforementioned topics.

In Chapter 2 ("Systems engineering, architecture, and simulation"), A. Tolk and T. K. Hughes propose the alignment of processes, methods, and means that are already used by both systems engineers and simulation practitioners. Aligning the processes and using system architecture artifacts as a common repository result in a consistent model that can be executed as a simulation, providing additional numerical insight and eventually resulting in high-quality, trustworthy, and cost-effective systems as required by the customer.

In Chapter 3 ("System modeling: Principled operationalization of social systems using Presage2"), S. Macbeth, D. Busquets, and J. Pitt first provide various reasons for modeling and simulating social systems (e.g., formalize human methods of problem solving as a source and inspiration for developing computational systems to address engineering challenges), and then describe the implementation of the multiagent-based animation and simulation platform, Presage2, which supports the process of principled operationalization, defined as the mapping between formal models of social systems and implemented systems.

In Chapter 4 ("Formal agent-based models of social systems"), I.-C. Moon introduces two approaches for formal agent-based modeling and simulation of social systems. The chapter first shows the meta-network model that is useful in describing the structure of social systems. Next, the discrete event systems (DEVS) formalism is applied to illustrate the behavior of social systems.

In Chapter 5 ("On the evolution toward computer-aided simulation"), L. F. Perrone addresses the problem of failures in simulation studies that arise from the complexity of the simulation workflow, which introduces several points of failure for those with less expertise in the field. He then proposes the introduction of *computer-aided simulation* systems, which include comprehensive support tools that guide users throughout most if not all the stages of the modeling and simulation endeavor, thus leading to highly credible results without overburdening the user.

In Chapter 6 ("Model-driven method to enable simulation-based analysis of complex systems"), P. Bocciarelli and A. D'Ambrogio address the use of distributed simulation techniques to model the inherently distributed architecture of complex systems. The chapter proposes a method that exploits principles, standards, and tools introduced in the model-driven engineering field and supports the automated generation of high-level-architecture-based distributed simulations from system models specified by the use of SysML, the UML-based general purpose modeling language for systems engineering.

In Chapter 7 ("Collaborative modeling and simulation in spacecraft design"), V. Schaus, D. Lüdtke, P. M. Fischer, and A. Gerndt deal with the collaborative and concurrent design of spacecrafts with respect to modeling and simulation. Looking at the systems engineering life cycle, they discuss expectations and advantages of model-driven design approaches, and highlight the potential of formalized design for continuous verification and validation. The chapter closes with a vision on promising ideas to integrate modeling and simulation as a fast and easy-to-use means of supporting the spacecraft design process.

In Chapter 8 ("Performance engineering of distributed simulation programs"), G. Iazeolla and A. Pieroni apply software performance engineering techniques to predict, at design time, the effect of program partitioning and network capabilities on the efficiency of distributed simulation systems, to establish whether the use of distributed simulation can lead to a speedup with respect to the use of conventional sequential simulation, and thus decide the convenience to proceed with the distributed simulation implementation.

In Chapter 9 ("Reshuffling PDES platforms for multi/many-core machines: A perspective with focus on load sharing"), F. Quaglia, A. Pellegrini, and R. Vitali address the use of parallel discrete event simulation (PDES) and discuss how rethinking the organization of PDES platforms to make them perfectly suited for exploiting the computing power offered by modern multi/many-core machines. The proposed approach is based on an innovative load-sharing paradigm suited for PDES systems run on top of multicore machines.

In Chapter 10 ("Layered architectural approach for distributed simulation systems: The SimArch case"), D. Gianni introduces a layered architecture named SimArch, for local and distributed simulation systems, based on the IEEE High Level Architecture standard. The chapter also outlines a mechanical procedure that informally proves that no extra know-how or effort is necessary to develop a distributed simulation system with regard to the equivalent local simulation one.

In Chapter 11 ("Reuse-centric simulation software architectures"), O. Dalle addresses the importance of reusability in simulation software, which is also motivated by business-specific reasons, such as providing a better reproducibility of simulation experiments, or avoiding a complex validation process, other than by more conventional reasons such as improved reliability and decreased development costs. The chapter considers reuse as a problem that may be considered in two opposite directions: reusing and being reused.

In Chapter 12 ("Conceptual models become alive with Vivid OPM: How can animated visualization render abstract ideas concrete?"), D. Dori, S. Bolshchikov, and N. Wengrowicz address the processes needed to define and coordinate activities in M&S-based systems engineering. The chapter describes OPM, an emerging ISO Standard 19450 for modeling complex systems using a compact set of concepts, and Vivid OPM, a software module that enables the modeler to create a moving cartoon, that is, a video clip that is driven by the OPM underlying conceptual model.

In Chapter 13 ("Processes to support the quality of M&S artifacts"), J. Himmelspach and S. Rybacki focus on well-defined processes as one of the preconditions for high-quality products. Based on an overview of work done before, they describe how M&S software can be created utilizing workflow management systems, methods, and technologies. This approach is leading to well-defined processes in the software and to well-defined experiments, by keeping all degrees of freedom for users.

In Chapter 14 ("Formal validation methods in model-based spacecraft systems engineering"), J.-P. Katoen, V. Y. Nguyen, and T. Noll introduce a novel modeling language and toolset for a (semi) automated validation approach in the context of spacecraft systems, in which validation methods are labor intensive, usually being based on manual analysis, review, and inspection. The proposed language enables engineers to express the spacecraft system design, the software design, and their reliability aspects in a single integrated model. The model can then be analyzed by the COMPASS toolset for checking requirements related to functional correctness, safety and dependability, and performance.

In Chapter 15 ("Modeling and simulation framework for systems engineering"), S. Diallo, A. Tolk, R. Gore, and J. Padilla introduce a modeling and simulation system development framework (MS-SDF) that can be effectively applied to successfully model, simulate, verify, and validate a model in support of systems engineering. Within the MS-SDF, the chapter helps the reader determine whether a model satisfies design requirements and adequately represents a real system without violating key assumptions.

In Chapter 16 ("Liquid business process model collections"), W. M. P van der Aalst, M. La Rosa, A. H. M. ter Hofstede, and M. T. Wynn address the problems due to the increasing amounts of data (*big data*) that many organizations need to deal with intelligently to turn data into valuable insights that can be used to improve business processes. The chapter proposes to manage large collections of process models and event data in an integrated manner. The collection should self-adapt to evolve organizational behavior and incorporate relevant execution data (e.g., process performance and resource utilization) extracted from the logs, thereby allowing insightful reports to be produced from factual organizational data.

Finally, in Chapter 17 ("Web-based Simulation using Cell-DEVS Modeling and GIS Visualization"), S. Wang and G. Wainer focus on web-based simulation as a means to integrate geographic information systems (GIS), modeling, simulation, and visualization, to support the simulation-based design process in system engineering. The chapter proposes a method to extract information from GIS, model with Cell-DEVS theory, run web-based simulation, and visualize results back in Google Earth. The web-based simulation is based on RESTful Interoperability Simulation Environment (RISE), which supports different simulation environments for both remote simulation on a single server and distributed simulation on multiple servers.

Conclusion

In his essays on life itself, the famous scientist and researcher Robert Rosen, who studied, in addition to biology, mathematics and the history of science, states (Rosen 1998):

> I have been, and remain, entirely committed to the idea that modeling is the essence of science and the habitat of all epistemology.

For M&S researchers, this is a powerful statement, as it makes the argument that models are the essential way to capture knowledge, and we can preserve this knowledge in the form of model repositories. Furthermore, we can make this knowledge applicable to all in the form of simulations.

Tolk et al. (2011) make the case that in order to capture intelligence for systems engineering support, three components are needed: (1) ontologies to provide the unambiguous structure for terms describing the concepts of the application domains as well as of the implementation domains;

(2) simulations to capture the knowledge in dynamic and executable form; and (3) software agents that support the process of identification of applicable models, selection of the optimal set of models, composition of these models into a usable solution, execution of these composition of models, and evaluation of the results in the light of the customers task. Again, simulation plays a pivotal role in capturing knowledge.

The contributions to this book show that we are making progress in all identified research domains, in architectures, collaboration, repositories, and theoretical foundations. M&S experts become aware of systems engineering processes that can support them and that they can support. Systems engineers recognize the potential of M&S to produce numerical insight into the behavior of complex systems—including emerging behavior that can be positive or negative in regard to the intended use of the system—in all life cycles of a system. MBSE opened the door for more collaboration, and the use of dynamic simulation, in addition to the static models, is a logical next step. We honestly hope that many young researchers will be motivated to contribute to the increasingly growing and overlapping domains of M&S and systems engineering so that an update of this handbook with many new insights will have to be published soon.

References

Cetinkaya, D. (2013). Model Driven Development of Simulation Models, Defining and Transforming Conceptual Models into Simulation Models by Using Metamodels and Model Transformations. PhD Thesis at the Technische Universiteit Delft, the Netherlands.

D'Ambrogio, A. and Petriu, D., *First International Workshop on Model-driven Approaches for Simulation Engineering (Mod4Sim'11)*, http://www.sel.uniroma2.it /mod4sim11, last accessed February 2014.

Dori, D. (1995). Object-process analysis: Maintaining the balance between system structure and behaviour. *Journal of Logic and Computation*, 5(2), 227–249.

Estefan, J. A. (2008). *Survey of Model-Based Systems Engineering (MBSE) Methodologies*. International Council on Systems Engineering, San Diego, CA.

Harrison, R. (2007). *TOGAF Version 8.1*. Van Haren Publishing, Zaltbommel, the Netherlands.

IEEE. (1998). *IEEE Standard 1220: Application and Management of the Systems Engineering Process*. IEEE Press, New York, NY.

Inmon, W. H., Zachman, J. A., and Geiger, J. G. (1997). *Data Stores, Data Warehousing and the Zachman Framework: Managing Enterprise Knowledge*. McGraw-Hill, Chicago, IL.

Kleppe, A. G., Warmer, J. B., and Bast, W. (2003). *MDA Explained: The Model Driven Architecture: Practice and Promise*. Addison-Wesley Professional, Boston, MA.

Mittal, S. and Martín, J. L. R. (2013). *Netcentric System of Systems Engineering with DEVS Unified Process*. CRC Press, Boca Raton, FL.

Rosen, R. 1998. *Essays on Life Itself*. Columbia University Press, New York.

Tolk, A., Adams, K. M., and Keating, C. B. (2011). Towards intelligence-based systems engineering and system of systems engineering. *Intelligence-Based Systems Engineering*. Edited by A. Tolk and L. C. Jain. Springer Verlag, Berlin/Heidelberg, Germany. ISRL Vol. 10, 1–22.

Tolk, A., Diallo, S. Y., Padilla, J. J., and Herencia-Zapana, H. (2013). Reference modelling in support of M&S—foundations and applications. *Journal of Simulation*, 7(2), 69–82.

U.S. Department of Defense (DoD) Deputy Chief Information Officer. (2014). *Web portal DoD Architecture Framework*, http://dodcio.defense.gov/dodaf20 .aspx, last accessed February 2014.

Wymore, A. W. (1993). *Model-Based Systems Engineering*. Series on Systems Engineering, Vol. 3., CRC Press, Boca Raton, FL.

Zeigler, B. P., Fulton, D., Hammonds, P., and Nutaro, J. (2005). Framework for M&S-based system development and testing in Net-centric environment. *ITEA Journal of Test and Evaluation*, 26(3), 21–34.

chapter two

Systems engineering, architecture, and simulation

Andreas Tolk and Taylor K. Hughes

Contents

Introduction

This chapter addresses system engineers who want to improve their work and products by using simulations. It also addresses simulationists who want to help improve systems developers to provide better or more cost-efficient solutions. The chapter proposes the alignment of processes, methods, and means that are already used by practitioners.

    Systems engineering (SE) and modeling and simulation (M&S) as engineering approaches are both helping in solving important problems in various application domains. The diversity and interdisciplinarity of SE and M&S, and the multiplicity of their application domains show the significance of both disciplines as solution providers for complex and highly relevant real-world problems. Systems architectures are built within the SE process to actually support the SE process in all phases. If the systems architectures are rich enough in the way they describe all components, functions, relations, processes, and status changes, they can be executed as a simulation of the envisioned system. This simulation is often referred to as an executable architecture. This chapter makes the case to tie SE, systems architecture, and simulation even closer together to maximize the synergy between these disciplines.

It should be pointed out that we are not limiting SE to software engineering or the development of simulation systems, but that we address SE as the discipline applied to successfully develop complex, real-world systems. Software and simulation systems are a subset, but not the focus. In the same sense systems architectures are defined as the structure of components, relationships, and rules governing their design and evolution over time. Software and simulation architectures are a subset of systems architectures as well, but not in the focus of this chapter. Finally, we apply simulation in support of systems engineering and architecture, not as the objective of it. As such, this chapter addresses a much broader community than the M&S experts, including everybody who is in the business to develop, maintain, upgrade, or retire a real-world system.

What is systems engineering?

Traditional systems engineering

According to Maier and Rechtin (2007), whenever the complex interrelationships among combined elements are beyond what the engineers' and builders' tool can handle, new methods for SE have to be introduced. As such, SE and architecting are not new disciplines, but they go back to the beginnings of the recorded history of classical architecting of more than 4000 years ago. However, the term "systems engineering" describing the academic discipline of interest in this chapter started to emerge at the end of World War II, in particular in the domain of aerospace developments. It is closely related with the development of the idea of systems thinking, as described by Churchman (1968).

Sage and Armstrong (2000) define SE as the management technology that controls the total systems lifecycle process, which involves the definition, development, deployment, and retirement of a system that is of high quality, trustworthy, and cost-effective in meeting the users' needs.

The International Council on Systems Engineering (INCOSE 2011) defines SE as follows:

SE is an interdisciplinary approach and means to enable the realization of successful systems. It focuses on defining customer needs and required functionality early in the development cycle, documenting requirements, and then proceeding with design synthesis and system validation while considering the complete problem: operations, performance, test, manufacturing, cost and schedule, training and support, and disposal. SE integrates all the disciplines and specialty groups into a team effort forming a structured development process that proceeds from concept to production to operation. SE considers both the business and the technical needs of all customers with the goal of providing a quality product that meets the user needs.

The NASA *Systems Engineering Handbook* (2007) gives a description of the role of SE in the context of an organization. Within this handbook, SE is understood as a methodical, disciplined approach for the design, realization, technical management, operations, and retirement of a system. Similar to the definition used by Maier and Rechtin (2007), a system is defined as a construct or collection of different elements that together produce results not obtainable by the elements alone. The elements of a system can include people, hardware, software, facilities, policies, and documents. The results provided by the system include qualities, properties, characteristics, functions, behavior, and performance. The value added by the system as a whole, beyond that contributed independently by the elements, is created by the relationship among the elements. SE is a way of looking at the big picture when making technical decisions. It is a way of achieving high-level stakeholder functional, physical, and operational performance requirements in the intended use environment over the planned life of the system(s). SE is a logical way of thinking, the art and science of developing an operable system capable of meeting requirements within all given constraints. To accomplish this objective, a holistic, integrative view must be applied, wherein the contributions of structural engineers, electrical engineers, mechanism designers, power engineers, human factors engineers, logistics engineers, reliability engineers, maintainability engineers, and many more disciplines are evaluated and balanced, one against another, to produce a coherent whole that is not dominated by the perspective of a single discipline. SE balances often competing requirements regarding system performance, available budgets, required resources, time constraints, life cycle cost, and overall perception of the final system. The role and responsibility of the systems engineer may vary, but in every project of sufficient complexity the SE functions must be performed. In smaller projects, the project manager must ensure that they are covered; in larger efforts, engineering managers and systems engineers will have to work hand in hand to accomplish this.

The NASA Handbook also states that the actual assignment of the roles and responsibilities of the named systems engineer may vary as well. The lead systems engineer ensures that the system technically fulfills the defined needs and requirements and that a proper SE approach is being followed. The systems engineer oversees the project's SE activities as performed by the technical team and directs, communicates, monitors, and coordinates tasks. The systems engineer reviews and evaluates the technical aspects of the project to ensure that the systems/subsystems engineering processes are functioning properly and evolve the system from concept to product. The entire technical team, which preferably includes customer representatives, is involved in the SE process.

In summary, the SE processes take the requirements for all stakeholders for all life cycle phases—development, manufacturing, deployment or fielding, training, operations and maintenance, refinement, and retirement and disposal—and engineers a system composed of existing and newly development components and support systems to satisfy these requirements.

System of systems engineering and infranomics

In the recent years, the idea of system of systems engineering (SoSE) was discussed intensively, and to what degree the ideas of traditional SE can be applied. Some of the main ideas have been summarized by Jamshidi (2009). Although the term *system of systems* has no widely accepted definition, Maier (1998) notes that the notion is widespread and generally recognized. As the topic of this book is focused on SE, we will just give a short overview of relevant ideas that are generally applicable in this section. Maier (1998) as well as Sage and Cuppan (2001) evaluated several definition attempts and proposed the following distinguishing characteristics:

1. *Operational independence of the individual systems*: A system of systems is composed of systems that are independent and useful in their own right. If a system of systems is disassembled into the component systems, these component systems are capable of performing useful operations independent of one another.
2. *Managerial independence of the systems*: The component systems not only can operate independently when operating on their own, they generally operate even then independently when they contribute to achieving an intended common purpose. The component systems are generally individually acquired and integrated, and they maintain a continuing operational existence that is independent of the system of systems.

3. *Geographic distribution*: Geographic dispersion of component systems is often large. Often, these systems can readily exchange only information and knowledge with one another, and not substantial quantities of physical mass or energy.
4. *Emergent behavior*: The system of systems performs functions and carries out purposes that do not reside in any component system. These behaviors are emergent properties of the entire system of systems and not the behavior of any component system. The principal purposes supporting engineering of these systems are fulfilled by these emergent behaviors.
5. *Evolutionary development*: Development of these systems is evolutionary over time and with structure, function, and purpose added, removed, and modified as experience with the system grows and evolves over time.

Although SE assumes some minimal form of common control in charge of reaching the common objective that the components alone could not accomplish, SoSE does not establish a common control. It accepts that each system already provides operationally relevant functionality in its own context. Also, each system has its own governance regulating its rules and methods. With this come independent business models for all systems. The aspect of emergence is an ongoing research topic and describes how behavior on the macro level of the system of systems can and will emerge from the interaction of systems. In any case, SoSE significantly widens the group of experts that have to be brought together, and therefore increases the need for alignment and synchronization.

A similar observation has been highlighted recently by Gheorghe and Masera (2010). They identify *Infranomics* as a crucial discipline for this century. Infranomics is the body of disciplines supporting the analysis and decision-making regarding the metasystem (e.g., the totality of the technical components, stakeholders, frame of mind, legal constraints, and other elements composing the set of infrastructures). It is the set of theories, assumptions, models, methods, and associated scientific and technical tools required for studying the conception, design, development, implementation, operation, administration, maintenance, service supply, and resilience of the metasystem. In this way, Infranomics will be the discipline-of-disciplines grouping all needed knowledge. Finally, it is tightly connected with the process of managing the SoSE processes, as only a metasystem can help to orchestrate the operational and managerial independent systems. As such, it has to address the technical aspects of SE as well as the managerial aspects of orchestrating governance and business models accordingly.

Alberts et al. (2010) introduced a concept for measuring the maturity of independent systems regarding their ability to support a common goal.

Within the North Atlantic Treaty Organization (NATO), nations remain sovereign but work together toward a common goal. Their command and control process—which can be seen as equivalencies to the SE process of the independent systems within a system of systems—has to mature accordingly. As shown by Tolk (2003), only providing a set of common technology is not sufficient to reach interoperability between organizations. The levels of interoperation maturity are defined as follows:

- Although currently the various systems are managed and governed independently, the plans are *conflicting*. Each system exclusively focuses on its own resources and capabilities to reach their own objectives as if no other participant were present. Therefore, conflicts between participants are the rule. Their objectives are potentially mutually exclusive. Plans and execution will compete with each other.
- Limiting the responsibility to each sector of the system results in *deconflicted operations*. It ensures that organizations avoid interfering with one another. However, this partition is artificial and suboptimal, as it excludes synergism. The orchestration of activities in each sector will be supported by special nodes in intersecting domains specifically designed for this task. Nonetheless, orchestration will be limited to synchronizing the execution of operations.
- *Coordinated operations* require joint planning driven by a shared intent. The synchronized plan allows decision-making on lower levels. However, each component is still acting on the participant's behalf, leaving the governance intact. The execution of the plan remains the responsibility of the local system. Coordinated operations require a common intent and a common awareness, supported by broader access to shared sensors for the sake of gathering and fusing information into a common picture. This is where sociability starts to make a real difference.
- *Collaboration* requires one to not only share the planning process, but also the execution process. Shared situational awareness supported by joint common operational pictures requires a unifying system of systems that integrates the heterogeneous contributions of all systems. Information shall be shared seamlessly. Although the execution process is still limited to the better component or better service in the coordinated approach, the collaborative approach synchronizes planning and execution vertically and horizontally.
- *Coherent operations* are characterized by rapid and agile decision-making processes based on seamless and transparent information sharing. Smart and social components of each system will have access to all information they need to make the decision, regardless of where they are, which components they use to gain access to the information, or where the information came from.

The methods applied to ensure this coherence in all efforts need to be well aligned, which requires a common view of the challenges that the team is facing. Similar to the traditional approach, SoSE has to cope with all requirements of all stakeholders for all life cycle phases, not only for one system, but also for a dynamically changing portfolio of independently managed and operated systems.

Systems architectures supporting systems engineering

The discussions so far have shown that multiple interest groups have to be synchronized and activities have to be orchestrated. This requires that the applied methods and tools are supporting such an objective. This is accomplished by using systems architectures to provide the structure, logic, syntax, and semantics necessary to arrive at a consistent understanding of the requirements for the system of systems being conceived. Chigani and Balci (2012) postulate that the process of architecting takes the problem specification and requirements specification as input and produces an architecture specification as an output work product. This shows the close relation to the SE process. Tolk et al. (2009) emphasize the need that the resulting architecture artifacts be used as inputs to build a repository that aligns the data, synchronizes the processes, and orchestrates the necessary activities to ensure consistency of understanding between all stakeholders of each life cycle as well as between all life cycles as well. This is also the general philosophy behind model-based SE approaches, such as summarized by Estefan (2008).

In practice, system architects, stakeholders, and system developers all work iteratively to generate an architectural concept that is both desired by the client and feasibly built by the developer. During this iterative process, the problem and the requirements may evolve and become clearer. The agreed upon architectural concepts lead to a more concise definition of the problem and a set of requirements and a solution strategy that will meet client needs while maintaining feasibility on the engineering side.

Modeling a systems architecture

The systems architecture must become a blueprint for all members of the technical and nontechnical team, including the customer, to support developing, training for, operating, maintaining, refining, and retiring the system.

Requirements

The community agrees on the prime of requirements, which means that every part of the system must either fulfill a requirement, or must provide necessary support for a part of the system fulfilling a requirement. If a requirement is not supported by the system, then it is not fulfilled, and the system does not provide all the characteristics that the stakeholders required. If a system part does not fulfill a requirement, either directly or supportively, this part is superfluous and has no justification in the requirement. Necessarily, requirements should be a direct focus and part of the systems architecture, in particular in agile environments that are characterized by often changing requirements. Buede (2009) recommends the following requirements taxonomy depicted in Figure 2.1:

- Input/output requirements characterize the overall functionality of the system.
  - Input requirements constrain the input parameters for the system or for the specific functions provided by the system.
  - Output requirements do the same for output parameters.
  - Functions requirements define the intended behavior and performance for functions.
  - External interfaces requirements address in particular the interfaces of the system to other external systems that are needed to fulfill the operation or mission, but that are not under the control of the SE team responsible for the system under evaluation.

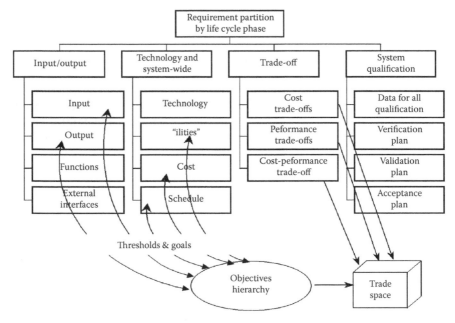

Figure 2.1 Requirements' taxonomy. (Buede, D.M.: *The Engineering Design of Systems: Models and Methods.* 2009. Copyright Wiley-VCH Verlag GmbH & Co. KGaA. Reproduced with permission.)

- Technology and system-wide requirements address the system holistically.
  - Technology requirements address system-wide technology constraints, such as the use of certain commercial and company standards, or even the use of certain implementation details, for example, to ensure the seamless integration into other solutions of the customer.
  - *Ilities* requirements address the system-wide capabilities regarding quality, safety, and usability but today spanning many more characteristics, such as maintainability, sustainability, interoperability, composability, reliability, durability, adaptability, flexibility, extensibility, and more. Overall, these define the desired performance of the system.
  - Cost requirements address the budgetary constraints.
  - Schedule requirements address the time constraints.
- Trade-off requirements define the flexibility of constraints regarding how much of one characteristic the user is willing to modify to gain more of another one.
  - Cost trade-off requirements address budgetary compromise constraints.
  - Performance trade-off requirements address performance compromises.
  - Cost-performance trade-off requirements allow for more mixed forms of cost and budgetary constraints.
  - Although not mentioned in Buede (2009), the third category of trade-offs is time.
- System qualification requirements define what needs to be observed and performed to ensure that requirements have been satisfied such that the user accepts the system. This involves a mix of experiments to be conducted and metrics to be applied.
  - Data requirements capture the need of data engineering to ensure that all data are accessible and can be collected. If planned in advance, this process can significantly reduce the cost and time required for operational testing support, data collection and management, test system support, and so forth.
  - The Verification Plan captures what needs to be done to ensure the transformation accuracies between different phase and artifacts used to model the architecture. It ensures that the system is built correctly.
  - The Validation Plan ensures that all requirements are understood as specified by the user and the system provides the functionality desired by the user. It ensures that the correct system is built.
  - The Acceptance Plan defines managerial and technical constraints the system has to meet to be accepted by the user.

Buede (2009) emphasizes the need to have these sets of requirements for each life cycle of the system, plus a set of possible trade-offs between the life cycles.

Functional and physical architecture viewpoints

The system provides for the customer the functionality that helps to solve the original problem of the user. The functional architecture comprises all functions, usually in the form of a hierarchy, that are needed to address all input/output requirements. Within a systems architecture, functions transform input parameters into output parameters, utilize resources, and comply with constraints. These functions can be interconnected, that is, the output parameters can be used as input parameters, control parameters, or resources in other functions. Loops and feedbacks are permitted. In addition, functions can be executed sequentially or in parallel, and under observance of certain flow parameters. The functional architecture therefore specifies which functions are provided, under what constraints they are executed and how their execution is orchestrated by call structures and flow control parameters, and what initial conditions and states have to be observed before, during, and after the execution. The Section "Enterprise architecture" will introduce several modeling languages that can be applied to support the systems engineer with this task.

Functions have to be executed somewhere. Each function must therefore be allocated to a component—or a group of components—of the system. When functions are related by input, output, resources, or control parameters, the hosting components have to be connected via links or internal interfaces. Functions that are operationally related, because they support, for example, a common important task that is pivotal for mission success, may benefit from being hosted on neighbored components. To ensure the effective availability of pivotal functions, the systems engineer may use several redundant implementations. If the reliability of the components is not as high as required, the systems engineer may implement redundancy of such functions as well to be able to execute several instances of the function in parallel: the correct result will hopefully show up more often than any defective outcomes. To ensure the efficient availability of pivotal functions, the systems engineer may apply a maintainability implementation that shortens the amount of time required to restore a failed function to a fully operational state. All these aspects and more have to be taken into account when the physical architecture is designed.

In the final step of the general architecture development, the allocation takes place. Buede (2009) refers to this as the operational architecture or allocated architecture. This is where validation (was the correct system built, i.e., are all requirements fulfilled as specified in the acceptance plan) and verification (was the system built correctly, accuracy of specification transformations and implementation) can be conducted. Also, this is the place where many of the final trade-offs will take place.

In summary, SE has to capture the behavioral modeling (the functional architecture) and the structural hierarchy (the physical architecture) of a system. Various stakeholders in all life cycles will have different viewpoints on and interests in the system. The system architecture must represent all these viewpoints based on a common and consistent representation. If two viewpoints represent the same thing, this thing should be represented only once in the architecture repository. The architecture must be the common model of the system that spans over all life cycles and between all stakeholders.

Enterprise architecture

Enterprise architectures address more than one system and include all the aspects mentioned in the discussion of system of systems and Infranomics. Within systems architectures, the technical challenges are addressed by systems engineers in close collaboration with other members of the technical team as well as with the customer. Enterprise architectures have to address all aspects around these technical challenges. They have to address organizational aspects of the enterprise, business models, general technological trends, and the underlying enterprise philosophy, and they have to take the political, cultural, and environmental constraints into account in which the enterprise is embedded. As such, an enterprise architecture is much more than just a set of systems architectures, as the enterprise architecture combines the business aspects with the technological perspectives on the strategic level. It needs to present enterprise-level functions and capabilities required to reach strategic goals and objectives with enabling technology, resources (including data), and business and governance regulations. This allows high-level decisions regarding risk management, change management, and knowledge management as well. All of these decisions are usually communicated as contextual constraints to the systems engineers, but having an enterprise architecture that is conceptually aligned with the systems architectures allows a similar degree of consistency and communication of assumptions and constraints in the unambiguous form of models, methods, and tools. The enterprise architecture provides different perspectives of stakeholders on the strategic level with different abstractions to cope with the interrogatives of who, what, where, why, and when. Systems architectures provide the how in this context.

Modeling languages and architecture frameworks

There are many modeling languages in use when it comes to modeling a system and creating a systems architecture or an enterprise architecture.

They all have in common that they support the unambiguous representation of requirements, assumptions and constraints, and derived solutions on various levels. They all are rooted in models that can be expressed with mathematics. All methods are artificial languages with syntax and semantics. Cetinkaya et al. (2011) emphasize in their model-driven development approach the necessity to ensure accurate transformations between different models used in different phases. Tolk et al. (2013) propose the use of mathematical Model Theory to support the validation of models and their transformation.

There is likely a good reason for this variety of modeling languages. Each one was developed to support a certain expert group, and if they are still applied, it is likely because these experts are satisfied with the methods and tools. This heterogeneity of model languages and methods is not in itself a disadvantage. Only if the homogeneity of concepts cannot be ensured, as the model languages and methods are conceptually not alignable, they create a problem. But as the modeling languages can be expressed as mathematical formal languages, Model Theory can again help to identify gaps and discrepancies. These observations are true on the systems architecture level as well as on the enterprise architecture level. In a series on options for command and control system modeling, George Mason University published results on process modeling (Levis and Wagenhals, 2000), structured analysis (Wagenhals et al., 2000), and object-oriented approaches (Bienvenu et al., 2000).

In the remaining part of the Section "Modeling languages and architecture frameworks," a short introduction to the often-used modeling languages and frameworks in practice will be given.

IDEF0 and EFFBD

In 1970, the U.S. Air Force worked on the Integrated Computer-Aided Manufacturing (ICAM) project to improve their processes. The resulting ICAM definition was referred to as Integrated Definition and Function Modeling, which is more popularly known today as IDEF. They focused on process and data modeling with the objective to support the cohesive development within several development teams. The efforts were successful enough to be considered as a general approach, so that in 1993 the National Institute of Standards and Technology (NIST) published the Federal Information Processing Standard (FIPS) Publication 183: *Integration Definition for Function Modeling* (IDEF). Although the function, process, and data modeling parts of IDEF are best known and applied in many projects, IDEF today addresses a multitude of standards that are enumerated in Table 2.1. Some are fully developed (emphasized by bold fonts in Table 2.1), some are partially developed, and some are recognized to be important but with no implementation driven forward so far. In any case with IDEF, it is possible to recognize the various necessary viewpoints that have to be supported.

Table 2.1 IDEF Standards

Number	Title	Description
IDEF0	**Function modeling**	Subset of structured analysis and design technique (SADT) with focus on functional/process model
IDEF1	**Information modeling**	Information model supporting functions of the system
IDEF1X	**Data modeling**	Semantic data model using relational theory
IDEF2	**Simulation model design**	Dynamic model of the system
IDEF3	**Process description capture**	Process and object state-transition model of the system
IDEF4	**Object-oriented design**	Object-oriented design method as promoted by the Object Management Group
IDEF5	**Ontology description capture**	Describing ontological structures/the ontological spectrum
IDEF6	**Design rationale capture**	Partially developed method to document the rationale (how and why) for the design of an information system
IDEF7	Information system auditing	Defined as a required method, but has not been developed yet
IDEF8	Human-system interaction modeling	Partially developed method used to specify how the system and the users of the system interact
IDEF9	Business constraint discovery	Partially developed method used to assist in the identification and analysis of business constraints on the business of an organization
IDEF10	Implementation architecture modeling	Defined as a required method, but has not been developed yet
IDEF11	Information artifact modeling	Defined as a required method, but has not been developed yet
IDEF12	Organization modeling	Defined as a required method, but has not been developed yet
IDEF13	Three-schema mapping design	Defined as a required method, but has not been developed yet
IDEF14	Network design	Partially developed method used to support the design of computer networks

Note: Some standards are fully developed (emphasized by bold fonts).

The focus in this section is on IDEF0, which provides a very commonly used language for modeling functions in systems architectures. To describe the items exchanged between functions in the case of information technology (IT) systems—that is, modeling data and parameters—IDEF1X is very often used, as it describes data elements. Alternatively, IDEF4 is becoming more dominant in the IT domain, as it addresses object-oriented approaches that combine data and functions into software objects.

In any case, it is recommended that a systems engineer should consider several options, including alternatives within the IDEF family. In this section, IDEF0 will be described as the general function-modeling language that originates from the IT domain, but is not limited to that domain. IDEF0 is a subset of the structured analysis and design technique. Kim and Jang (2002) show the general applicability, including enterprise modeling challenges.

IDEF0 starts with the top-level function provided for the system that is designed or described. The function in IDEF0 is a black box that is described by a verb–noun phrase. Flow of items between functions—which can be material, personal, or data—is represented by arrows. The syntax of IDEF0 clearly defines how the flow is connected to the functions. A function transforms input into output, using resources, and observing constraints provided by the control parameters. Input always enters the function block, represented as a rectangle named with the verb–noun phrase describing the function, from the left. Output flows out of the block from the right. Mechanisms, often also called resources, enter the block from below, and control parameters enter from above. Figure 2.2 depicts the syntax of an IDEF0 function and the input, control, output, and mechanism (ICOM) arrows.

In recent versions, the call parameter was introduced as a special arrow under the mechanism category. A call leaves the function and

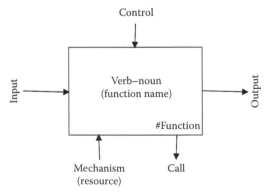

Figure 2.2 IDEF0 function.

calls another function. With a call, a function provides information for other functions. In particular in systems that mix IT functions and other system functions, that is, IT provides a supporting role when producing something else, systems engineers prefer to use calls to communicate data and outputs to exchange items between functions.

For the highest level function, the incoming parameters are provided by the context of external systems, and the outgoing output is used by an external system (it makes no sense that output is generated without knowing for what and for whom, that is why one cannot generate output for the context of a system). The highest function is next broken down into subfunctions. If the upper level function was, for example, to prepare a barbeque dish, proper subfunctions would be to grill the meat, prepare a side dish, and take care of salad and dressing. Of importance here is that items can neither be created out of nothing nor can they disappear into nothing: whatever is needed must either be brought in from the outside or must be produced by an internal function.

IDEF0 defines a numbering scheme for functions—as well as pages where these functions are specified—that gives the systems engineer an immediate idea of where the function plays a role. The system function is referred to as A0. The next level of function that results from decomposing A0 consists of functions A1, A2, A3, and so on. Each of these functions can be decomposed as well. The decomposition of A1, for example, leads to the functions A11, A12, A13, and so on. Although not required by the standards, it is a good practice to use a similar number schema for the ICOM parameters as well.

IDEF0 is a very powerful tool for handling the functional complexity by decomposition. However, it only addresses the static hierarchy of functions, not the dynamic behavior of functions. This behavioral aspect is very often supported by using the functional flow block diagram (FFBD) or the extended functional flow block diagram (EFFBD) as a complementary modeling method. An EFFBD diagram contains all information of an IDEF0 diagram, but both have unique perspectives supporting different kinds of users. While IDEF0 builds a taxonomy of functions within the system, EFFBD focuses on the execution logic. It is therefore a good practice to support both as different viewpoints within a common, consistent repository. EFFBD connects functions that are executed on the same level with arrows as well, but adds control structures to allow the expression of whether functions are executed as a series or concurrently. If they are executed concurrently, the systems engineer can define whether all functions are executed in parallel (AND), or whether only a selection of functions is executed (OR). Finally, loops (LP) and iterations (IT) are introduced as well. EFFBD supports the same naming convention and numbering schema as IDEF0 does. In addition to the logic, parameters that flow between the functions can be shown as well, although EFFBD does not differentiate between input, control, or control parameters, they all flow into the function. EFFBD has sufficient information to support the automatic generation of a discrete event simulation, although internal timing of the functions needs to be assumed or generated. More details will be discussed in the Section "Simulation" of this chapter.

Similar to the numbering scheme introduced for the functions of IDEF0, a clear scheme for numbering the pages and sheets describing the functions and the system is standardized as well. The pages also compose standardized form of the metadata information that communicates the maturity of the solution, responsible engineers, and so on. The resulting IDEF diagram form therefore minimizes the structure and constraints supporting only the functions that are important to the discipline of structured analysis, these being the establishment of context, cross-referencing between pages, and notes about the content of each page or sheet. The diagram form is a single standard size for ease of filing and copying. It includes the working information on top, the message field in the center, and the identification fields at the bottom. Everything written in an IDEF0 project follows this standard so that every systems engineer knows immediately what he or she is looking at when receiving a part of the system architecture documentation.

In summary, IDEF0 and EFFBD provide a well-understood method and language for describing functional decomposition and dynamic behavior. Augmented by an unambiguous description of the items that flow between the functions (such as IDEF1X when these are data entities) it is a well-established way to define and describe systems.

Unified Modeling Language and System Modeling Language

This section looks at three modeling concepts driven by object-oriented approaches to systems architecture methods and languages. The Unified Modeling Language (UML) and the System Modeling Language (SysML) are both administered by the Object Management Group (OMG).

The traditional view of systems modeling distinguishes between static views describing the structures and functional views describing the behavior exposed. These views describe what entities are modeled (as classes describing the types and objects describing the instances of things), how entities react when they receive input (as the state changes, which can be broken down in case the entity is a composite made up of several smaller entities), and how entities interact with each other (in the form of activities in which more than one entity work together and exchange messages with each other). UML extends these categories for object-oriented software engineering resulting in the diagrams shown in Figure 2.3.

The diagrams of UML are used to either explain the structure or the behavior of a system. UML supports the object-oriented paradigm and,

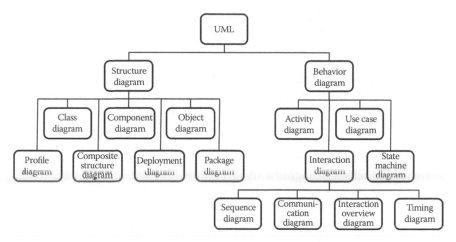

Figure 2.3 UML diagram hierarchy.

as such, distinguishes between classes that represent type information (properties shared by all things of this type) and objects that represent instance information (properties exposed by the individual instantiations of a thing of a given type).

The structure diagrams deal with modeling the objects representing the entities that make up the system. The atomic entity is an object that has only attributes and methods assigned. Using such objects, composites can be built that combine *atomic* objects and potentially already existing composites. More diagrams are used to add more ideas to facilitate the work of object-oriented software engineers, and many of them can be used by systems engineers as well. Here is an overview of the structure diagrams:

- Class diagrams describe the entity types including their attributes relationships between the types.
- Object diagrams show views of the system with instantiated objects based on their types.
- Composite structure diagrams describe internal structure of composite objects, that is, objects that have at least one object among their attributes. This allows the systems engineer to treat the composite as one structure from the outside and describe how the composite works on the inside.
- Component diagrams describe how the system is split up into components. These diagrams show interdependencies and interactions between these components.
- Package diagrams describe logical groupings and their interdependencies and define software packages.

- Deployment diagrams describe the allocation of software to hardware components: what software package is executed on which hardware component.
- Another diagram type that has been added recently is named the *profile diagram*. This profile diagram operates at the meta-model level and shows stereotypes. Each object that fulfills the requirements of the stereotype can be used in the operation in the respective role.

The most important three diagrams of the functional view of the systems are activity diagrams (how entities interact), state machine diagrams (how entities react), and use case diagrams (how the overall system behaves). These three core diagrams are supported by more detailed diagrams that often focus on software engineering details. The behavior diagrams are described as follows:

- Use case diagrams describe the interactions of actors with the main functionality of the simulation system in terms of the user. These diagrams are often used to communicate what the system does and why it does it with the users.
- Activity diagrams can be best described as workflows between the entities of the simulation system. They show the overall flow of control, including synchronization between activities, interim results, and so on.
- State machine diagrams show how the objects or entities react to certain inputs they receive. While activities focus on the entity, external communication state machine diagrams focus on the internal changes.
- Interaction diagrams use four subtypes to better describe the details of interactions within the system: (1) interaction overview diagrams provide the big picture and build the frame for the other three diagram types to fill in the details; (2) sequence diagrams visualize causal orders of an interaction and define the lifespan of a temporary object used within such interactions; (3) timing diagrams focus on timing constraints of sequences to define the temporal orders; and (4) communication diagrams specify the interactions between entities that are part of the modeled sequence. They are very often used as system internal use cases, as they show several aspects of which entities come together how and when.

UML has been applied in a variety of communities and has become something like the lingua franca between modelers that specify software intensive systems. As such, the systems engineer needs to be able to read and interpret the associated artifacts. The main disadvantage of UML, however, is the focus on object-oriented software systems. Although UML easily can be interpreted to support other systems as well, the software engineering roots are apparent and ubiquitous.

These observations led to the development of SysML, also under the umbrella of the OMG. SysML is a general modeling language for systems and is receiving more and more support from systems engineers. There is a huge overlap between UML and SysML, but the focus is on the system, not on the software. As such, SysML started by stripping all software-specific diagrams from UML, modifying the others in case of need, and adding system-specific new diagrams where needed. The resulting diagram types are shown in Figure 2.4.

In Figure 2.4, the diagrams with the standard frame are identical to the UML diagrams with the same name; those with the bold frame were modified and those with the dashed bold frame are new. The differences between UML and SysML can be summarized as follows:

- Requirement diagrams represent a visual modeling of system requirements, which are pivotal for SE.
- Parametric diagrams show the relations between parameters for system components in all levels.
- Block definition diagrams are based on the class diagrams, but they use system blocks instead of classes. System block diagrams are well known by systems engineers and are easy to map to class representations.
- Internal block diagrams extend the composite structure diagrams of UML by the application of restrictions and extensions that govern system component interactions and interrelations.
- Activity diagrams were extended to allow modeling of continuous flows, a capability which is often needed in systems. Finally, logical operators are introduced to allow the use of EFFBD.

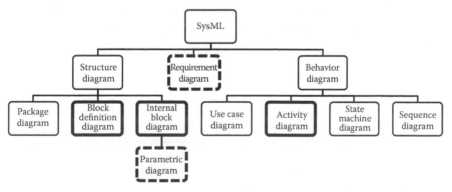

Figure 2.4 SysML diagram hierarchy.

Of particular interest are the requirement and parametric diagrams. They allow one to integrate traceability of requirements into the architecture. Furthermore, the parametric diagram allows one to define metrics that have to be applied to evaluate the performance of the system as well as to define thresholds that have to be met when validating and verifying the system. Integrating this idea of the prime of requirements into the language is one of the biggest advantages of SysML when compared to other approaches.

Another aspect of SysML is the use of system diagram frames that are part of every artifact. Similar to the diagram forms described for IDEF0, they describe the metainformation needed to better understand what the diagram shows in the context of the whole system. The header specifies the version, what kind of diagram is used, what kind of model element is used (block, activity, interaction), whether the diagram is complete or has elements elided, and so on. In addition, there is also a systematic standardized numbering scheme that allows the engineer to understand with one number on which system level the observed artifact is located. In summary, SysML utilizes the best of both worlds, namely the rigor of IDEF0 and the flexibility of UML.

Discrete event systems specification

There are numerous efforts in play on the discrete event systems specification (DEVS), in particular the foundational work of Zeigler et al. (2000). It is well known as a simulation formalism and also used to define software architectures for simulation systems, but not many systems engineers are aware of the system architecture roots of DEVS. Nonetheless, DEVS can be used to define and describe systems architectures, not only for virtual systems. Mittal and Risco Martin (2013) only recently addressed the opportunities to support SE as well as SoSE with the DEVS Unified Process (DUNIP) that adds a general SE process to the well-known formalism.

The DEVS formalism builds models from the bottom up. The most basic behavior of system components is captured in atomic DEVS components that are defined by the following sets and functions:

- A set of input events X
- A set of output events Y
- A set of states S (states in which the atomic DEVS component can be in)
- The time advance function t_a (that defines how long the component remains in a state)
- The external transition function δ_{ext} (that defines how an input event changes the state of the system)

- The internal transition function δ_{int} (that defines how a state of the system changes internally)
- The output function λ (that defines how a state of the system generates an output event)

All state transitions are not only governed by the input, but they are also governed by the time advance function. Therefore, the atomic DEVS component is well defined by the seven-tuple of the four sets and the three functions: $< X, Y, S, t_a, \delta_{ext}, \delta_{int}, \lambda >$. For parallel execution, authors also include an additional confluent transition function that decides the next state in cases of collision between external and internal events or multiple external events. Figure 2.5 depicts an atomic DEVS component.

Coupled DEVS use DEVS components and couple them to represent the structure of the system. The lowest layer must be atomic DEVS components, but several layers are allowed. In general, coupled DEVS is defined by an eight-tuple:

- A set of input events X
- A set of output events Y
- A set of names for the subcomponents D (that is used for unambiguous identification of subcomponents)
- A set of subcomponents C_i (that can be atomic DEVS components or coupled DEVS components and that are named by D)
- Mapping M_{EiIi} of external inputs to internal inputs (all external input events must be mapped to internal input events via ports of the components)
- Mapping M_{IoIi} of internal outputs to internal inputs (as the result of one internal component can become the input for another internal component)
- Mapping M_{IoEo} of internal outputs to external outputs (all results must be produced by an internal component)
- A tie-breaking function that defines which event to select in case of simultaneous events, which is equivalent to the confluent function of the atomic DEVS component.

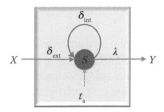

Figure 2.5 Atomic DEVS.

The behavior is exclusively defined by the state changes and the related time on the atomic DEVS level. If these state changes are not modeled deterministically but are modeled stochastically instead, DEVS becomes stochastic as well. The change of input parameter to output parameters as well as the required time for the execution of this activity is exclusively done by the atomic DEVS components. Atomic and coupled DEVS together build the classic sequential DEVS approach. With the advancements in computer technology, there have also been numerous extensions, explained with examples in Wainer (2009), such as parallel DEVS, dynamic DEVS, cellular DEVS, and more. Although not as mature as IDEF0 or SysML, DEVS has the potential to be used as a systems architecture language, in particular when supported by the DUNIP (Mittal and Risco Martin, 2013). As several simulation development frameworks have been proposed based on the DEVS formalism, a bridge between systems architecture and simulation is in particular easy for DEVS-based system architecture approaches.

Architecture frameworks

To go into the details of architecture frameworks goes far beyond the scope of this chapter, but at least the basic ideas need to be addressed in the context of SE, architecture, and simulation. So far, the focus has been on methods and languages. Architecture frameworks add an additional aspect to these efforts by providing *blueprint* schemas that guide the systems engineer regarding what needs to be specified, not just how to specify it.

The U.S. defense domain uses the Department of Defense Architecture Framework (DoDAF) in support of their SE efforts. Several other nations are using slight variations of this framework; for example, NATO has the NATO Architecture Framework (NAF), the United Kingdom has the Ministry of Defense Architecture Framework (MoDAF), and Canada uses the Department of National Defense and Canadian Forces Architecture Framework (DNDAF). To develop a common modeling method in support of these main frameworks, OMG defined the Unified Profile for DoDAF/MODAF (UPDM) mainly using UML products as the common denominator. The definition for DoDAF is freely distributed via the Web portal of the DoD deputy chief information officer (http://dodcio .defense.gov/dodaf20.aspx).

DoDAF describes concepts and models defining the six core processes within its metamodel (DM2). The supported core DoD processes are (1) capabilities integration and development; (2) planning, programming, budgeting, and execution; (3) acquisition system; (4) SE; (5) operations planning; and (6) capabilities portfolio management. To support the various technical team members, stakeholders, and other partners, DoDAF defines viewpoints that each provides a specific perspective on components, data, and processes. The following viewpoints are recommended by the DoDAF documentation, but additional user specific viewpoints can be created to make the support fit for its purpose:

- *All Viewpoint* describes the overarching aspects of architecture context that relate to all viewpoints.
- *Capability Viewpoint* articulates the capability requirements, the delivery timing, and the capability deployment.
- *Data and Information Viewpoint* articulates the data relationships and alignment structures in the architecture content for the capability and operational requirements, SE processes, systems, and services.
- *Operational Viewpoint* includes the operational scenarios, activities, and requirements that support capabilities.
- *Project Viewpoint* describes the relationships between operational and capability requirements and the various projects being implemented. The project viewpoint also details dependencies among capability and operational requirements, SE processes, systems design, and services design within the defense acquisition system process.
- *Services Viewpoint* articulates the performers, activities, services, and their exchanges, providing for or supporting operational and capability functions. Services are the backbone for netcentric solutions.
- *Standards Viewpoint* articulates the applicable operational, business, technical, and industry policies, standards, guidance, constraints, and forecasts that apply to capability and operational requirements, SE processes, systems, and services.
- *Systems Viewpoint* is the description for solutions articulating the systems, their composition, interconnectivity, and context providing for or supporting operational and capability functions. The systems viewpoints complement the services viewpoint to better integrate legacy descriptions and systems that are not netcentric.

On the civilian side, at least four frameworks have to be mentioned that are widely distributed and applied. Sessions (2007) describes them in his whitepaper as follows:

- The Zachman Framework for Enterprise Architectures: Although self-described as a framework, it is actually more accurately defined as a taxonomy.
- The Open Group Architectural Framework (TOGAF): Although called a framework, it is actually more accurately defined as a process.

- The Federal Enterprise Architecture (FEA): It can be viewed as either an implemented enterprise architecture or a proscriptive methodology for creating an enterprise architecture.
- The Gartner Methodology: It can be best described as an enterprise architectural practice.

As the core processes that have to be supported are not as easily characterized as is the case for the defense domain, these frameworks are all more generic and address the need to create viewpoints in support of business, governance, maintenance, IT support, data support, and more from various perspectives. However, at the end all have the same objective: To make sure that all business relevant aspects are supported by appropriate viewpoints derived from the common and consistent repository captured in the form of an architecture that drives the viewpoints for the various team members of all phases of the life cycle of a relevant system.

Simulation

There are several selected papers that address the need to look at SE and M&S as mutually supportive disciplines, going back to early vision papers, such as those by Ören (1971) and Heil (1985). It is also possible to include papers addressing the issue of mutual support directly, such as those by Ören and Yilmaz (2006). Other papers address the need to apply the SE process and systems architecture for simulation as well, in particular in the domain of defense and security (Gallant and Gaughan, 2010). A third group is looking at simulation support for SE, such as Lackey et al. (2007). All of these are valid viewpoints. However, the focus in this chapter is the direct use of systems architecture artifacts that are developed within the SE process to be used in simulation, also known as executable architectures, such as that earlier envisioned by Pawlowski et al. (2004), and recently implemented as described by Mittal (2006), Wagenhals and Levis (2009), Helle et al. (2013), or Garcia and Tolk (2013).

Types of executable architectures

There is no generally agreed definition of executable architecture in the current literature, but three types emerge in the various descriptions. The following three types all use a simulation to execute an architecture, but they differ in the degree of automation and use of the systems architecture artifacts to do so:

- *Native Executable Architectures* use executable languages, such as executable UML, to describe the architecture. Architecture development environments with an integrated *simulation* option fall into the same category. They execute entities, relations, events, and behavior exactly as specified in the architecture. No additional software or integration effort is necessary, but as the context of the architected system is not simulated, only the function within the system can be evaluated (measure of performance [MOP]: how well are the system functions executed?). It is not possible to evaluate the system in the context of an operation or within a portfolio of systems (measure of effectiveness [MOE]: how well does the system contribute to the overall mission success?).

* *Federated Executable Architectures* use the information defined in the system architecture to generate an executable federate with external interfaces. While all internal elements represent the architectural definition one by one, the federate can be embedded into a simulation environment that provides the valid context for the system. Although this approach allows one to determine MOPs and MOEs, it is also more complex and challenging, as the systems engineer has to understand all components as well as how to federate them and evaluate them afterwards.

* *Calibrated Executable Architectures* use an existing simulation system that already provides all entities, relations, events, and behavior specified in the systems architecture so that only the parameters have to be calibrated accordingly to the architecture specifications. The degree of automation can reach from fully automated to the need to set up all scenario data and systems parameters manually. This allows one to collect MOPs and MOEs as provided by the simulation system without having to support a federation, and this simplifies the data collection for the evaluation. The disadvantage is that the simulation system concepts constrain the systems engineer, as only what is built into the simulation by its developers can be used, and there may be significant differences in scope, structure, and resolution of entities, relations, and processes.

Federated and calibrated executable architectures require a simulation to represent the context and eventually the specified system as well. Which will be the best simulation system for this effort needs to be answered in the context of the customers' question. The decision criteria can be derived from the *Guide for Understanding and Implementing Defense Experimentation* (GUIDEx) (TTCP 2006). Although the application domain is defense and security, the general principles are applicable to all other application domains as well. Guiding questions regarding the simulation system in comparison with the system architecture specifications are as follows:

* Are the systems of interest represented in sufficient detail?
  * Characteristic attributes
  * Appropriate resolution

- Are the events/effects represented in sufficient detail?
  - Events/effects created by the system (actor)?
  - Events/effects directed at the system (victim)?
- Is the expected behavior represented in sufficient detail?
  - State changes when an event is created?
  - State changes when an event is received?

In addition to these questions, the GUIDEx further defines the following four requirements for valid experiments that have to be fulfilled by a suitable simulation system as well:

1. The new or alternative capability must be modeled in the system.
2. It must be possible to detect a change in effects when the new or alternative capability is used.
3. It must be possible to isolate the reason for the change in effect that is observed.
4. It must be possible to relate the simulation results to actual operations.

As a good practice for the systems engineer, it is helpful to enumerate the entities with attributes contained in the scenarios, the relations, the processes and behaviors, the MOP for the system functions, and the MOE for the operational effectiveness of the system and simply compare those with the modeled properties in the simulation system. If they are sufficiently similar, the system is likely to be a good fit. If this is not the case, the system may still be applied by the systems engineer to support the work, but some engineering work will be necessary.

Examples of executable architectures

The easiest way to reach an executable architecture is to use an executable language and method to model the system resulting in a system architecture. If, for example, UML is used the systems engineer can take advantage of the work toward executable UML (xUML), as described—among others—by Meller and Balce (2002). The OMG is working on a standardized version, so that systems architectures modeled with xUML will inherently be executable.

Regarding Dickerson and Mavris (2010), the composable object (COB) knowledge presentation was developed at the Georgia Institute of Technology originally representing and integrating design models with diverse analysis models. This method can be used to enrich SysML to make it executable as well, as the COB can represent the blocks in SysML and be calibrated using the other diagrams accordingly. The mapping of the various SysML diagrams has not been standardized, but the first prototypes are promising to support native executable architectures even better than xUML, as the parametric diagram allows for a native representation of MOPs within the system architecture.

Shuman (2011) evaluated several system architecture languages and methods regarding their contribution to executable architectures by populating a multidimensional analysis framework. He evaluated architecture elements, which were mainly derived from the UPDM, that leverage modeling language descriptions, conform to M&S formalisms, and that form a baseline for an executable architecture specification. The modeling language descriptions, which were derived from decomposition and which represent the results in a common format of IDEF, UML, SysML, and BPMN, instantiate architecture elements, conform to M&S formalisms, and inform the executable architecture specification. The M&S formalism DEVS informs the modeling language descriptions, informs the architecture elements, and validates the executable specifications. The resulting architecture specification leverages the modeling specifications, expands the baseline architecture elements, and conforms to M&S formalism. The research produced a detailed metamodel for executable architectures, which is based on architecture elements and relationships derived from the UPDM architecture framework, and elements and relationships from key modeling languages (UML, SysML, BPMN, and IDEF). The metamodel was subjected to a plausibility analysis through comparison with M&S formalisms (DEVS and colored Petri nets) elements and relationships. A result of the analysis of practical interest was that no modeling approach covered all required components of Shuman's framework, but SysML comprised most of the elements required, so that this can be a valuable baseline for a comprehensive definition of executable architectures.

The Framework for Assessing Cost and Technology (FACT) is a web-based tool which utilizes SysML-based standards for structuring and presenting content for analysis. Ender et al. (2012) describe the framework that can contribute to as well as benefit from the proposal given in this chapter, as all artifacts captured in this chapter can become elements of FACT as a general integration framework. In addition, FACT can provide all other components it uses to the following methods discussed.

Simulation and architecture validation

Simulation as a means to validate system requirements and the resulting system architecture could easily fill a whole chapter if not a book. In the context of this chapter we only address some general issues, which in no sense shall diminish the importance of this topic.

It is good practice to use a simulation to validate the system requirements before going on with design and production. As stated in the examples in this section, executable architectures help to identify logical loops in the system's functionality, they can be used to identify dead locks, and when executed within an operational context even conceptual challenges. In some cases, simulation may be the only way to validate requirements, such as for nuclear systems or missile defense systems in which it is prohibitive to use or test the real system itself. Whenever the test environment for the real system is too dangerous, too costly, or physically not accessible, simulation is the only way to test the validity of requirements and the resulting system architecture. Heesbeen et al. (2003) give an example on using a traffic simulation to test new air traffic management system concepts. Maturana et al. (2005) describe the use of an agent-based simulation environment to validate the architecture for a distributed manufacturing organization.

Summary and conclusions

Within this chapter, we showed that the SE processes need to be aligned and synchronized to support a variety of technical team members and stakeholders and all phases of the life cycle of a system. The variety of interested and affected groups increases with the complexity of the system as well as with the complexity of the context, such as SoSE and Infranomics.

The proposed solution is the seamless integration of SE process knowledge within the systems architecture. The architecture becomes the common and consistent knowledge repository. Model-based systems engineering is already recommending the use of models instead of documentation to support the SE processes. We support these ideas with the proposal in this chapter, but recommend furthermore to use the system architecture framework used by the community as the foundation for this repository instead of introducing new methods not yet established among practitioners of the field.

Various modeling languages and methods can be used, and if the artifacts are rich enough, they can be executed. Executable architectures can help to identify shortcomings of systems in dynamic situations before the systems are built, making the process safer and less costly.

In summary, the alignment of processes ensures collaboration and synergy between the groups of team members and stake holders, complementing instead of contradicting each other. The use of a common system architecture as the system knowledge repository ensures knowledge transfer between the phases and groups. The data collected by the aligned SE processes in the common system architecture will be rich enough to be executed to deliver numerical insights into the dynamic behavior of the systems and its integration into the operational environment, integrating all viewpoints provided by the different team members and stake holders: technical, managerial, financial, operational, and so on.

Such a comprehensive approach synthesizing SE, architecture, and simulation should be integrated into academic curricula as well as in practitioners' continuous education.

References

Alberts, D. S., Huber, R. K., and Moffat, J. (2010). *NATO NEC C2 Maturity Model.* Command and Control Research Program (CCRP) Press, Washington, DC. Available at http://www.dodccrp.org/files/N2C2M2_web_optimized.pdf; last accessed September 2013.

Bienvenu, M. P., Shin, I., and Levis, A. H. (2000). C4ISR architectures: III. An object-oriented approach for architecture design. *Systems Engineering* 3(4): 288–312.

Buede, D. M. (2009). *The Engineering Design of Systems: Models and Methods*, 2nd Edition. Wiley, New York.

Cetinkaya, D., Verbraeck, A., and Seck, M. D. (2011). MDD4MS: A model driven development framework for modeling and simulation. *Proceedings of the Summer Computer Simulation Conference*, pp. 113–121. SCS Press, San Diego, CA.

Chigani, A. and Balci, O. (2012). The process of architecting for software/system engineering. *International Journal of System of Systems Engineering* 3(1): 1–23.

Churchman, C. W. (1968). *The Systems Approach*. Delacorte Press, New York.

Dickerson, C. E. and Mavris, D. N. (2010). *Architecture and Principles of Systems Engineering*. CRC Press, Taylor & Francis Group, Boca Raton, FL.

Ender, T. R., Browne, D. C., Yates, W. W., and O'Neal, M. (2012). FACT: An M&S Framework for Systems Engineering. *Proceedings of the Interservice/Industry Training, Simulation and Education Conference* (I/ITSEC 2012), Paper 12115, 12 pages.

Estefan, J. A. (2008). Survey of model-based systems engineering (MBSE) methodologies. *International Council on Systems Engineering*, San Diego, CA.

Gallant, S., and Gaughan, C. (2010). Systems engineering for distributed live, virtual, and constructive (LVC) simulation. *Proceedings of the Winter Simulation Conference*, pp. 1501–1511. IEEE Press, Piscataway, NJ.

Garcia, J. J. and Tolk, A. (2013). Executable architectures in executable context enabling fit-for-purpose and portfolio assessment. *Journal of Defense Modeling and Simulation*; DOI: 10.1177/1548512913491340.

Gheorghe, A. V. and Masera, M. (2010). Infranomics. *Critical Infrastructures* 6(4): 421–427.

Heesbeen, W. W. M., Hoekstra, J. M., and Clari, M. V. (2003). Traffic manager: Traffic simulation for validation of future ATM concepts. *AIAA Modeling and Simulation Technologies Conference & Exhibit*, Austin, TX.

Heil, D. E. (1985). Simulation and systems engineering: An affirmation. *Proceedings of the 18th Annual Symposium on Simulation* (ANSS'85), pp. 29–47. IEEE Press, Piscataway, NJ.

Helle, P. S., Giblett, I., and Levier, P. (2013). An integrated executable architecture framework for system of systems development. *Journal of Defense Modeling and Simulation*; DOI: 10.1177/1548512913477259.

International Council on Systems Engineering (2011). *INCOSE Systems Engineering Handbook* (Version 3.2.2). Report INCOSE-TP-2003-002-03.2.2. San Diego, CA. Available at http://www.incose.org/ProductsPubs/products/sehandbook.aspx; last accessed September 2013.

Jamshidi, M. (2009). *System of Systems Engineering: Innovations for the 21st Century*. Wiley, New York.

Kim, S.-H. and Jang, K.-J. (2002). Designing performance analysis and IDEF0 for enterprise modelling in BPR. *International Journal of Production Economics* 76(2): 121–133.

Lackey, S. J., Harris, J. T., Malone, L. C., and Nicholson, D. M. (2007). Blending systems engineering principles and simulation-based design techniques to facilitate military prototype development. *Proceedings of the Winter Simulation Conference*, pp. 1403–1409. IEEE Press, Piscataway, NJ.

Levis, A. H. and Wagenhals, L. W. (2000). C4ISR architectures: I. Developing a process for C4ISR architecture design. *Systems Engineering* 3(4): 225–247.

Maier, M. (1998). Architecting principles for systems-of-systems. *Systems Engineering* 1(4): 267–284.

Maier, M. W. and Rechtin, E. (2007). *The Art of Systems Architecting*, 3rd Edition. CRC Press, Taylor & Francis Group, Boca Raton, FL.

Maturana, F. P., Staron, R., Hall, K., Tichý, P., Šlechta, P., and Mařík, V. (2005). An intelligent agent validation architecture for distributed manufacturing organizations. *Emerging Solutions for Future Manufacturing Systems*, Springer Series on International Federation for Information Processing 159, pp. 81–90. Springer-Verlag, Berlin/Heidelberg, Germany.

Meller, S. J. and Balce, M. J. (2002). *Executable UML: A Foundation for Model-Driven Architecture*. Addison-Wesley, Menlo Park, CA.

Mittal, S. (2006). Extending DoDAF to allow integrated DEVS-based modeling and simulation. *Journal of Defense Modeling and Simulation* 3(2): 95–123.

Mittal, S. and Risco Martin, J. L. (2013). *Netcentric System of Systems Engineering with DEVS Unified Process*. CRC Press, Taylor & Francis Group, Boca Raton, FL.

National Aeronautics and Space Administration (NASA). (2007). *Systems Engineering Handbook*. NASA/SP-2007-6105 Rev 1. NASA Headquarters, Washington, DC.

Ören, T. I. (1971). GEST: A combined digital simulation language for large-scale systems. *Proceedings of the Association Internationale pour le Calcul Analogique*; Symposium on Simulation of Complex Systems, pp. B-1/1–B-1/4, Tokyo, Japan.

Ören, T. I. and Yilmaz, L. (2006). Synergy of systems engineering and modeling and simulation. *Proceedings of the SCS International Conference on Modeling and Simulation—Methodology, Tools, Software Applications* (M&S-MTSA'06), pp. 10–17, Calgary, Canada.

Pawlowski, T., Barr, P., Ring, S., Williams, M., and Segarra, S. (2004). Executable Architecture Methodology for Analysis. MITRE Technical Report MTR 03W0000081. Washington, DC.

Sage, A. P. and Armstrong Jr., J. E. (2000). *Introduction to Systems Engineering*. Wiley, New York.

Sage, A. P. and Cuppan, C. D. (2001). On the systems engineering and management of systems of systems and federations of systems. *Information, Knowledge, Systems Management* 2(4): 325–345.

Sessions, R. (2007). A Comparison of the Top Four Enterprise-Architecture Methodologies. Microsoft Developer Network Library. Available at http://msdn .microsoft.com/en-us/library/bb466232.aspx; last accessed September 2013.

Shuman IV, E. A. (2011). Understanding the Elements of Executable Architectures through a Multi-Dimensional Analysis Framework. PhD Dissertation. Batten College of Engineering, Old Dominion University, Norfolk, VA.

Technical Cooperation Program (TTCP). (2006). *Guide for Understanding and Implementing Defense Experimentation* (GUIDEx). TTCP Report JSA AG-12. Ottawa, Canada.

Tolk, A. (2003). Beyond technical interoperability: Introducing a reference model for measures of merit for coalition interoperability. *Proceedings of the 8th International Command and Control Research and Technology Symposium.* Command and Control Research Program (CCRP) Press, Washington, DC.

Tolk, A., Diallo, S. Y., King, R. D., and Turnitsa, C. D. (2009). A Layered Approach to Composition and Interoperation in Complex Systems. *Complex Systems in Knowledge-based Environments: Theory, Models and Applications*, edited by Tolk, A. and Jain, L. J., SCI 168, pp. 41–74. Springer Verlag, Berlin, Germany.

Tolk, A., Diallo, S. Y., Padilla, J. J., and Herencia-Zapana, H. (2013). Reference modeling in support of M&S—Foundations and applications. *Journal of Simulation* 7(2): 69–82.

Wagenhals, L. W. and Levis, A. H. (2009). Service oriented architectures, the DoD architecture framework 1.5, and executable architectures. *Systems Engineering* 12(4): 1520–6858.

Wagenhals, L. W., Shin, I., Kim, D. and Levis, A. H. (2000). C4ISR architectures: II. A structured analysis approach for architecture design. *Systems Engineering* 3(4): 248–866.

Wainer, G. A. (2009). *Discrete-Event Modeling and Simulation: A Practitioner's Approach.* CRC Press, Taylor & Francis Group, Boca Raton, FL.

Zeigler, B. P., Praehofer, H., and Kim, T. G. (2000). *Theory of Modeling and Simulation*, 2nd Edition. Academic Press, San Diego, CA.

chapter three

System modeling
Principled operationalization of social systems using Presage2

Sam Macbeth, Dídac Busquets, and Jeremy Pitt

Contents

Introduction

There are (at least) three reasons to model and simulate a social system. First, to understand and explain how the social system works, for example, to investigate the effect of policy change on a large population, or to understand some other property or phenomenon, such as a financial system (Christensen et al. 2002), or the spread of information in a social network (Draief and Ganesh 2010). Second, to formalize human methods of problem-solving as a source and inspiration for developing computational systems to address engineering challenges (Edmonds et al. 2005), an example being trust in open multiagent systems (Castelfranchi

43

and Falcone 1998). Third, to represent and reason about the sociological concepts involved, as the basis for implementing sociotechnical or cyber-physical systems (Ferscha et al. 2012; Hoffman et al. 2009), in which the technical components understand the same concepts as the human ones (commonly required in workflow, legal systems, serious games, etc.).

In this chapter, we describe the implementation of the multiagent-based animation and simulation platform Presage2. This tool has been developed as an experimental platform to support the process of *principled operationalization*, defined here as the mapping between formal models of social systems and implemented systems. The use of Presage2 is demonstrated through two applications of modeling and simulating social systems, one involving Rescher's theory of Distributive Justice (Rescher 1966), and the other the Kyoto Protocol (United Nations [UN] 1998). These examples provide a demonstration of the first two reasons for modeling and simulating social systems, and we also briefly discuss the third reason, modeling and simulating sociotechnical systems.

We begin with a brief description of methodology for engineering artificial social systems, called sociologically inspired computing. Following that, our two examples of social systems, Distributive Justice and the Kyoto Protocol, are given. The system Presage2 is introduced, and we discuss its architecture and operational basis. A computational model of each formal model of the two social systems is then implemented in Presage2: this is the process of principled operationalization. Finally, we summarize and draw conclusions on our methodology.

Methodology

The methodology we employ is called *sociologically inspired computing* and is illustrated in Figure 3.1 (Jones 2013). This method builds on the synthetic method underlying research in artificial societies and artificial life (Steels

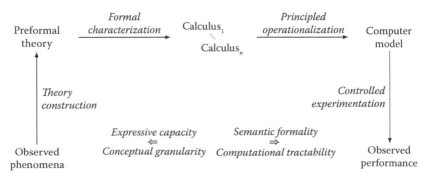

Figure 3.1 Methodology for sociologically inspired computing.

and Brooks 1994), the formalization of social relations (such as trust) in multiagent systems (Neville and Pitt 2004), and other attempts to apply ideas from the social sciences to the design of computational systems (Edmonds et al. 2005).

To apply the methodology, we start from an observed phenomenon, for example, from the social sciences, a human social, legal, or organizational system. The process of theory construction creates a preformal *theory*, usually specified in a natural language, although there may be some formal analysis. The process of formal characterization represents such a theory in a calculus of some kind, whereby calculus we mean any system of calculation or computation that is based on symbolic representation and manipulation. This representation can be at different levels of abstraction depending on the intended role of the calculus: expressive capacity or conceptual granularity with regard to theory; computational tractability or declarative semantics with regard to implementation. The step of principled operationalization embeds such formal representations in simulations that can include both a detailed implementation of individual agents, and/or a treatment of large populations. The computer model can be animated or executed in a series of controlled experiments, and the performance of the model can be observed.

The method is also similarly structured to the CosMos methodology (Andrews et al. 2010) for biologically inspired computing. In biologically inspired computing, the aim is to use biological sciences as the source of inspiration for algorithms to address engineering challenges. For example, insect swarms, or a process such as genetics, embryonics, or the auto-immune response system, have provided inspiration for much research in network traffic management, swarm robotics, evolutionary computing, and autonomic computing. The CosMos methodology also identifies three phases: discovery, development, and exploration. This corresponds to the steps of theory construction, formal characterization, and principled operationalization, which in our methodology includes controlled experimentation.

In the following, we apply our general sociologically inspired computing methodology to two specific examples of social systems, first defining a formal model, then introducing a tool with which to implement an operational model.

Examples of social systems

In this section, we give two examples of the kind of social system that we propose to model formally. We first present a system where we take a human method of problem-solving, in the form of Rescher's canons of Distributive Justice (Rescher 1966) and use it to develop a computational system to address the problem of resource allocation in open systems.

We then present an example of modeling a social system, in this case the Kyoto Protocol, to understand and predict its properties and effects.

Distributive Justice: Formal model

In the work of Pitt et al. (2011), we addressed the problem of resource allocation in open systems, in which we used a formal axiomatization of Ostrom's principles of self-governing the commons (Ostrom 1990) to define a self-organizing electronic institution. The members (agents) of such an institution used conventional rules to both select and self-organize the roles, rules, and parameters for implementing a resource allocation method. However, given a set of agents A, a set of divisible resources R, and mapping from A to (some subset of) R, a natural question is to ask, is it *fair*?

To address this question, there are many different models of Distributive Justice from various fields of social science (economics, psychology, philosophy), including Rescher's model (Rescher 1966). Rescher reviews many canons (principles) of Distributive Justice, all of which have strengths and weaknesses. His position instead is one of legitimate claims: that justice consists of determining which of these canons are appropriate in context, how they are accommodated in case of plurality, and how they are reconciled in case of conflict. Using the method of sociologically inspired computing, we can define a formal characterization of Rescher's model of legitimate claims, which can then be given a principled operationalization.

To model this resource allocation problem, we use a variant of the linear public good (LPG) game (Gaechter 2006) called *LPG'* (Pitt et al. 2012). In this situation, the agents (players) form clusters (institutions), and in consecutive rounds of the LPG' game, they provision resources, an allocation is computed according to the legitimate claims, and the agents make an appropriation. Note that in an open system, agents may not comply with the specifications, so the appropriation may be greater than the allocation (one of the variants in LPG' from LPG). Agents calculate their utility, satisfaction and evaluate the legitimate claims, and the next round starts.

Therefore, we define a formal model of the system as a 4-tuple <A, C, L, G> where

- A is set of agents.
- C is a set of clusters.
- L is a norm-governed system specification.
- G is the LPG' game.

Note that L is a specification of the operational-choice and collective-choice rules that are used to compute the resource allocation.

Each cluster $c \in C$ is a 3-tuple: $C = <M, l, e>$ where

- M is a set of members, subset of A.
- l is a specification instance of L.
- e is an environment of institutional and brute facts.

In particular, e, the environment, specifies which legitimate claims are being used in the computation of the resource allocation defined by the rules in l, and the weight which is attached to each of them.

Each agent $a \in A$ is a 9-tuple $<q, g, p, d, r, r', s, t, u>$ where

- q is resources needed.
- g is resources generated.
- p is resources provisioned.
- d is resources demanded.
- r is resources allocated.
- r' is resources appropriated.
- s is satisfaction.
- t is threshold for staying in or leaving a cluster.
- u is its utility.

In each round of the game

- Each player
  - Determines the resources it has available and its need for resources.
  - Makes a demand for resources and a provision of its resources to the common pool.
- The allocation to each agent is made according to legitimate claims.
- Each player
  - Makes an appropriation of resources.
  - Updates its satisfaction and utility.
  - Leaves the cluster if $s < t$ for more than N rounds.
- The weights on legitimate claims are recomputed.

The set of legitimate claims that prioritize resource allocation to agents will be according to a set of criteria, for example, their average allocation in previous rounds, number of rounds they have participated in the game, and so on. Plurality is accommodated by treating each claim as a *voter* in a Borda count protocol that implements the resource allocation policy. Conflict is reconciled by attaching a weight to each claim, and at the end of each round, the agents self-organize the weights on each claim. This method is outlined in Model 3.1. For a more detailed description of these functions, see Pitt et al. (2012).

---

input : $A \leftarrow$ set of n agents
input : $F \leftarrow$ set of m voting functions f_* each with weight w_*
output: allocation $R : A \rightarrow \mathbb{R}$
1 **foreach** *agent* $i \in A$ **do**
2      $d_i \leftarrow i$.demand ;
3      $p_i \leftarrow i$.provision ;
4      $r_i \leftarrow 0$;
5 **end**
6 $P \leftarrow \sum_{i=1}^{n} p_i$;
7 *rank_orders* $\leftarrow$ [] ;
8 **foreach** *function* $f_* \in F$ **do**
9   *rank_orders* $\leftarrow$ *rank_orders* $\cup f_*(A)$;
10 **end**
11 $Borda\_ptq \leftarrow$ Borda_count(*rank_orders*, F)
12 **repeat**
13      $i \leftarrow$ head($Borda\_ptq$) ;
14      $Borda\_ptq \leftarrow$ tail($Borda\_ptq$) ;
15   **if** $P \geqslant d_i$ **then**
16          $r_i \leftarrow d_i$;
17          $P \leftarrow P - d_i$;
18   **else**
19          $r_i \leftarrow P$;
20          $P \leftarrow 0$;
21   **endif**
22 **until** $P == 0$;

Model 3.1 Resource allocation with legitimate claims.

Kyoto Protocol: Formal model

Having recognized that developed countries are primarily responsible for the high levels of greenhouse gas emissions in the atmosphere, the United Nations Framework Convention on Climate Change formed the Kyoto Protocol to not only encourage countries to reduce emissions, but to commit to that reduction (UN 1998). To make the Protocol more appealing, carbon taxation was avoided in favor of other carbon-reduction mechanisms.

Signed in 1997, and coming into force in 2005, the Protocol sets mandatory targets on greenhouse gas emissions for participating nations with the aim of reducing global emissions. The Protocol aims to address the differing capabilities of countries to reduce emissions by classifying them according to current emissions, wealth, and capacity. Each group has different responsibilities, thus allowing developing countries room for economic expansion and putting a higher burden on nations able to afford the cost of emissions reductions. These membership levels are as follows:

- Annex I: Industrialized countries and economies in transition. These countries commit to reducing their emissions levels.
- Annex II: Developed countries. This is a subset of Annex I that are encouraged to invest in developing countries.

- Non-Annex: Developing countries which are not required to reduce emission levels.

While participant countries must meet their emissions targets primarily through domestic carbon reduction methods, flexible mechanisms were also put in place to make targets more attainable and affordable. Three mechanisms were defined in the protocol: emissions trading, where members may trade a commodity representing the surplus created if their emissions are below their assigned target; clean development mechanism, allowing Annex I and II countries to invest in sustainable greenhouse gas reduction projects in non-Annex countries in exchange for emissions reductions; and joint implementation, where Annex I countries can invest in sustainable greenhouse gas reduction projects in other Annex I countries as an alternative to reducing their own emissions.

We discuss in this section how the Kyoto Protocol and its different Annex countries can be mapped to a common-pool resource (CPR) problem. Through targets, greenhouse gas emissions become a CPR where countries may only emit as much as they are allowed by their allocated target. Unlike traditional CPR problems, such as water appropriation from reservoirs, there is no physical depletion of the resource in the case of countries missing their targets as the resource is artificially constrained.

We define a formal model of the system at a discrete time t as a 4-tuple $<A, M, K, e>$ where

- A is set of agents, each representing a single country.
- M is the set of membership levels in the Kyoto Protocol.
- K is the Kyoto Protocol rules.
- e is the system's environment.

Each membership level $m \in M$ is a 2-tuple $m = <C, k>$ where

- C is the set of countries with this membership level, subset of A.
- k is the Kyoto Protocol rules and obligations applying to these members. For example, for Annex I countries reducing emissions, this specifies how their targets are set as well as members' obligations regarding reporting of emissions.

Each country $c \in A$ is an 8-tuple of the following:

- Total land area
- Arable land area
- Carbon output
- Energy output
- Gross domestic product (GDP)

- Emissions target
- Member level
- Budget

Each time *t*, every country may choose a course of action to meet their goals. The actions available are the following:

- Invest in carbon absorption: Reduces carbon output and arable land area; cost increases as arable land area reduces.
- Reduce energy output: Reduces energy output and carbon output; will affect the rate of GDP change.
- Ask for clean development mechanism investment: Reduces emissions target.
- Invest in carbon reduction: Reduces carbon output; cost increases as the proportion of remaining *dirty* industry falls (*dirty* industry proportion is measured as the ratio of carbon output to energy output).
- Invest in carbon industry: Increases carbon output, energy output, and rate of GDP change.
- Trade in carbon offset: Changes emissions target and budget according to the value of trades made.
- Join/leave Kyoto: Change member level to new value.
- No action.

Each year (defined as a fixed number of time increments) the environment defines how each country's GDP and budget are updated. This function takes the state of the global economy, each country's GDP rate and energy output into account. In addition, the Kyoto Protocol dictates that countries must report their carbon emissions as well as contribute to the cost of monitoring. A random subset of reporting countries are monitored to ensure they are truthful. In the case of a false report, sanctions are imposed in the form of a fine, reducing the country's budget.

At the end of each session of the Kyoto Protocol (defined as a fixed number of years), countries' emissions targets are assessed and new targets created. Model 3.2 shows the entire process executed each discrete time step.

Presage2

To make the transition from formal models, as we have described, into operational models that can be queried for experimental results, we use the platform Presage2. In this section, we will describe the platform's design and implementation, the foundation for our subsequent demonstration of its use on the Kyoto Protocol and Distributive Justice systems.

---

input : $A \leftarrow$ set of n country agents
1 $t \leftarrow 0$;
2 **repeat**
3   **if** $endOfYear(t)$ **then**
4        $updateGDP(A)$;
5        $updateBudgets(A)$;
6        $M \leftarrow kyotoMembers(A)$;
7        $monitoringTax(M)$;
8        $updateCarbonOffsets(M)$;
9        $sanction(monitored(M) \cap cheated(i, t))$;
10       $setYearEmissionsTargets(M)$;
11       $reportCarbonEmissions(A)$;
12   **endif**
13   **if** $endOfSession(t)$ **then**
14       $M \leftarrow kyotoMembers(A)$;
15       $resetCarbonOffsets(M)$;
16       $setSessionEmissionsTargets(M)$;
17   **endif**
18    $doCountryBehaviour(A)$;
19    $t \leftarrow t + 1$;
20 **until** $t > t_{finish}$;

Model 3.2 Kyoto Protocol simulation loop.

Presage2 is designed as prototyping platform to aid the systematic modeling of systems and generation of simulation results. To this end, we require that the platform can simulate computationally intensive agent algorithms with large populations of agents; simulate the physical environment in which the agents interact, including dynamic external events; reason about social relationships between agents including their powers, permissions, and obligations; systematic experimentation; and aggregation and animation of simulation results.

Presage2 is one of many software platforms that would be suitable for this process of principled operationalization. A general survey of the basic features of agent-based modeling tools allows some evaluation of which platforms are suitable (Nikolai and Madey 2009). Popular tools that would also be suitable for the systems described in this chapter are NetLogo (Wilensky 1999), Repast (Collier 2009), and MASON (Luke et al. 2005). The feature sets of these platforms are quite similar; NetLogo being more assessable to nonprogrammers, with usage being largely graphical user interface based; MASON and Presage2 sit at the other end of the scale, with powerful capabilities for skilled programmers and all functionality code based; Repast lies somewhere in between these two points. Beyond these differences, the choice between platforms comes down to user preferences such as familiarity with the platform language, aptitude of the platform's model or framework to the target problem, and any differentiating features that will ease the implementation of a system. Presage2 was chosen due to features that ease the implementation of institutional and conventional structures prevalent in the social systems we are implementing.

Architecture

The platform is composed of the following packages that work together to control the execution of a simulation as illustrated in Figure 3.2:

- Core—controls and executes the main simulation loop and core functions.
- State engine—stores and updates simulation state.
- Environment and agent libraries—implementations of common-use cases that can be used in the environment and/or agent specifications.
- Communication network simulator—emulates a dynamic, inter-agent communication network.
- Database—enables storage of simulation data and results or analysis.
- Batch executor—tools to automate the execution of batches of simulations.

The Presage2 simulator's core controls the main simulation loop as well as the initialization of a simulation from parameter sets. The simulation uses discrete time, with each loop being a single time step in the simulation. Each time step, each agent is given a chance to perform physical and communicative actions, and the simulation state is updated according to these actions as well as to any external events. Agent architecture and computational complexity are not limited (except by the limitations of the computer running the simulation); the platform will wait for every agent's function to terminate before moving to the next time step.

A time-driven approach is used over an event-driven for simplicity and performance. Although event-driven approaches are theoretically more computationally efficient as they can ignore periods of inactivity, the reactive nature of agent behavior largely negates this. In addition, we gain performance within a time-driven simulation through parallelization.

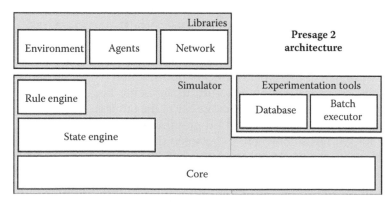

Figure 3.2 Presage2 architectural block diagram.

We exploit this with our multithreaded implementation. Any processes that can be run in parallel can do so to utilize all available computing resources. For example, the execution of agent code within a single time step can be done in parallel, gaining significant performance when agents are computationally complex.

A key concept of multiagent simulation is that the agents share a common environment. The state engine package simulates this environment as a state space. This package allows the user to control two important functions: the observability of state for each agent, and the effect of an agent's action on the state. The former specifies what state an agent can read given the current environment state, while the latter determines a state change given the current state and the set of all actions performed in the last time step. The user defines these functions from a set of modules. Each module can be seen as an independent set of rules regarding observability and/or state changes. This method allows for behavioral modules to be slotted into a system without conflict, building up complex system rules from the composition of modules.

As well as allowing agents to observe raw environment states, environment modules can provide processed data to agents. This abstraction layer enables agents to ask simple questions such as "who's near me?" rather than manually querying x and y coordinates of other agents. With this feature, we can model complex phenomena such as dynamic relationships between agents. Figure 3.3 illustrates how environment modules mediate agents' access to the environment state.

We have developed another implementation to extend the basic state engine implementation described using the Drools rule engine to offer an extended feature set. Drools is a production rule engine based on the Rete algorithm and written in Java.* Our optional implementation uses this rule engine as the underlying state storage engine, encapsulating all features of the initial implementation plus additional benefits afforded by Drools. Using Drools, we are able to have a much richer state representation. While our initial version uses raw data points with text strings to reference each one, with Drools we can store structured and relational data in the state. Drools also features a forward chaining rule engine. We are

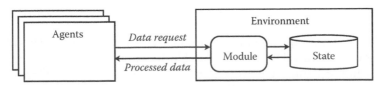

Figure 3.3 Interaction between agents and their environment via environment modules.

* http://www.jboss.com/drools

thus able to use declarative rules to modify the state in the engine. Users can create a set of rules that the platform will trigger each time step. This can be used to add additional functionality to the environment. For example, we can use these rules to describe complex state changes, which are triggered from actions that agents have performed. These rules are often easier to understand and verify than blocks of procedural Java code. Their declarative nature is close to that of action languages such as the Event Calculus. Therefore, a dynamic specification in the Event Calculus can be almost directly ported into Presage2 via Drools. Work has been done to allow Event Calculus predicates to be used in Drools directly (Bragaglia et al. 2012).

The environment and agent libraries provide abstractions for creating agents and environments for the platform. The libraries include environment modules for common use cases such as mobile agents as well as other useful code snippets such as a finite state machine implementation.

Similarly, the communication network simulator includes implementations for common use cases in this domain, from a simple fully connected network to dynamic ad-hoc networks. The network allows agents to send messages containing structured data to each other.

Presage2's abstractions and separation between agents and their environment as illustrated in Figure 3.3 enable animation of multiagent systems. This process involves rewiring data requests from agents such that the module instead is able to retrieve data from the real-world environment. The same is done for the actions agents perform. We then are able to animate the real-world system with the same code we used in the simulated system.

To accommodate data visualization and analysis, the platform includes a data storage package. Raw data can be stored during simulation and can be associated with specific agents and/or time steps. Drivers for several different database management systems (DBMSs) are included so that the data can be sent to different database types depending on the user's needs. In the early prototyping phases, storing data in flat files is usually sufficient to verify behaviors. When the system is ready for experimentation, a simple configuration is all that is needed to start storing data in a much higher performance DBMS such as MySQL. Once raw data has been stored, then it can be analyzed and visualized with third-party tools such as MATLAB® or Python.

The primary implementation language of Presage2 is Java. This allows simulation specifications to be portable to multiple execution environments while still having performance comparable to compiled languages. A simulation specification will run on any Java compatible system, thus facilitating the exchange of models between users. Java can also aid the deployment of simulations on high-performance computing systems as there is no need for recompilation on each target machine.

Operational basis

In Figure 3.4, we show how a simulation is created in Presage2. In the design and implementation phase, the user selects from available modules, as well as creating their own, to meet the requirements of the system specification. These selections include which network topology to use for interagent communication and modules to control the environment state engine. There are no constraints on the number or composition of modules used for a simulation. In addition, the user must implement any agent types to be used in the simulation. Together these create an executable simulation specification that can be invoked with a set of simulation parameters.

Once the simulation specification has been created, we can generate experimental data by instantiating the specification with a parameter set. A database configuration can be provided to specify where simulation data is stored. Having generated a series of data sets, external tools can be used to analyze and visualize it.

When dealing with large batches of simulations, running each individual set of parameters including repeats would be tedious. Presage2 is able to automate many parts of this process by generating parameter set permutations and executing batches of simulations. This batch executor is able to use a user-specified set of available machines on the network, or interface

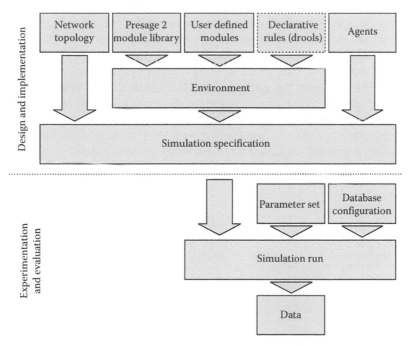

Figure 3.4 Simulation creation process in Presage2.

with a job scheduler as is often available in high-performance computing facilities, and run simulations in parallel. These tools streamline the
experimentation process by enabling automated, unsupervised running of
simulation batches and utilizing all available computing resources without
the need for dedicated high-performance computing infrastructure.

Worked examples

Previously we described a formal model for social systems, one modeling the Kyoto Protocol, the other resource allocation using Distributive
Justice. In this section we describe, for each of these examples, the process of principled operationalization, translating the formal model to a
Presage2 simulation, and controlled experimentation, using this simulation specification to run simulation and then observe its performance.

Distributive Justice: Operational model in Presage2

The formal model of the Rescher's Distributive Justice and the model of
the LPG' game have been presented earlier. We follow the simulation creating process outlined in Figure 3.4 to explain how to achieve a queryable
computer model using Presage2.

In this example, we utilized the Drools rule engine to model the rules
of the LPG' game. The environment processes these rules in response to
actions the agents perform. As all of the environment behavior is encapsulated in these rules, we do not need to define any environment modules.

There are six actions in the game: *generate, need, demand, provision, allocate,* and *appropriate.* Each of these is represented in the environment state
as a tuple containing the game round, the agent that is associated with
the action, and the quantity of resource. *Generate* and *need* are values we
will randomly generate for each agent; *demand, provision,* and *appropriate*
are actions the agents perform themselves. We treat *allocate* as being done
by the environment, though it could be an agent with a distinguished role
performing these actions. Figure 3.5 illustrates how this translates into the

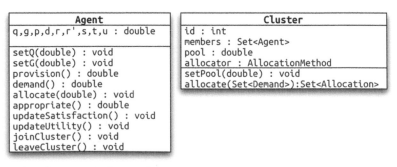

Figure 3.5 Agent and cluster class diagrams.

implementations for both an agent and the clusters. A sequence of these actions creates a game narrative that the rule engine processes to modify the environment state and generate more actions in response to the narrative. For example, a series of *demands* and *provisions* will create a pool of available resources and an *allocation* for each demanding agent.

To interface this agent implementation with Presage2, we must create a mapping from the LPG′ game to an agent function that is invoked each discrete time step by the simulator. This is done simply by providing an abstraction that keeps track of the current state of the game and then calls the appropriate method on the subclass agent implementation.

Each agent represents a single player in the LPG′ game with the following behavior. Agents read their randomly generated *generate* and *need* then decide how much to *demand* and *provision*. Once they have an allocation, they choose how much to *appropriate*. Compliant agents will demand what they need, provision what they have generated, and appropriate what they have been allocated. Noncompliant agents have some probability of demanding, provisioning, or appropriating more or less than they should. Agents may also join and leave clusters. Figure 3.6 shows this sequence of interactions during one game round.

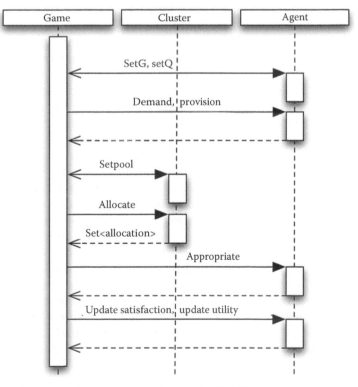

Figure 3.6 Sequence diagram of single round of LPG′.

With the core game functionality in place, we add different allocation methods. For each cluster, the rule engine will determine the quantity of resources available from the *provision* actions of cluster members, and then pass this pool plus the set of agents to the allocator. The allocator then creates an allocation for each agent in the set. For memory-less allocation methods such as ration and random, this is trivial. However, those based on legitimate claims require knowledge of agents' past actions to rank them. As we store actions in the environment state as a narrative, we can look back on each agent's history of actions.

The simulation specification is completed by creating an initial game state. We create a set of available clusters each with a designated allocation method and a set of agents to participate in these clusters. This implementation supports controlled experimentation with different agent behavior profiles, population sizes, population compositions, cluster configurations, and allocation policies, as well as different sets of legitimate claims, different weights, and so on. Data logged from the experiments allow us to verify the performance of the different allocation methods as well as different agent strategies.

From our experiments, we were able to verify that the use of legitimate claims with self-organization for resource allocation in an economy of scarcity improved the endurance of a cluster over ration and random allocation methods and benefitted compliant agents. In addition, this method improved the fairness for agents in the cluster using the Gini index as a fairness measure. This index measures the statistical dispersion of values in a frequency distribution, and it is widely used to measure income inequality. It is computed as half the relative mean difference of the values of the distribution, and has a value of 0 for complete equality and a value of 1 for complete inequality. We found that compliant agents preferred to participate in a cluster using legitimate claims with self-organization. Finally, our results showed that, with our chosen fairness measure, allocations are quite unfair in the short term, but this leads to a very fair overall allocation in the long term, as shown in Figure 3.7. It shows the Gini index of each allocation performed at each time step, as well as the Gini index of the accumulated allocations, when using legitimate claims in a self-organized cluster. It can be observed that while each individual allocation is not fair (with all indices being around 0.32 ± 0.4), the accumulated index reaches values indicating a fair distribution (with values around 0.014 ± 0.013). Moreover, as the figure shows, such fair value is reached very quickly, just after 20 or 30 rounds of the game. Actually, if we do not take into account these first 30 rounds, the average accumulated Gini index is 0.012 ± 0.002.

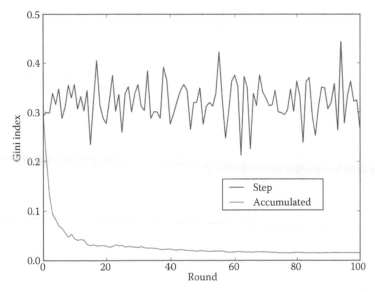

Figure 3.7 Evolution of Gini indices: index of individual allocations at each game step (round) in dark gray (top line) and index of accumulated allocations over time in light gray (bottom line).

Kyoto Protocol: Operational model in Presage2

As with the previous example, we take the formal model of the Kyoto Protocol as a CPR from the Section "Examples of social systems" and translate it into a computer model using Presage2.

The agents in the simulation each represent a single country. To model the economic and physical properties of the environment, the user must define several environment modules. These will describe the state of the environment, what state is readable by whom, and what actions agents can perform and their consequences on the environment.

A country has several values to describe its state at any point in the simulation: land area, arable land area, carbon output, energy output, GDP, emissions target, member level, and budget. Some of these values are modified directly by country actions, for example, building carbon sinks on arable land will reduce carbon output, while others are controlled by our economic and emissions models. An agent's state is only observable to itself, any state of other countries may be accessed via interagent communication. We allow all countries to communicate with each other, using a fully connected communication network topology. Countries can perform several different actions that affect their carbon emissions. In addition, they can engage in carbon trading, report carbon output, and participate in other Kyoto Protocol schemes. The modules define the state change as a result of each action according to our emissions and economic models. To handle proactive phenomena in the simulation, such as economic fluctuations that affect countries' growth and budget, we have other state changes that trigger every time step.

Several rules are embedded into the environment, for example, each country has an obligation to report its carbon emissions every year. These emissions are then compared to the target level and failure to meet a level will result in a sanction. We implement sanctions with a monetary fine, reducing the country's available budget. Countries may also falsify their carbon emissions report to avoid a potential sanction. However, a monitoring agent randomly checks reports against actual emissions to detect this cheating.

As Presage2 is a discrete time simulator, we must compose a time scale for these actions to occur, which maps onto real time. In our simulation specification, we split a year into a number of *ticks*, 100 by default, which were the smallest unit of time. Each *tick* corresponds to a single time step in Presage2. In the Kyoto Protocol, we also have a *Session* that is composed of multiple years. The length of a Session is mutable depending on global economic and political climate.

We define how an agent determines actions to perform depending on its state and the environment. In every simulation tick, each country must decide a course of action (or inaction). Given their member level (Annex I [Sustain], Annex I [Reduce], Non-Annex, etc.) they will have an emissions target to reach. Agents can spend budget in several ways to help meet this target. Investment in clean industry directly reduces carbon output (with diminishing returns), while one can also invest in forestation of arable land to absorb some carbon output. If actual carbon output falls above or below the emissions target, carbon trading can be used to prevent sanctions or to raise more cash in each case. Countries may choose to simply invest in industry to increase GDP at a cost of increased carbon output. They also have the option to leave the protocol if they wish.

Given all the possible actions listed previously, we can implement an algorithm to find optimal executions of these actions given certain constraints and goals. Because of the nature of how emissions targets are set for each member level in the Kyoto Protocol, and the nature of those countries, this gives differing behaviors for countries in a simulation.

Finally, to complete the simulation specification, we define how to create the system's initial state from a set of parameters. The environment is initialized with several parameters for our economic models, and each country is initialized with real-world data. From this specification, we are able to instantiate various scenarios and observe the system behavior.

Having initialized our simulation with the emissions and economic data from 1998, around the time the Kyoto Protocol was adopted, it was run to simulate the effect of the protocol during 40 years. The result from the first 15 years can be directly compared to the real-world data allowing the model's

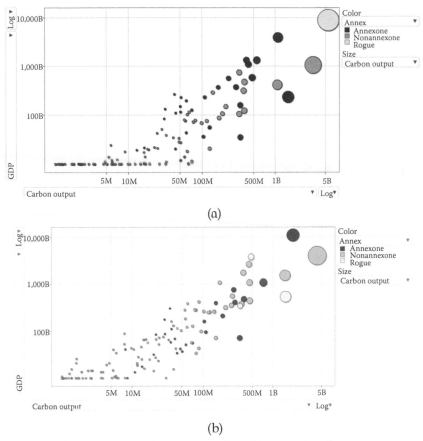

(a)

(b)

Figure 3.8 Projected gross domestic product (GDP) and carbon emissions per country in (a) 1998 and (b) 2038.

validity to be accessed. Extrapolating into the future may allow us to predict how these policies affect greenhouse gas emissions. Figure 3.8 shows each country's GDP and emissions at the beginning and end of a simulation, illustrating that in general, countries have managed to increase GDP without a rise in emissions over this period. We present these results merely as a demonstration of how one can model policies in social systems using our methodology and Presage2. Our simplifying assumptions for this model, a basic economic model, and certain assumptions about the behavior of countries, render this system as a thought experiment on the construction of system models.

Simulating sociotechnical systems

The third reason we gave for modeling and simulation of social systems is for the development sociotechnical systems, that is, systems that involve the active participation of both human and computational *components*. In recent work (Bourazeri et al. 2012), we have examined the use of Serious Games (Marsh 2011) to provide a virtual environment as a user–infrastructure interface for SmartGrids.

Serious Games are digital games, simulations, and virtual environments whose purpose is not just to entertain, but also to assist learning and develop skills such as decision-making, long-term engagement, and collaboration. Serious Games can be used for modeling and simulating new complex social systems, especially those where there are agents operating on behalf of human users.

For example, there is increasing attention being paid to demand-side management in SmartGrids (Strbac 2008) and the use of SmartMeters in domestic residences. We propose to use the SmartMeter as an agent that is acting on behalf of the user, hiding complex calculations and negotiations, but drawing their attention to issues that need human intervention, especially issues that require collective action.

For example, during peak times it is possible for transformers in the electricity grid to be overloaded, and if its temperature rises too high, a protector switch trips and the transformer is disconnected. This of course causes a power outage for domestic consumers, who care not that the transformer was overloaded but that their energy supply was interrupted, and they ring the generator to complain. However, if they are motivated enough to be reactive in this way, perhaps they could also be motivated to be *proactive*, and take action to prevent the overload from happening in the first place. However, if one consumer reduces their consumption, it will not have any effect, but if a *critical mass* of consumers takes positive (collective) action, then the trip can be avoided.

To experiment with this kind of system, we have been developing Serious Games with user interfaces on top of Presage2 (Bourazeri et al. 2012). In future work with Presage2, for modeling and simulation purposes for this kind of system, we intend to address the question of how to integrate a Presage2 simulation also with a virtual environment for a Serious Game. In part, this would involve linking the plug-ins to the agent library to virtual characters in the virtual environment, and plug-ins to the environment library to locations in the virtual environment. In this way, the human player becomes an *animation-by-proxy* (i.e., the virtual character) and could participate in and interact with a large-scale energy grid simulation powered by Presage2. This should make it possible to simulate sociotechnical systems with direct engagement of people, that is, without having to create computationally intensive, and possibly inaccurate, cognitive models of people. The results of such simulations can inform the actual development and deployment of the sociotechnical system itself.

Summary and conclusions

There are myriad agent-based social simulation systems, ranging from dynamic social psychology (Nowak and Vallacher 1998), to analytical sociology (Hedström and Bearman 2009), geography and ecology (Matthews et al. 2007), or public health (Noble et al. 2012), among many others. However, despite the variety of topics they address, there is a common need for tools and methodologies for simulating and analyzing complex systems involving a large number of individual and heterogeneous agents.

This chapter has discussed the design, modeling, and principled operationalization of social systems using the Presage2 multiagent system animation and simulation platform. In particular, we have described the operationalization of self-organizing resource allocation in a collective, adaptive system as exemplified by the Kyoto Protocol and Distributive Justice as exemplars. These examples demonstrate the first two reasons for modeling and simulating social systems as we have mentioned: understanding and explaining how a social system works, and formalizing human methods of problem-solving to address engineering challenges.

There are many reasons to use a software platform for the simulation of social systems, as evidenced by the proliferation of platforms created for this purpose (Bellifemine et al. 2003; Luke et al. 2005; North et al. 2007). Primarily, a platform allows us to speed up the implementation process, to ensure results can be verified, and to potentially aid collaboration through the exchange of code. The trade-off when choosing whether to use a simulation platform is between these benefits and the one-off time investment learning to use the software. We have shown Presage2's capabilities in these respects and how it can be used for the simulation and animation of sociotechnical systems.

As we mentioned when we introduced our choice of software platform for principled operationalization, the decision is largely made based on subjective factors. In fact, objective evaluation of platforms in respect to each other is very difficult to do. In Railsback et al. (2006), the authors compare five simulation platforms by following a methodology where they invent a simple model and then incrementally add more complex. Although they do generate some empirical data to compare the platforms by execution time, their evaluation focuses on these subjective issues, noting that a clear model design process and framework, and good documentation are the most important requirements.

When observing sociotechnical systems, we often see that the processes and interactions within the systems are governed by rules. Our methodology of socially inspired computing preserves the rule syntax, first through the use of an appropriate calculus such as the Event Calculus, and second by using a rule engine in the computer model, such as the

Drools rule engine in Presage2. One can then directly compare the implemented rules with their corresponding rule in the sociotechnical system and their effect, that is, comparing observed phenomena and observed performance.

We often encounter large parameter spaces when implementing these sociotechnical systems. Exploring these spaces is a problem of effectively utilizing available computing resources to find cases or trends of interest. We have shown that Presage2's architecture allows it to excel at this task, running single simulations quickly through multithreading, as well as running large batches of simulations across compute clusters. This batch simulation process has the advantage that it requires no prior installation or configuration on individual nodes.

However, there are some potential pitfalls in this method. One must be aware of implicit and explicit assumptions gained when following this method. For example, when generating a computer model with Presage2, we assume that time is discrete and therefore actions always have the same duration. This granularity of time is particularly important when dealing with real-world systems with time constraints, like in the Kyoto Protocol example. There the number of *ticks* we slice a year into will affect the outcomes. In addition, when attempting to model a real-world phenomenon, there is always a limit to the number of factors that it is possible to simulate, and are worthwhile simulating. In doing this, we are making assumptions that the effects of the factors we ignore are inconsequential to what we want to observe. When interpreting the simulation results, we must take these limitations into account.

Finally, this chapter has given case studies of modeling and simulation of two systems using Presage2, for two of the reasons we gave in the introduction: modeling a *technical* system (composed of autonomous computing entities) and modeling a social system (composed of autonomous social entities, e.g., people, organizations, countries, etc.). For the third reason, we have also briefly considered modeling and simulation of a sociotechnical system through the use of Presage2 and Serious Games.

Presage2 is open source and available under the LGPL license from http://www.presage2.info.

References

Andrews, P. S., Polack, F. A. C., Sampson, A. T., Stepney, S. et al. (2010). The CoSMoS Process Version 0.1: A Process for the Modelling and Simulation of Complex Systems. *Technical Report YCS-2010–453*, University of York, United Kingdom.

Bellifemine, F., Caire, G., Poggi, A., and Rimassa, G. (2003). Jade-a white paper. *EXP in Search of Innovation*. 3(3), 6–19.

Bourazeri, A., Pitt, J., Almajano, P., Rodriguez, I. et al. (2012). Meet the Meter: Visualising SmartGrids using Self-Organising Electronic Institutions and Serious Games. In: *2nd AWARE Workshop on Challenges for Achieving Self-Awareness in Autonomic Systems.* SASO 2012, Lyon, France.

Bragaglia, S., Chesani, F., Mello, P., and Sottara, D. (2012). A rule-based calculus and processing of complex events. In: *Rules on the Web: Research and Applications.* Springer, Berlin/Heidelberg, Germany, pp. 151–166.

Castelfranchi, C. and Falcone, R. (1998). Principles of trust for MAS: cognitive anatomy, social importance, and quantification. In: *Proceedings International Conference on Multi Agent Systems,* Paris, France, pp. 72–79.

Christensen, K., di Collobiano, S. A., Hall, M., and Jensen, H. J. (2002). Tangled nature: A model of evolutionary ecology. *Journal of Theoretical Biology.* 216(1), 73–84.

Collier, N. (2009). RePast : An extensible framework for agent simulation. *Natural Resources and Environmental Issues.* 8(4).

Draief, M. and Ganesh, A. (2010). A random walk model for infection on graphs: Spread of epidemics & rumours with mobile agents. *Discrete Event Dynamic Systems.* 21(1), 41–61.

Edmonds, B., Gilbert, N., Gustafson, S., Hales, D. et al. (2005). AISB'05: Social intelligence and interaction in Animals, Robots and agents. *Proceedings of the Joint Symposium on Socially Inspired Computing.* B. Edmonds, N. Gilbert, S. Gustafson, D. Hales et al. (eds.). AISB, Denmark Hill, London, United Kingdom.

Ferscha, A., Farrahi, K., Hoven, J., Hales, D. et al. (2012). Socio-inspired ICT. *The European Physical Journal Special Topics.* 214(1), 401–434.

Gaechter, S. (2006). Conditional cooperation: Behavioral regularities from the lab and the field and their policy implications. *Discussion Papers 2006-03.* The Centre for Decision Research and Experimental Economics, School of Economics, University of Nottingham.

Hedström, P. and Bearman, P. (2009). *The Oxford Handbook of Analytical Sociology.* P. Hedström and P. Bearman (eds.). Oxford University Press, United Kingdom.

Hoffman, R. R., Norman, D. O., and Vagners, J. (2009). Complex Sociotechnical Joint Cognitive Work Systems. *IEEE Intelligent Systems.* 24(3), 82–89.

Jones, A. J. I., Artikis, A., and Pitt, J. (2013). The design of intelligent socio-technical systems. *Artificial Intelligence Review.* 39(1), 5–20.

Luke, S., Cioffi-Revilla, C., and Panait, L. (2005). MASON: A multiagent simulation environment. *Simulation.* 81(7), 517–527.

Marsh, T. (2011). Serious games continuum: Between games for purpose and experiential environments for purpose. *Entertainment Computing.* 2(2), 61–68.

Matthews, R. B., Gilbert, N. G., Roach, A., Polhill, J. G. et al. (2007). Agent-based land-use models: A review of applications. *Landscape Ecology.* 22(10), 1447–1459.

Neville, B. and Pitt, J. (2004). A computational framework for social agents in agent mediated E-commerce. In: A. Omicini, P. Petta, and J. Pitt (eds.). *Engineering Societies in the Agents World IV.* Springer, Berlin/Heidelberg, Germany, pp. 376–391.

Nikolai, C. and Madey, G. (2009). Tools of the trade: A survey of various agent based modeling platforms. *Journal of Artificial Societies and Social Simulation.* 12(2), 2.

Noble, J., Silverman, E., Bijak, J., Rossiter, S. et al. (2012). Linked lives: the utility of an agent-based approach to modeling partnership and household formation in the context of social care. In: *Proceedings of the Winter Simulation Conference 2012 (WSC2012)*. December 9, 2012, Winter Simulation Conference, Berlin, Germany, p. 93.

North, M. J., Howe, T. R., Collier, N. T., and Vos, J. R. (2007). A declarative model assembly infrastructure for verification and validation. In: S. Takahashi, D. L. Sallach, and J. Rouchier (eds.). *Advancing Social Simulation: The First World Congress*. Springer, Heidelberg, Germany, pp. 129–140.

Nowak, A. and Vallacher, R. (1998). *Dynamical Social Psychology*. Guilford Press, New York.

Ostrom, E. (1990). *Governing the Commons*. Cambridge University Press, United Kingdom.

Pitt, J., Schaumeier, J., and Artikis, A. (2011). The Axiomatisation of Socio-Economic Principles for Self-Organising Systems. *2011 IEEE Fifth International Conference on Self-Adaptive and Self-Organizing Systems*, Ann Arbor, MI, pp. 138–147.

Pitt, J., Schaumeier, J., Busquets, D., and Macbeth, S. (2012). Self-Organising Common-Pool Resource Allocation and Canons of Distributive Justice. *Sixth International Conference on Self-Adaptive and Self-Organising Systems*, Lyon, France, pp. 120–128.

Railsback, S. F., Lytinen, S. L., and Jackson, S. K. (2006). Agent-based Simulation Platforms: Review and Development Recommendations. *Simulation*. 82(9), 609–623.

Rescher, N. (1966). *Distributive Justice*. Bobbs-Merrill, Indianapolis, IN.

Steels, L. and Brooks, R. (1994). *The Artificial Life Route to Artificial Intelligence: Building Situated Embodied Agents*. Lawrence Erlbaum Associates, New Haven, CT.

Strbac, G. (2008). Demand side management: Benefits and challenges. *Energy Policy*. 36, 4419–4426.

United Nations. (1998). *Kyoto Protocol to the United Nations Framework Convention on Climate Change*. Available at http://unfccc.int/resource/docs/convkp/kpeng.pdf.

Wilensky, U. (1999). NetLogo. Center for Connected Learning and Computer-Based Modeling, Northwestern University, Evanston, IL.

chapter four

Formal agent-based models of social systems

Il-Chul Moon

Contents

Introduction

Social systems are ubiquitous and include the following types: traffic systems (Balmer et al. 2004; Wang 2005), market systems (Tesfatsion 2002), government systems (Comfort 2007; Oh and Moon 2008), and military systems (Pew and Mavor 1998), among others. These systems have human and societal entities put in place to achieve a goal that neither a single individual nor organization can achieve (Blanchard and Fabrycky 2010).

As all systems require to be understood for improvements, social systems, too, need to be understood. For example, the government system is constantly checked and balanced through various mechanisms: voting, auditing, and assessment of policies. Because of checking and balancing the systems, the government structures and behavior believe to be changing over time to better achieve the system objectives, that is, improving the efficiency of social welfare through implementing the policy. However, an interesting question is that whether such changes are aligned to the best way to achieve its system's objectives. Often, it is very difficult to estimate the following consequence of changes, and there are occasions that changes might invoke unintended secondary consequences. Such unintended consequences from interactions between various social elements are only visible after the actual implementation of changes, and we need a theory, a technique, and a tool to see a glimpse of the consequences.

Another difficulty in understanding and influencing the social systems, particularly compared to the other systems, (e.g., physical engineering systems) (Forrester 1971), is that social systems are difficult to observe and assess. If the systems are physical in nature, it is likely that they have limited degrees of freedom. For example, a robot arm may have six degrees of freedom in the arm movement, and maintaining the arm will be focused on observing and assessing the movement in the six dimensions. However, if multiple individuals involve in a collaborative task execution, the execution has countless outcomes with various behavior exhibited by the individuals. As opposed to physical systems, social systems sometimes have (1) unclear boundaries with difficulties in identifying the membership and the group identity (Weber 1947), (2) not fully observable social entities (Carley and Newell 1994), and (3) infinite degrees of freedom regarding system behavior (Carley 2002). On top of these difficulties, the recent evolution of social systems has created more challenges in understanding the systems. The recent evolution of the systems is characterized by the following implementations: (1) heterogeneous compositions of technical elements and social entities (Krackhardt and Carley 1998), (2) a nonhierarchical structure of system elements (Guetzkowm and Simon 1955), and (3) dynamic changes of the structure and the roles of the systems (Moon and Carley 2007). These difficulties eventually call for new analysis approaches of social systems, and one such approach entails modeling and simulating social systems.

The modeling and simulating of social systems is an overlapping area of modeling and simulation (Zeigler et al. 2000), as well as analyzing social systems (Schweitzer 2002). In the 1960s and 1970s, researchers began modeling and simulating social systems after simplifying the systems. Particularly, some researchers viewed the problem of social systems as a problem that occurred due to interactions between multiple social entities (Bonabeau 2002). Therefore, they invented a modeling approach to regenerate the behavior and interactions of social entities in a social system to ensure that the model could eventually illustrate the observable social phenomena present in the social system. This particular modeling approach has several monikers, such as multiagent simulation, agent-based simulation, and entity-based simulation, among others, and this chapter will call this modeling approach as the agent-based modeling. For example, a group of researchers simplified the concept of social systems to being like ant colonies by assuming that the social entities in the system share the basic characteristics; that is, feeding, security, and incremental colony construction, that the ants exhibit (Nakamura and Kurumatani 1997). This ant colony modeling is one example of an agent-based model in the early days. Similarly, other researchers created hypothetical entities and their systems (e.g., Sugarscape, in which a hypothetical animal ate sugarcanes on a hypothetical plane [Epstein and Axtell 1996]). The researchers modeled these simplified entities and systems with basic models, and they performed diverse virtual experiments by changing the settings of the simplified hypothetical worlds that they created. At the beginning, this simplification of the systems was enough to provide a surprising insight into various issues of social systems (e.g., racial discrimination between social entities and the resulting segregation areas, based on Schelling's model [Schelling 1971]). After they began the initial practice of utilizing simple models, researchers started building more detailed models to represent the social systems more accurately and usefully.

Strengthening agent-based models by providing more details used to better regenerate social systems seems to be a natural direction leading to further development of this modeling approach. Nevertheless, this direction resulted in a major challenge for determining how to describe these complex models. Adding more details inherently increases the complexity of the models, which leads model users to have troubles in fully understanding the model. This lack of understanding of the models invokes in users a lack of trust in the process of applying the model's results to real-world solutions. This is not a simple validation request that a simulation model is generally evaluated upon. The agent-based models are fundamentally expected to regenerate social systems to tell the process of social phenomena within such a system, and this constructed narrative is expected to provide an insight into understanding these social systems. However, without understanding the complex model in detail, the model users cannot produce this important insight. Complex models have many states of entities and the system itself, in addition to various transitions between the states. Further, the complex models include many heterogeneous features across social entities; therefore, a systematic approach is needed to describe the details of the models.

This chapter introduces two formal modeling approaches of social systems in the agent-based approach. Currently, some dominant modeling environments (North et al. 2006; Sklar 2007) of agent-based models provide technical implementation methods, but these technical implementation, that is, program source code, is not clear enough to be understood by other modelers. This leads the problem of model transparency and model reusability. On the other hand, formal specification approaches developed in the system's engineering field are more adapted to composing heterogeneous models and to representing models as black boxes, as well as white boxes. The formal specification approaches are grounded by a set-theory-based mathematics, or formalism (Goodman 1979). Hence, the main question of formal specification of agent-based models is how is it possible to apply formalism to these agent-based models? To introduce these formal modeling approaches in the agent-based models, this chapter first reviews the existing formal specification approaches. Then, the chapter introduces two formal modeling approaches to formally describe an agent-based model. The two approaches are meta-network modeling (Carley 2002; Krackhardt and Carley 1998) and discrete event systems (DEVS) modeling (Zeigler 1972; Zeigler et al. 2000). This research has illustrated the approaches by providing theoretical background and a practical case study. The meta-network model provides a structural template to formally specify how social entities, resources, and information pieces are assigned, linked, and owned. This model is particularly useful when diverse types of entities are included to build an agent-based model. The DEVS model formally illustrates the detailed behavior of models by events and state transition functions. This model can be hierarchically composed as well as reused in other models.

Reviews of agent-based models and formal modeling

This chapter focuses on modeling and simulating social systems with formally specified agent-based models. Therefore, the chapter discusses two specific aspects of the modeling and simulation. First, the chapter argues what should be modeled in social systems—as well as how these entities should be modeled—from the perspective of agent-based models. Second, the chapter reviews how to build and describe an agent-based model with formal descriptions, so the models are appreciated, expanded, and utilized by other modelers.

Key features and characteristics in agent-based models of social systems

Social systems can be modeled from either macro or micro perspective. When modeling a system at the macro level, one of the most frequently utilized models is the system dynamics model (see Figure 4.1) (Forrester et al. 1976; Wolstenholme 2004). In a system dynamics model, the major features are a collection of factors in the modeled social system, and the

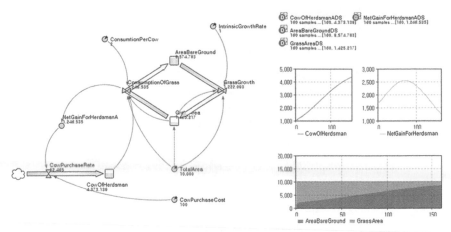

Figure 4.1 Example visualization of system dynamics models. Variables of interests are linked with associated functions. The changes of the observed variables are charted.

features will be selected by assessing whether the features have an influence power to determine a variable to observe and manage. For example, the mortality rate of influenza would be determined by analyzing over-the-counter drug sales, the number of doctors, and the number of children and elderly in the population (Satsuma et al. 2004; Shulgin et al. 1998). The values of the factors either positively or negatively influence each other; eventually, these factors collectively determine the final outcome variable. Particularly, the influence between factors can be modeled as partial derivatives; this macro model would result in continuous changes of the final outcomes over time. System dynamics models rarely describe individual entities, so it is impossible to model heterogeneity between the social entities involved. On the contrary, the system dynamics models aggregate individual entities to multiple variables of diverse aspects. Because of this aggregation, individual attributes that are not homogeneous are not aggregated, so these are not represented in the system dynamics models. Such individual attributes include the interactions between social entities that are inherently different across individuals.

The microlevel models of social systems are the focus of this chapter, and the dominant modeling approach is to utilize agent-based models (see Figure 4.2) (Bonabeau 2002; Davidsson 2002; Nikolai and Madey 2009). The agent-based models describe individual social entities, in addition to the relation between entities. The individual entity has its own specific attributes and behavior. The attributes are the heterogeneous features that capture key characteristics of the entities, and the behavior is the action that the entities execute based on the entities' decision-making processes and their attributes. The behavior affects the environment in which the

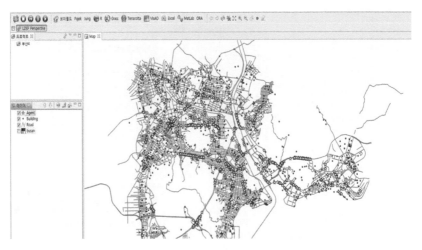

Figure 4.2 Example visualization of agent-based models. Each dot represents a vehicle in a city, and the individual model of vehicles collectively consists of the city traffic model.

entities reside, as well as the other agents in the system. Therefore, unlike the aggregative nature of system dynamics models, agent-based models concentrate on describing heterogeneous individuals with their own attributes and behaviors. Moreover, the agent-based model specifies interactions between the individual entities; often, the interactions become the network of the social system structure. The individual models and the interaction models are the building blocks of the agent-based models, so specifying the building blocks would be the direction to modeling and simulating social systems at the micro level.

Building agent-based models with formal descriptions

Agent-based modeling has been widely used from the domain research, that is, biology, ecology, sociology, and so on, as well as the methodological research, that is, systems engineering, artificial intelligence, and so on. The broad range of users made the model description diverse, and each field has developed distinct methods to explain their models to other modelers. In addition, the agent-based models might have complex behavior routines in agents, diverse parameters in environment, and complex interactions between them, so the descriptions are often difficult to be fully appreciated by model users. Therefore, there has been an effort to clearly represent an agent-based model, so that modelers can appreciate models from others and build upon them. This section introduces such efforts.

As an effort of standardizing the agent-based model, Railsback and Grimm suggested a standard protocol of describing an agent-based model—Overview, Design concepts, and Details (ODD) (Grimm et al. 2006, 2010). This protocol was introduced initially to the ecology modeling community, and now, it is a recommended protocol of *Journal of Artificial Society and Social Simulations*, one of the major venues of agent-based modeling research. The ODD is essentially a structured description of agent-based models by stating (1) modeling purpose, (2) state variables and their scales, (3) model process overview and scheduling of the processes, (4) design concepts of the model, (5) initialization information, (6) input description, and (7) submodels, that is, detailed formula, used by components in the agent-based models. This protocol is a comprehensive list that is not just limited to the model description itself, but this includes the background of the model and the modeling focus from the modeler. The actual descriptions in the steps can be textual, formula and numeric variables, and flowcharts.

In the systems engineering field, the DEVS formalism (Zeigler et al. 2000) is known to be a complete specification of the DEVS, which agent-based models belong to. Therefore, applying the DEVS formalism would be a natural step to describe an agent-based model in the systems engineering field. Moreover, after the classic DEVS formalism is introduced, many variants are invented with the basis of the classic formalism. Cell-DEVS (Wainer 2006), a variant of the DEVS formalism, integrates the designs of cellular automata, which is a basis of agent-based models, into the formalism. Two variant formalisms, dynamic structure discrete event (DSDE) (Barros 1998) and dynDEVS (Uhrmacher 2001), are introduced by strengthening the dynamic structuring of the DEVS formalism, and these two formalisms can be applied to the agent-based model specification because the dynamic interactions between the models are the keys of evolutions that are expected to see in the agent-based models. While both extended formalisms have a transition function for structural changes, the difference between DSDE and dynDEVS is the place of the transition function. DSDE invokes the structure change at the network model corresponding to the DEVS coupled model, whereas dynDEVS changes structure at the individual model level corresponding to the DEVS atomic model (Shang and Wainer 2006). The DEVS coupled model and the DEVS atomic model will be introduced in depth in the Section "Two formal model specifications: Meta-network and DEVS formalism." Assuming that agent-based models are inherently decentralized models, dynDEVS might be more suitable to apply, yet this decentralized structure change might cause conflicts in structure changes between agents.

Systems engineering and software engineering, which are two relevant fields of agent-based modeling, have utilized Unified Modeling Language (UML) (Medvidovic and Rosenblum 2002) to describe the designed system. Therefore, there have been suggestions to apply UML to agent-based modeling. One approach in this line of works is using the original UML without modification. Rather than introducing new notations to UML, Bersini introduces the design patterns of components in agent-based models (Bersini 2012). Another approach is introducing new notations to UML, just like SysML (Friedenthal et al. 2008) for the systems engineering domain. AgentUML (Bauer et al. 2001) is one example of the new notations. For instance, AgentUML specifies a new *agent class diagram* that is expanded from the *class diagram*. The agent class diagram has (1) an agent name; (2) state descriptions; (3) actions; (4) methods; (5) capabilities, service descriptions, and supported protocols; (6) society names; and (7) related automata, and this is much expanded from the class diagram that has only three information pieces: a class name, a property set, and a method set. Similarly, the state diagram and the sequence diagram in UML are expanded to accommodate the agent-based modeling description.

Another set of model description standards has a format of logical statements, like the first-order logic. Strictly declarative modeling language is a programming language that specifies agent behaviors with antecedents and consequents that are specified by predicate clauses (Moss et al. 1998). Similarly, planning domain definition language has an agent action definition that has precondition clauses and postcondition clauses (Fox and Long 2003). Soar (Carley et al. 1992; Laird et al. 1987) is another programming language as well as simulation environment that specifies agent actions with a sequence of *if-then* statements, and the condition clauses and the block statements within the *if* statement consist of logical clauses to alter the states of components in the agent-based model by conditions. Unlike the DEVS formalism and the UML, these logic-based modeling languages require modelers to infer agent behaviors with a set of conditions and consequences.

Besides the formal description of models, the modelers can implement the model with a simulation programming environment, and then the modelers make the model available to the public. This has been a recent trend in the publication of the agent-based modeling field because this practice helps others review, expand, and experiment with the published model. Furthermore, there is a chance of gap between a formal description of the model and actual program codes of the model, so in a sense, this might be the most accurate description of the model though this might not be the description with clarity. Frequently, the agent-based modelers utilize modeling environment and software libraries, specialized to build agent-based models, to code their models. The implementation of models with generic programming language has much lower clarity than the other implementations. Notable environments are Repast (North et al. 2006), NetLogo (Sklar 2007), Mason (Luke et al. 2004), AnyLogic (Borshchev et al. 2002), and so on. There is a comprehensive list of environments in a paper (Nikolai and Madey 2009) as well as a Wikipedia entry (Wikipedia 2011).

Two formal model specifications: Meta-network and DEVS formalism

Formalism (Goodman 1979) is the theory that statements of logics can be thought of as statements of sequences of string manipulations. This theory emphasizes a formal statement and inductive reasoning to rationalize phenomena. The formal statements are specifically important in analyzing systems because the systems will be logically reasoned by mathematical tools if the systems have been formally stated. Here, the formal statements would include the statements of mathematical sets, matrices, and functions. This section presents two particular formal models that are useful in describing social systems models.

Meta-network model

A meta-network model (Carley 2002; Krackhardt and Carley 1998) is a multimode and multirelation network model used to represent the structure of a social system. We might describe it using a matrix of relations, as is demonstrated in Table 4.1. From a task performance perspective of social systems, there are four basic types of nodes of interest: agents, knowledge, resources, and tasks. In addition, other extensive types of nodes (e.g., location, belief, event, and organization) can be included. These node types can vary depending on the modeling objectives. The relations among these included the following: who interacted with whom, who has access to what resources, what has what knowledge or expertise, who can or has done what task, what resources are needed for what task, and what knowledge is needed for completing what task or for using what resource. Each of these can be observed, albeit with some level of uncertainty, but for many social systems, this can only be done in an *after the fact* fashion. In Table 4.1, for example, we define a possible network for each of the cells.

Table 4.1 Meta-Network Component Networks

	Agent (A)	Knowledge (K)	Resources (R)	Tasks (T)
Agent (A)	Social network (AA)	Knowledge network (AK)	Resource network (AR)	Assignment network (AT)
Knowledge (K)		Information network (KK)	Skills network (KR)	Knowledge needs network (KT)
Resources (R)			Substitution network (RR)	Resource needs network (RT)
Tasks (T)				Task precedence network (TT)

From the viewpoint of formal mathematical notations, each type of social entity is represented as a vector (e.g., a collection of agents is identified as a vector A). The interactions between entities in two different types can be represented as an adjacency matrix whose basis is the two vectors corresponding to the types. For example, the assignment of an agent to a task can be represented by AT_{ij}. Then, an agent A_i is the agent assigned to a task T_j. This formal model of a social system structure is useful in the further reasoning of social systems. If we look for an agent assigned to the same task, we can perform a matrix multiplication of $(AT \cdot AT^T)_{ij}$, which is a multiplication of AT with the transpose matrix of itself, or AT^T. After the multiplication, the matrix will indicate the shared task relation between agents: the agent A_i and the agent A_j share a task if $(AT \cdot AT^T)_{ij}$ is one. In the agent-based models, the dynamic changes of social system structures can be explicitly stored by following the meta-network model, and modelers can further observe and investigate the simulated social systems by utilizing this formal network model.

The meta-network is not just limited to a social network. Rather, the meta-network contains multiple networks together. The meta-network covers the broader concepts related to a social system's structure. These concepts are task precedence, task assignment, resource distribution, information diffusion, and resource/information requirements for tasks. In the meta-network, each of these concepts is formally represented, and there is a tacit acknowledgment that each of these subnetworks influences the other. From a representational perspective, the meta-network is a representation scheme that modelers can use to store any of their knowledge regarding how the social system is structured for, prepared for, and capable of executing various functions. Clearly, each of these networks can be—and, in some cases, has already been—modeled on its own. For instance, the task precedence network in Table 4.1 is often analyzed by analysts in the operations research field, whereas the information network in Table 4.1 is a frequent topic for information scientists researching a knowledge management system or knowledge map. However, collectively, these networks are necessary for understanding how a social system operates. This meta-network formulation supports the integrated modeling of social systems from a multidisciplinary focus.

Discrete event system model

Introduced by Zeigler (Zeigler and Vahie 1993; Zeigler et al. 2000), DEVS formalism is a formal specification of a DEVS. DEVS formalism is applicable to social systems modeling because many social entities exhibit the characteristics of the discrete event system. For example, an individual en route to a certain point stays at the *driving* state until the individual arrives at the destination. This state transition of an agent can be represented as a discrete event system. Moreover, this formalism enables specifying a large discrete event system by hierarchically decomposing the system into modules. Continuing the presentation of the previous example, many driving agents represented as state transitions exist, so a modeler might decompose the interactions between the driving agents into co-locating groups of agents. In other words, the models, both agents and environment, can be represented as atomic models, and their interactions can be represented by coupled models. Each coupled model has identical inputs and outputs as shown in atomic models, so the coupled model can be treated as models to be coupled in the higher hierarchy of models, which is called *closed under coupling*. This idea is represented in Figure 4.3.

After the decomposition, we have two types of specifications: (1) the hierarchy structure of the system modules and (2) the internal structure, or the state transition, of the system modules. These two specifications of the decomposed DEVS correspond to the coupled model and the atomic model in the DEVS formalism, respectively. First, a coupled model is specified as the following 7-tuple in the set-theoretic context:

$$CM = <X, Y, M, EIC, EOC, IC, SELECT>, \text{where}$$

X: a set of input events

Y: a set of output events

M: a set of component models

$$EIC: \subset CM.X \times \bigcup_i M_i.X, \text{external input coupling}$$

$$EOC: \subset \bigcup_i M_i.Y \times CM.X, \text{external output coupling}$$

$$IC: \subset \bigcup_i M_i.Y \times \bigcup_i M_i.X, \text{internal coupling}$$

$$SELECT: 2^M - \phi \rightarrow M, \text{tie-breaking function}$$

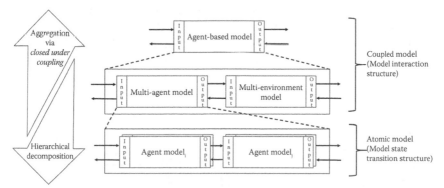

Figure 4.3 Hierarchical structuring of agent-based models. This is one potential scenario of configuring an agent-based model when we compose the model with the DEVS formalism.

Second, under the assumptions of the hierarchical structure of the coupled model, we identify an atomic model representing an individual social entity. According to DEVS formalism, an atomic model is specified as the following 7-tuple:

AM = <X, Y, S, δ_{ext}, δ_{int}, λ, ta>, where

X: a set of input events

Y: a set of output events

S: a set of sequential states

δ_{ext}: Q×X → S, an external transistion function,

where Q = {(s, e)| s ∈ S, and 0 ≤ e ≤ ta(s)}, *total state set of AM*

δ_{int}: S → S, an internal transistion function

λ: S → Y, an output function

ta: S → $R_{0,\infty}$, a time advance function

The previous specification is the DEVS formalism in the algebra format, and this specification style can be visualized with the DEVS diagram (Song and Kim 2010) illustrated in Figure 4.4. The DEVS atomic model has a single box to represent the model itself and multiple ellipses to enumerate the states of the atomic models. In an ellipse, the upper indicates the state name, and the lower specifies the time advance function value of the state. The dotted line between the ellipses is an internal transition function, and the solid line is an external transition function between two states. On top of the lines, the annotation with a question mark specifies the triggering event of the transition, and the annotation with an exclamation mark indicates the output event of the model. The bold outlined ellipse is the initial state of the model. The input and the output events are tagged to the box. The DEVS coupled model has a similar diagram with a few differences. First, instead of states, the diagram for the DEVS coupled model has multiple boxes representing the DEVS models that can be either atomic or coupled, by virtue of closed under coupling. A line links (1) models, (2) a model and an output event, and (3) an input event

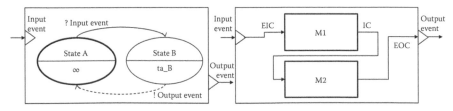

Figure 4.4 DEVS diagrams (left) for an atomic model and (right) a coupled model.

and a model, and these lines represent internal couplings, external output couplings, and external input couplings, respectively. This section uses the DEVS diagram instead of the algebra format.

Comparison of two proposed formal representations

Agent-based models generally require two specifications: a model and an input to the model. The previous two formal approaches correspond to the two specifications. First, the meta-network is a formal matrix representation of the interaction structure between agents and environment, if we include the locations, the resources, the information pieces, and so on, in the scope of the environment. Hence, this formal representation is applicable when we survey the target system and represent the system in the simulation model. While there are multinary definitions of interactions, the binary relation can be the fundamental basis of diverse interaction types. The meta-network is a collection of such binary relations between multiple entities that varies in the ontology of focused social situation. Moreover, the meta-network shares a matrix structure to infer the shared relations and the multihop relations through matrix algebra, that is, transpose and matrix multiplication. If a model requires such information, the meta-network is an easy form to manipulate. In addition, this network-based representation of modeled population is frequently utilized particularly in sociology, that is, social network analysis.

While the input dataset uses the meta-network, the agent behavior needs a different formal representation. Often, an agent makes an autonomous decision by combining the internal states and the external events, so the basis of the agent behavior is a finite state machine from the perspective of systems engineering. The finite state machines can formally be specified by petri-net, DEVS, event graph, and so on. This chapter chose the DEVS formalism because this formalism provides the time advance and the internal transition, which are essential in simulating agent behavior with state machines. Furthermore, a common problem of state machines, the growing number of states in complex systems, is partially resolved by adapting the hierarchical structure.

These two formal representations are complementary: the meta-network supports the input dataset representation, and the DEVS formalism strengthens the model description. In the agent-based models, the agents and the environment dynamically interact and change the interaction structure, so the model interaction would be a common issue at the input and the model level at the same time. The meta-network is a network model with a matrix structure, so the interactions are the key data in the representation. The DEVS formalism also has a network structure between models, which can be either agents or environment, by defining coupled models. This creates a partial overlapping between the two formal representations. For example, the social network among agents would be the basis of the coupling structure between agent models. However, the meta-network model does not specify which events should be directed from which model to which model, so the DEVS formalism should provide further details on the model interaction. In addition, the DEVS formalism gives a full specification of behavior of agents and environment as discrete event system models. However, the meta-network model only provides the abstract status of societal structure. This chapter provides an example of applying the meta-network model to the input dataset, and applying the DEVS formalism to the model descriptions in the Section "Case study: Agent-based model in disaster response."

This section presented two formal approaches to specify the features of agent-based models. The first formal approach was the meta-network model, which formally describes the interactions between diverse types of social entities. This model does not specify the behavior of social entities, but it is very applicable to modeling the structure of social entities. The second formal approach was the DEVS formalism, which specifies the detailed state transition mechanisms of social entities. In addition, the DEVS formalism supports the formal specification of model structures, but the structure was limited to representing hierarchically composed simulation models, rather than social entities in general. In the next section, the chapter presents a case study of agent-based modeling of a disaster response scenario, and the case study illustrates how to apply the meta-network model to the input data description, and how to apply the DEVS formalism to the model behavior description.

Case study: Agent-based model in disaster response

Disasters such as earthquakes (Beavan et al. 2011), hurricanes (Comfort et al. 2010), tsunami (Matanle 2011), and massive landslides (Lee and Jung 2011), need well-coordinated responses to minimize the damage to our society. The damage is not limited to one facet of our society. Disasters affect our traffic infrastructure, health care services, communication lines, energy supply, shelters, and security (Tierney et al. 2001). Even if we respond to a single dimension, this single response calls upon the supports from the full spectrum. Getting food to evacuees, a single response, requires traffic routes for delivery, communication between provider and shelter, and more. To guarantee this general support, an organization would need authority over the multiple dimensions of our society. However, this idea is impractical in our society. Many societal infrastructures, resources, and information are managed by separate organizations, to ensure technical specialization and task delegation. Under this decentralized structure, organizations should cooperate as a single body to respond to disasters effectively (Harrald 2006).

Structures between social entities, including disaster response organizations, have evolved from hierarchies to networks (Simon 1964; Weber 1947). Traditionally, a tree-shaped hierarchy was enough to handle the complexity of organizational cooperation. Cooperation was limited to a simple subgroup of organizations, with no demands crossing group affiliations. However, disaster response makes simple hierarchy impractical for several reasons. First, it is a collective response of organizations from multiple layers and multiple branches (Comfort et al. 2010). Second, disaster's characteristics require collaboration crossing affiliations. The disaster scene is often a local area not well known to federal organizations and nongovernmental organizations (NGOs). The local residents frequently require outside support from government, military, and NGOs, which calls upon cooperation and the flow of information to and from the local area (Schneider 2005). Third, disaster response takes time and shifts its cooperation structure within that time. For these reasons, disaster response organizations instead collaborate in a network structure. That network grows complex by coupling between subgroups, changing demands of information and resources, and the dynamism of participations. Diverse techniques have sought to answer it, including social network analysis (Wasserman and Faust 1994), agent-based modeling and simulations (Bonabeau 2002), and text analysis (Diesner and Carley 2004). Previous research analyzes potential organizational networks in the domain with computational approaches. Many works evaluate the potential performance of typical structures like cellular, hierarchy, or scale-free (Airoldi and Carley 2005) with agent-based models and social network metrics to gauge information propagation speed, task accuracy, and completeness (Moon and Carley 2007), and so on.

This case study is presented because this study utilizes the two formal representations of social systems: the meta-network of an input dataset and the DEVS formalism of model descriptions. First, the disaster response organizations have complex interactions, and these interactions are not just cooperation without context, but the interactions are motivated by sharing resources of one organization with another. Hence, the interorganizational structure with background context is a collection of organizations, resources, and communication messages, which becomes simulation models and exchanged objects. To efficiently show this structure, the study applies the meta-network. Second, the disaster response organizations are dynamic entity with autonomous behavior. This is formally modeled by utilizing the DEVS formalism. The decision-making process of an organization is specified by the state transition model. This study assembles the two formal specifications and results in the performance of the interorganizational structure over the course of disaster responses.

Meta-network for interorganizational structure description in disaster response

This subsection describes the scenario and dataset of this study. The case study starts from the logs of the disaster response communication during Hurricane Katrina in 2005. Specifically, the communication logs come from situation reports by the Louisiana Office of Homeland Security and Emergency Preparedness, refined by the Center for Disaster Management at the University of Pittsburgh. These logs illustrate communicative links between organizations during disaster response in the context of resource request. Descriptive statistics of the dataset are found in Table 4.2 and Figures 4.5 and 4.6.

From the log dataset, we recovered a meta-network structure. The network includes three node classes. The log specifies what to be delivered to whom, so the network has a set of organization nodes and a set of resource nodes. One more node class is included, the communication message node, not from the dataset, but from the simulations. This node

Table 4.2 Summary of Simulation Dataset from the Katrina Situation

Category	Descriptive statistics
Number of events	2576
Collection period	Aug.27th ~ Sep.6th (250 hr)
Involved organizations	169
Exchanged services and resources	123
Number of communication links	616
Number of status type	8

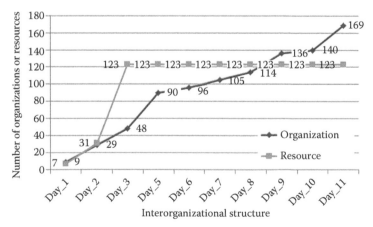

Figure 4.5 Number of involved organizations and delivered resources over period.

represents every message exchanged in a simulation. The links of the network have six types of semantics: interorganizational structure, resource ownership, resource requirement, received message, sent message, and resource delivery message. These uncovered networks comprise a meta-network that represents the scenario of interest throughout the simulation. This particular meta-network setting is illustrated in Table 4.3, and this becomes a major input dataset to the simulator, excluding the messages set that are dynamically added in the simulation.

DEVS formalism for interorganizational structure description in disaster response

To evaluate the organized disaster response efficiency, an agent-based model is utilized with the given input dataset. The model is a simplified description of message propagations through an interorganizational structure, and the message is created when an organization demands a resource

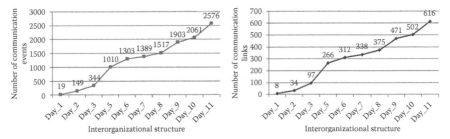

Figure 4.6 Descriptive statistic of dataset: (left) the number of communication events and (right) the number of communication links.

Table 4.3 Meta-Network of the Dataset

	Organizations (O)	Resources (R)	Messages (K)
Organizations (O) (169 organizations)	Interorganizational Structure (OO) (Density = 0.022)	Resource Ownership (OR^S) (Density = 0.030)	Received Message (OK^S) (Density = 0.006)
		Resource Requirement (OR^D) (Density = 0.051)	Sent Message (OK^D) (Density = 0.006)
Resources (R) (123 resources)	—	Not Used	Resource Delivery Message (RK) (Density = 0.008)
Messages (K) (2576 messages)	—	—	Not Used

that the organization does not have. Once another organization with that resource receives the message, it will respond to. This is a simplified process of single-resource delivery, simulated with multiple deliveries. We model the organizational structure into two layers. The first layer is the interorganizational structure, which is the network structure between organizations, and this is modeled by a coupled model in the DEVS formalism. The second layer is the intraorganizational structure, which is an agent model corresponding to a single organization, and this is modeled by using a coupled model that includes four atomic models in the DEVS formalism.

Interorganizational structure

An interorganizational structure describes communication links among organizations (O), where a single organization requests, broadcasts, and responds to messages (K) demanding resources (R) through the interorganizational structure. The interactions have two types: broadcast links for message broadcasts and response links for resource delivery. In this chapter, we assume the broadcast links are a directed network connecting pairs of organizations as defined in the dataset, and a message broadcasts through the interorganizational structure. On the other hand, the response links comprise a complete network of organizations. Figure 4.7 shows a conceptual interorganizational structure, with broadcast links as described in Figure 4.8 in the simulations using the interorganizational structure specified in the meta-network, specifically the network, OO in Table 4.3. This is actually the overlapping area of the meta-network and the DEVS formalism. A network in the meta-network becomes a design of model coupling in the DEVS formalism.

Intraorganizational structure

An intraorganizational structure describes a single organization as an agent model and single node in the interorganizational structure. We hypothesized that each organization consists of a decision-making model and three information buffer models: event buffer, broadcast buffer, and response buffer. These submodels divide a real-world disaster-response organization

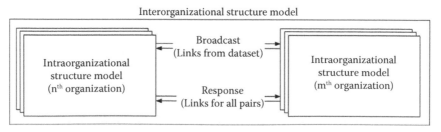

Figure 4.7 A sample of interorganizational structure model.

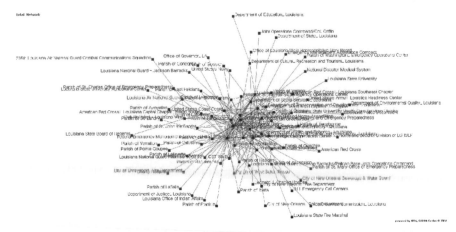

Figure 4.8 Visualized topology of the interorganizational structure model.

into functional subgroups. The decision-making model corresponds to the leadership of the disaster response organization where key decisions are made, such as when to send a resource to other organizations. The event, broadcast, and response buffers correspond to the field, communication, and response units, respectively. The field unit gathers the resource demand on the ground, and sends the demands to the leadership, also known as the decision-making model, as an event message object. The communication unit receives resource request messages from other organizations, and delivers that message to the decision-making model. Finally, the response unit, under the control of the decision-making model, responds to the requesting organization. Figure 4.9 captures these processes and structures in the DEVS formalism.

The decision-making model describes various decision operations: sending a message that requests a resource, broadcasting a received message, and responding to a message by providing a resource. We formally describe the decision-making process with meta-network notation in the Section "Meta-network model." When an organization (o_i) finds the needs of a resource (r_j) in a message (k_k) from the request buffer, the decision-making model checks whether o_i has r_j or not, which can be easily confirmed by the meta-network. If o_i has r_j, the decision-making model does not broadcast or respond k_k because this request was solved internally. However, if o_i does not have r_j, the decision-making model decides to broadcast k_k to the linked organizations through the interorganizational structure (OO). When o_i takes k_k from the broadcast buffer, the decision-making model decides to respond and transfer r_j (when $OR_{(i,j)}{}^S>0$, $OK_{(i,k)}{}^S = 1$, and $RK_{(j,k)} = 1$) or to broadcast k_k to other linked organizations (when $OR_{(i,j)}{}^S = 0$, $OK_{(i,k)}{}^S = 0$, or $RK_{(j,k)} = 0$). When o_i takes r_j for k_k from the response buffer, the decision-making model assumes that the message

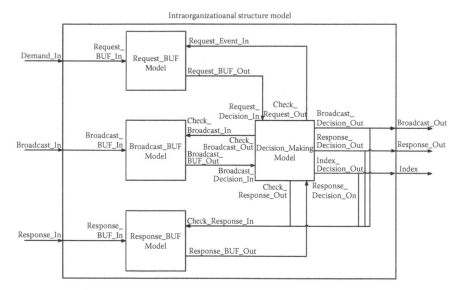

Figure 4.9 Model diagram for intraorganizational structure described in the DEVS coupled diagram.

(k_k) is resolved. These decision-making cases are encoded in the states of the DEVS atomic model illustrated in Figure 4.10.

Summary of agent-based model and simulation result

We summarized inputs, outputs, and auxiliary parameters for our agent-based model in Table 4.4. The input set is the past communication logs, including the communication links, resource demands, and responses, over the course of disaster responses, and this input organization is specified by the meta-network. The resource requirement network (OR^D) consists of relationships between demanding organizations and demanded resources in the meta-network. Following the simulation timeline, we send each simulation message (OK^D) from a demanding organization to request a demanded resource. The resource requirement network is independent of the other network in the meta-network, because the resource requirement information comes from the demanding organization at a disaster scene. The outputs are four different aspects of organizational performances. The auxiliary parameters are a set of waiting time specifications for an agent's behaviors.

Figure 4.11 shows the simulation results of the model. This result shows a simple resource exchange records over the simulation period. One piece of information from the result is that the number of resource exchange counts converges to a plateau from Day 3. This suggests that there is a limitation on the resource delivery frequencies. On the other hand, the delivery latency, which can only be estimated by the simulation because this particular information

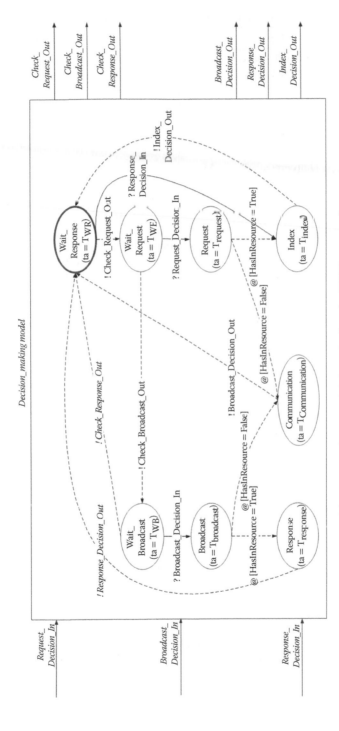

Figure 4.10 Operational behavior of decision-making model described in the DEVS atomic diagram.

Table 4.4 Parameter Table for Simulation Model of Crisis Management

Type	Name	Value	Implications
Input	Network source (OO network)	Fixed from the past records	Communication records between disaster response organizations
	Simulation scenario of resource demands	Fixed from the past records	Resource demand records from organizations in the crisis period
Output	Delivered resource count	Measured from simulations	Number of resources delivered as requested
	Delivered resource latency	Measured from simulations	Time interval between a request and a response of a resource
Parameters	T_{out} in three buffer models	Default = 0.1	Buffer's delay time in responding to the decision-making model's message checking
	T_{WR}, T_{WB}, and T_{WE} in decision-making model	Default = 0.1	Decision-making model's delay time in message checking
	$T_{communication}$ in decision-making model	Default = 0.1	The lambda value of the Poisson distribution for the decision-making model's delay time in broadcasting the message to linked organizations
	$T_{broadcast}$ and $T_{request}$ in decision-making model	Default = 1	The lambda value of the Poisson distribution for the decision-making model's delay time in deciding whether a message should be rebroadcast or responded by itself
	$T_{response}$ in decision-making model	Default = 1	The lambda value of the Poisson distribution for the decision-making model's delay time in transferring the resources for responses
	T_{index} in decision-making model	Default = 1	The lambda value of the Poisson distribution for the decision-making model's delay time in finishing the delivery of the resources internally

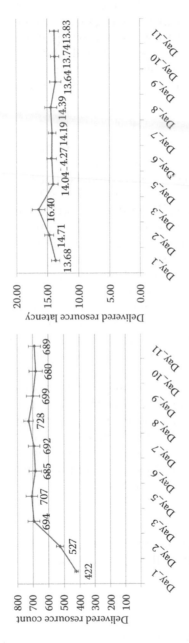

Figure 4.11 Virtual experiment results: (left) delivered resource count and (right) delivered resource latency.

was absent in the dataset, does not improve over the time. Finally, Figures 4.5 and 4.6 illustrating the dataset implies that the number of responding organizations is increasing. In short, this interorganizational structure has limited resources to deliver; the responding organizations are arriving at the scene, but the speed of the delivery is not improving. From this short simulation example, it is reasonable to conjecture that the interorganizational structure, which determines the cooperation between organizations, should be examined further to increase the performance of this social system.

Conclusion

This chapter introduced the method for formally specifying the structure and behavior of social systems. The formal specification is applied to the description of agent-based models that have suffered from the model transparency and reusability issues. When applying a model to an analysis case in the real world, the model should be trusted by the users, as well as expanded enough to fully capture the scenario of interests. To trust a model and its results, the users should see through the model inside, yet many agent-based modeling environments depend on technical specifications, such as programming languages, rather than a formal approach, which the system's engineering field utilizes. This technical-level specification becomes a significant burden when the model is intended to be reused and reapplied to a modified environment with a changed model composition. Therefore, this chapter proposes to provide the formal specifications of agent-based models to potential model users. The chapter suggests two formal models that modelers can use to specify their agent-based models and scenarios. One formal model is the meta-network model, and the other formal model is the DEVS model. The meta-network model is useful in representing the structure of the social systems, whereas the DEVS model is applicable when modelers illustrate the behavior of the social systems. Finally, this chapter presents a case study demonstrating how to apply such formal methods to an agent-based model of a social system.

Acknowledgments

I gratefully acknowledge the support that I received from the Public Welfare & Safety Research Program through the National Research Foundation of Korea (NRF) (2012-0029881).

References

Airoldi, E. and Carley, K. M. (2005). Sampling algorithms for pure network topologies: Stability and separability of metric embeddings. *ACM KDD Explorations (Special Issue on Link Mining)*, 7, 13–22.

Balmer, M., Cetin, N., Nagel, K., and Raney, B. (2004). Towards Truly Agent-Based Traffic and Mobility Simulations. In *AAMAS '04 Proceedings of the Third International Joint Conference on Autonomous Agents and Multiagent Systems* (pp. 60–67). New York.

Barros, F. (1998). Abstract simulators for the DSDE formalism. *Simulation Conference Proceedings, 1998*. Winter (Vol 1, pp. 407–412). Washington, DC.

Bauer, B., Müller, J., and Odell, J. (2001). Agent UML: A formalism for specifying multiagent software systems. *International Journal of Software Engineering and Knowledge Engineering*, 11, 207.

Beavan, J., Fielding, E., Motagh, M., Samsonov, S., and Donnelly, N. (2011). Fault location and slip distribution of the 22 February 2011 Mw 6.2 Christchurch, New Zealand, earthquake from Geodetic data. *Seismological Research Letters*, 82(6), 789–799. doi:10.1785/gssrl.82.6.789.

Bersini, H. (2012). UML for ABM. *Journal of Artificial Societies and Social Simulation*, 15(1), 9.

Blanchard, B. S. and Fabrycky, W. J. (2010). *Systems Engineering and Analysis* (5th ed.). New York: Prentice Hall.

Bonabeau, E. (2002). Agent-based modeling: Methods and techniques for simulating human systems. *Proceedings of National Academy of Sciences*, 99(3), 7280–7287.

Borshchev, A., Karpov, Y., and Kharitonov, V. (2002). Distributed simulation of hybrid systems with AnyLogic and HLA. *Future Generation Computer Systems*, 18(6), 829–839.

Carley, K. M. (2002). Smart agents and organizations of the future. In *The Handbook of New Media: Social Shaping and Consequences of ICTs* (pp. 206–220). Thousand Oaks, CA: Sage Publications Inc.

Carley, K. M., Kjaer-Hansen, J., Newell, A., and Prietula, M. (1992). Plural-Soar: A prolegomenon to artificial agents and organizational behavior. In *Artificial Intelligence in Organization and Management Theory* (pp. 87–118). Amsterdam, the Netherlands: North-Holland.

Carley, K. M. and Newell, A. (1994). The nature of the social agent. *Journal of Mathematical Sociology*, 19(4), 221–262.

Comfort, L. K. (2007). Crisis management in hindsight: Cognition, communication, coordination, and control. *Public Administration Review*, 67(1), 189–197.

Comfort, L. K., Oh, N., Ertan, G., and Scheinert, S. (2010). Designing adaptive systems for disaster mitigation and response: The role of structure. In *Designing Resilience: Preparing for Extreme Events* (1st ed., pp. 59–97). Pittsburgh: University of Pittsburgh Press.

Davidsson, P. (2002). Agent based social simulation: A computer science view. *Journal of Artificial Societies and Social Simulation*, 5(1), 7.

Diesner, J. and Carley, K. M. (2004). Using Network Text Analysis to Detect the Organizational Structure of Covert Networks. In *North American Association for Computational Social and Organizational Science Conference (NAACSOS)*. Pittsburgh, PA.

Epstein, J. M. and Axtell, R. (1996). *Growing Aritificial Societies* (pp. 1–20), Washington, DC: The Brookings Institution.

Forrester, J. (1971). Counterintuitive behavior of social systems. *Theory and Decision*, 2(2), 109–140.

Forrester, J. W., Mass, N. J., and Ryan, C. J. (1976). The system dynamics national model: Understanding socio-economic behavior and policy alternatives. *Technological Forecasting and Social Change*, 9(1–2), 51–68.

Fox, M. and Long, D. (2003). PDDL2. 1: An extension to PDDL for expressing temporal planning domains. *Journal of Artificial Intelligence Research*, 20, 61–124.

Friedenthal, S., Moore, A., and Steiner, R. (2008). *A Practical Guide to SysML: Systems Model Language*. Waltham, MA: Morgan Kaufmann.

Goodman, N. D. (1979). Mathematics as an Objective Science. *American Mathematical Monthly*, 86(7), 540–551.

Grimm, V., Berger, U., Bastiansen, F., Eliassen, S., Ginot, V., Giske, J., Gross-Custard, J. et al. (2006). A standard protocol for describing individual-based and agent-based models. *Ecological Modelling*, 198(1), 115–126.

Grimm, V., Berger, U., DeAngelis, D. L., Polhill, J. G., Giske, J., and Railsback, S. F. (2010). The ODD protocol: A review and first update. *Ecological Modelling*, 221(23), 2760–2768.

Guetzkowm, H. and Simon, H. A. (1955). The impact of certain communication nets upon organization and performance in task-oriented groups. *Management Science*, 1(3), 233–250.

Harrald, J. R. (2006). Agility and discipline: Critical success factors for disaster response. *ANNALS of the American Academy of Political and Social Science*, 604(1), 256–272.

Krackhardt, D. and Carley, K. M. (1998). A PCANS model of structure in organizations. In *1998 International Symposium on Command and Control Research and Technology* (pp. 113–119). Conference held in June. Monterray, CA. Vienna, VA: Evidence Based Research.

Laird, J. E., Newell, A., and Rosenbloom, P. S. (1987). SOAR: An architecture for general intelligence. *Artificial Intelligence*, 33(1), 1–64.

Lee, G. and Jung, K. (2011). Analysis of flood and landslide due to torrential rain in Seoul. In *Korean Society of Industrial and Applied Mathematics*. (pp. 47–48).

Luke, S., Cioffi-Revilla, C., Panait, L., and Sullivan, K. (2004). Mason: A new multi-agent simulation toolkit. In *Proceedings of the 2004 SwarmFest Workshop* (Vol. 8). Ann Arbor, MI.

Matanle, P. (2011). The Great East Japan Earthquake, tsunami, and nuclear meltdown: towards the (re) construction of a safe, sustainable, and compassionate society in Japan's shrinking regions. *Local Environment*, 16(9), 823–847.

Medvidovic, N., Rosenblum, D. S., Redmiles, D. F., and Robbins, J. E. (2002). Modeling software architectures in the Unified Modeling Language. *ACM Transactions on Software Engineering and Methodology*, 11(1), 2–57.

Moon, I. C., & Carley, K. M. (2007). Modeling and Simulating Terrorist Networks in Social and Geospatial Dimensions. *IEEE Intelligent Systems*, 22(5), 40–49.

Moss, S., Gaylard, H., Wallis, S., and Edmonds, B. (1998). SDML: A multi-agent language for organizational modelling. *Computational & Mathematical Organization Theory*, 4(1), 43–69.

Nakamura, M. and Kurumatani, K. (1997). Formation mechanism of pheromone pattern and control of foraging behavior in an ant colony model. *Artificial Life V*, 67. Cambridge, MA: The MIT Press.

Nikolai, C. and Madey, G. (2009). Tools of the trade: A survey of various agent based modeling Platforms. *Journal of Artificial Societies and Social Simulation*, 12(2).

North, M. J., Collier, N. T., and Vos, J. R. (2006). Experiences creating three implementations of the repast agent modeling toolkit. *ACM Transactions on Modeling and Computer Simulation (TOMACS)*, 16(1), 1–25. doi:10.1145/1122012.1122013.

Oh, N. and Moon, I. C. (2008). Searching New Structure for the Effective Disaster Management System. In *69th American Society of Public Administration Annual Conference (ASPA'08)*. Dallas, TX.

Pew, R. W. and Mavor, A. S. (1998). *Modeling Human and Organizational Behavior: Application to Military Simulations.* Washington, DC: National Academies Press.

Satsuma, J., Willox, R., and Ramani, A. (2004). Extending the SIR epidemic model. *Physica A: Statistical Mechanics and its Applications,* 336(3–4), 369–375.

Schelling, T. C. (1971). Dynamic models of segregation. *Journal of Mathematical Sociology,* 1(2), 143–186.

Schneider, S. K. (2005). Administrative breakdowns in the governmental response to Hurricane Katrina. *Public Administration Review,* 65(5), 515–516.

Schweitzer, F. (2002). *Modeling Complexity in Economic and Social Systems.* 394. Singapore: World Scientific.

Shang, H. and Wainer, G. (2006). A simulation algorithm for dynamic structure DEVS modeling. *Proceedings of the 38th Simulation Conference on Winter.* Monterey, CA.

Shulgin, B., Stone, L., and Agur, Z. (1998). Pulse vaccination strategy in the SIR epidemic model. *Bulletin of Mathematical Biology,* 60(6)1123–1148.

Simon, H. A. (1964). On the concept of organizational goal. *Administrative Science Quarterly,* 9(1), 1–22.

Sklar, E. (2007). NetLogo, a multi-agent simulation environment. *Artificial Life,* 13(3), 303–311. doi:10.1162/artl.2007.13.3.303.

Song, H. and Kim, T. (2010). DEVS diagram revised: A structured approach for DEVS modeling. In *Proc. European Simulation Conference. Eurosis,* Hasselt, Belgium.

Tesfatsion, L. (2002). Agent-based computational economics: Growing economies from the bottom up. *Artificial Life,* 8(1), 55–82. doi:10.1162/106454602753694765.

Tierney, K. J., Lindell, M. K., and Perry, R. W. (2001). *Facing the Unexpected: Disaster Preparedness and Response in the United States.* Washington, DC: Joseph Henry Press.

Uhrmacher, A. (2001). Dynamic structures in modeling and simulation: A reflective approach. *ACM Transactions on Modeling and Computer Simulation,* 11(2), 206–232.

Wainer, G. (2006). Applying Cell-DEVS methodology for modeling the environment. *Simulation,* 82(10), 635–660.

Wang, F. (2005). Agent-based control for networked traffic management systems. *IEEE Intelligent Systems2,* 20(5), 92–96.

Wasserman, S. and Faust, K. (1994). *Social Network Analysis: Methods and Applications* (Vol. 8, p. 827). New York: Cambridge University Press.

Weber, M. (1947). *Theory of Social and Economic Organization.* New York: Oxford University Press.

Wikipedia. (2011). Comparison of agent-based modeling software. *Wikipedia.* Retrieved from http://en.wikipedia.org/wiki/Comparison_of_agent-based _modeling_software.

Wolstenholme, E. (2004). Using generic system archetypes to support thinking and modelling. *System Dynamics Review,* 20(4), 341–356.

Zeigler, B. P. (1972). Toward a formal theory of modeling and simulation: Structure preserving morphisms. *Journal of the ACM,* 19(4), 742–764. doi:10.1145/321724. 321737.

Zeigler, B. P., Praehofer, H., and Kim, T. G. (2000). *Theory of Modeling and Simulation* (Vol. 100). New York: Academic Press.

Zeigler, B. P. and Vahie, S. (1993). DEVS formalism and methodology: unity of conception/diversity of application. In G. W. Evans, M. Mollaghasemi, E. C. Russell, and W. E. Biles (Eds.), *Proceedings of the 25th Conference on Winter Simulation* (pp. 573–579). Piscataway, NJ: Institute of Electrical and Electronics Engineers, Inc.

chapter five

On the evolution toward computer-aided simulation

Luiz Felipe Perrone

Contents

Introduction

As soon as the first computers became available, scientists in the most varied disciplines started using them to run mathematical models to predict the behavior and the performance of physical systems. As computational capabilities evolved, simulation became an essential tool in the support of activities in science and systems engineering. Models grew in size and in complexity, and became widely accessible with the advent of free, open-source simulators.

In spite of significant, positive advances, the general area of simulation has witnessed growing concerns with the fidelity of results published in research literature. The so-called "crisis of credibility" in the field of network simulation is emblematic of this trend. Through years of investigation, scholars have identified a multiplicity of issues that have conspired against the scientific rigor of network simulation results. The simulation community learned that many of the failures in simulation studies arise from the complexity of the simulation workflow, which introduces several points of failure for those with less expertise in the field and in the application area. As this realization sank in, it became apparent that simulation tools could be augmented with functionality to keep users on track to produce rigorous results. What ensued was a growing investment in creating tools to provide high levels of support to the simulation experimenter. The simulation support tools started to embody more and more responsibilities in streamlining the workflow of the experimental process; some do so with the explicit intent of guaranteeing the rigor and the quality of results.

It would not be unreasonable for the community to start viewing these tools as *computer-aided simulation* (CAS) systems, even if the term has been used with different meanings over the years. In medical sciences, this term goes back 42 years. Harless et al. (1971) applied it in the context the computer system playing the role of a virtual patient in a textual conversation with medical students as they learn to develop diagnoses from medical histories. Later on, as computer graphics advanced, the application of computer simulation in the training of medical professionals became a common aid for the training of surgeons (Caponetti and Fanelli 1993; Xia et al. 1995; Murphy et al. 1986). In several engineering disciplines, in which computer simulation met computer-aided design, the term CAS took on yet a different meaning. Takano et al. (1997) remind us that many authors have used the term CAS to mean the application of computing techniques to the analysis of engineering problems, particularly those requiring numerical solutions.

At the same time that the older meaning of *computer-aided* has been used in various fields of *applied* computer simulation, the community that develops modeling and simulation methodology and tools, suggests that a different trend has emerged. This shift may have started with Luna (1993) who proposed the notion of doing for simulation model construction something similar to what had already been done for *computer-aided software engineering*. The concept of *computer-aided simulation model engineering* (CASME) called for the creation of tools that enable the modeler to work at a high level of abstraction to build entirely new models, to leverage existing models stored in a repository in the creation of more complex ones, and finally, to use those models to generate executable code automatically to support empirical simulation studies. Although the concept of CASME may once have seemed ambitious, there have been advances from industry and academia that indicate that it is viable and that the modeling and simulation (M&S) community is progressing toward tools that include capabilities that go well beyond those proposed originally.

The current state of the art for M&S tools indicates that many of the discrete components embodying capabilities proposed by CASME already exist. This chapter discusses how the integration of these components into a cohesive whole is likely to give rise to what might become known as CAS systems. These systems will include comprehensive support tools that guide users throughout most, if not all, stages of the modeling and simulation endeavor. Ultimately, the use of automation in this process will ensure the rigorous application of methodology throughout the entire workflow and, therefore, leads to highly credible results without overburdening the user.

The structure of the remainder of this chapter is as follows. The Section "Building a simulation study" discusses the general methodology for a simulation study establishing a foundation for discussion in the remainder of the chapter. The Section "Automating the simulation workflow" explores the context of automation by relating the advances in various stages of the simulation workflow to a sample of existing tools. This material highlights how various accomplishments in the state of the art are paving the way for more ambitious and comprehensive tools. Finally, the Section "Conclusion" concludes the chapter by summarizing the lessons that can be used to guide the development of CAS systems, which are deployable in applied computer simulation communities.

Building a simulation study

Computer simulation goes well beyond "writing some code to test a few ideas." The rigor and the credibility of a simulation study depend on following a well-established process or methodology. Although the complete process appears in simulation textbooks such as the one by Law (2007), as well as in many other books and conference and journal articles, this section presents a brief summary to illustrate how complex simulation is and to identify where software support can benefit systems engineers in their use of simulation. Grouping into two complementary activities the steps in a simulation study by Law (2007), one identifies larger structural elements as a *modeling workflow* that is followed by a *simulation workflow*.

The steps in the modeling workflow, shown in Figure 5.1 comprise the following sequence. A simulation study starts with the clear definition of objectives, which can be expressed as hypotheses or as simple questions for which the answer hinges on the analysis of numerical results. The importance of this step is paramount as it defines the level of abstraction for the construction of models and the broader scenario for experiments with those models. The next most important task in the process is the

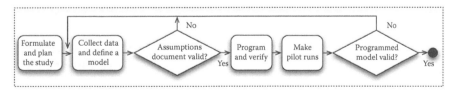

Figure 5.1 Steps of the modeling workflow.

construction or the definition of *simulation models,* which are *abstractions* that capture the essential features of a physical system for the specific goals of the study. This process may involve the collection and the analysis of a body of numerical data and/or detailed descriptions of the interactions between system components at a behavioral level.

The success and the credibility of a simulation study depend on the quality of its models, the faithfulness of their computational counterparts, and the rigor of the output data analysis methods it uses. Each model must be *validated,* that is, understood as a faithful representation of a component of the real system, and later *verified,* that is, understood as a correct implementation of that representation. Pilot runs are useful in this stage: they can help the modeler determine whether the models capture what is needed to meet the objectives of the study and whether their computational implementations are free of errors. Often enough, the results from pilot runs cause the modeler to go back to the proverbial drawing board or, possibly, to refine models and their implementations in an iterative process.

Once models have been validated and verified, the modeling workflow is complete and the process moves on to the simulation workflow, illustrated in Figure 5.2. The first step is the *design of experiments* (DOE) stage. Taking guidance from the objectives of the simulation study, the experimenter identifies a set of model parameters (*factors*) that will be assigned a range of different values (*levels*) in the production runs. Each unique combination of levels constitutes a *design point* in the DOE space and corresponds to one or more runs of the computer simulation. To avoid a combinatorial explosion of design points, which would lengthen the experimental process unnecessarily, one can apply one of the several DOE methodologies, such as those identified by Law (2007) and Sanchez (2007).

Once a DOE methodology is chosen, one or more simulation runs must be executed for each design point. The termination condition for the entire simulation experiment and/or for individual runs depends on whether the simulation is *terminating* or *nonterminating.* A *terminating* simulation executes until the occurrence of a particular event, such as the simulation clock reaching a particular value or the collection of a given number of output data samples. A *nonterminating* (aka. *steady-state*) simulation, on the other hand, does not have a clear-cut termination time: it

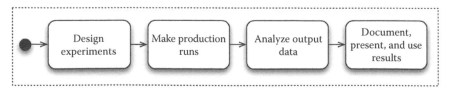

Figure 5.2 Steps of the simulation workflow.

must run until enough data has been collected to produce estimates of "the behavior of the system in the long run" (Pawlikowski 2003a; Law 2007). The type of termination condition for a simulation determines the correct method for output data analysis. As the literature indicates, many published simulation studies have either failed to discuss the details of the simulation experiment or applied the wrong output data analysis method for the type of simulation at hand, which compromises reproducibility and credibility (Pawlikowski 2003a; Camp et al. 2005).

Enumerating the relevant design points for the chosen DOE methodology and knowing the type of simulations enable the experimenter to stage the right production runs, to collect the appropriate body data for statistical processing, and ultimately to meet the objectives of the study. Executing simulation runs for each design point, collecting and organizing output data can be laborious and tedious, but is essential to the simulation workflow. This data is the input for statistical analysis, which provides quantitative metrics for making design decisions in a systems engineering activity, creating insight, answering questions, and confirming or denying hypotheses.

At the conclusion of a simulation study, the experimenter collects the results of the analysis and reports conclusions in some form of media. This may involve the construction of plots, tables, and documents to communicate what was learned to high-level decision makers. Results, however, are unreliable in the absence of convincing evidence that models were validated and verified and/or without records of the assumptions and tools used in their elaboration.

Automating the simulation workflow

The combined workflow for M&S is often not rigorously followed, even though it is well understood and carefully explained in a large variety of simulation publications that range from research articles to entire textbooks. In the field of computer network simulation, the literature is rife with indicators that this lack of adherence to methodology lies at the epicenter of a crisis of credibility. Pawlikowski et al. (2002), Pawlikowski (2003a), and Camp et al. (2005) are among the most notable articles covering this discussion.

Years of first-hand experience with obstacles and pitfalls toward the application of solid simulation methodology drove the community to an important conclusion: *Although some of the steps in the M&S workflow require the application of high-order cognitive functions of which only the human brain is capable, several other steps can be automated by software* (Perrone et al. 2009; Dalle 2012; Uhrmacher 2012). As a matter of fact, this same conclusion has been reached in fields of natural sciences that rely on the management of computerized data and processes, where it spurred vigorous efforts in the research and development of *scientific workflow systems*. The desiderata proposed by McPhillips et al. (2009) illustrate that the needs of researchers in natural sciences and in M&S are closely related. Just as workflows that support natural sciences should be well formed, clear, predictable, recordable, reportable, and reusable, so should be the workflows that support any computer science. Fortunately, similar problems often have similar solutions. As Rybacki et al. (2010) demonstrate, the lessons from scientific workflows are starting to have an impact on the M&S community. The remainder of this section discusses evidence from the literature that show the progress that has been made in automating the M&S workflow over the last couple of decades.

Supporting modeling activities

Law (2007) states that "most real-world systems, however, are quite complex, and coding them without supporting software can be a difficult and time-consuming task." The creation, the validation, and the verification of models are tasks of the highest importance because the quality of the models determines the quality of the simulation study. For many years, there have been many software tools created to support the development of mathematical and behavioral models, which is evidenced by the number of vendors in the exhibit space annually at the Winter Simulation Conference. There exist today several software tools that support the extraction of essential features from data collected from real systems and that fit probability distributions to data sets. Other tools offer sophisticated graphical user interfaces (GUIs) for the construction of process-oriented models and animate the interactions between model components, which are a good aid in verification.

The advances in modeling methodologies hint at the potential for reality to go significantly farther the tools that we see today. Olson et al. (2007) exemplify formalism for model description that enables the automation of analysis and diagnosis that "supports model verification and validation in the early stages of the model development process." This type of formalism raises the level of abstraction in model description and enables the *static* and the *dynamic* analysis discussed by Olson and Overstreet (2011), which are described as follows: *Static analysis* can be performed before the execution of any simulation code to provide insight into whether the relationships between model subcomponents meet the expectations of the modeler and correspond to the stated constraints and specifications. *Dynamic analysis* can be performed during the simulation experiment using data collected from simulation runs to verify whether the causal relationships between the events observed correspond to the modelers' expectations. Formalisms for model specifications can also be powerful when they can be automatically translated into the code used in simulation experiments. Examples of such formalisms have been presented by Röhl and Uhrmacher (2006), Olson et al. (2007), and Brumbulli and Fischer (2012).

An expressive, high-level formalism for model description enables the rich collection of functionalities necessary to realize Luna's (1993) vision of CASME, which includes the means to create, select, delete, and modify model components, the means to express relationships between model components such as composition, dependence, constraints, inheritance, and communication, and the means to store models in repositories, which enables reuse and sharing. This type of formalism is only a *syntactical* tool and as such it is nothing more than notation that describes what is known about model components and their relationships. The actual knowledge described in this notation is derived from the modeler's personal expertise or extracted from an authoritative source such as ontology. The use of ontologies is proving to be important because they support the development of credible models even by those less experienced. Ontologies serve as modeling oracles: well-organized, searchable resources that aggregate and organize *semantic* knowledge related to model components. In addition, the information codified in ontologies can drive the automatic verification of consistency and completeness in component-based models, a functionality desirable to users of any level of expertise (Benjamin et al. 2006). Area-specific ontologies shared with the M&S community in virtual repositories can escalate substantially the potential for rich functionalities in modeling tools. Developing ontologies is challenging, however; it is a time-consuming, laborious process that requires deep knowledge of model components that only experts can provide.

The discussion in the Section "Facilitating experiment control and output data analysis" explores the stages of the simulation workflow, which are predicated on two important assumptions. First, the models used in the experimental process are complete, consistent, and support the goals of the study. Second, the models have been validated and verified.

Facilitating experiment control and output data analysis

The *Akaroa2* project was among the first to demonstrate the potential of automating the simulation workflow and stands out as a significant turning point toward automating the simulation experimental process to address issues of credibility in network simulation (Ewing et al. 1999; Pawlikowski 2003b). At first glance, one might mistakenly see Akaroa2 only as a framework for controlling the execution of simulation experiments over a collection of computers. Its functionalities extend in more important directions, however, and can substantially improve the quality of results.

When multiple computers are available, Akaroa2 can use the *multiple replications in parallel* (MRIP) paradigm to accelerate execution time. For steady-state simulations, Akaroa2 uses a *local data analyzer* to detect the length of transients in the model metrics observed in each replication. This component automatically deletes samples generated during the transient, which avoid statistical biases in output data analysis. The *good samples* of each metric are sent to a corresponding *global data analyzer* that aggregates data from all replications. When the global analyzer has collected enough samples to reach a predetermined statistical precision for its estimates, it terminates all replications. By automatically determining run length to meet clearly defined criteria of estimation, Akaroa2 generates more credible results. Alternatively, the system allows the experimenter to specify directly the length of simulation runs, for instance, when the experiment uses terminating simulations. The system produces a confidence interval with the correct statistical coverage and displays the relative precision of the estimators for each metric through its GUI.

An important additional feature provided by Akaroa2 is the generation of high-quality pseudorandom numbers. Not every simulator is up-to-date with the best and most recent developments in mathematical techniques for pseudorandom number generation. Although some simulators may implement these advances as custom libraries of their own, it is safer and more productive to leverage the work of the experts. Akaroa2 follows this approach, which simplifies the code of the simulator and, most importantly, guarantees that the random number streams are validated, verified, and meet the desired statistical rigor. This type of solution allows one to avoid one of the problems that commonly undermine the rigor of published simulation studies (Pawlikowski 2003a; Camp et al. 2005).

Akaroa2 has a well-conceived architecture that allowed for its integration with various simulators such as ns-2 (http://www.isi.edu/nsnam/ns/), OMNeT++ (http://www.omnetpp.org), and OPNET Modeler (http://www.opnet.com). Although it can be used free of charge in teaching and nonprofit research, Akaroa2 is not distributed under a flexible free software licensing model that allows copying, modification, and redistribution, such as General Public Licenses GPLv2 and GPLv3.

A second framework that implements steady-state and run-length analyses and multiple replications is AutoSimOA (Hoad et al. 2010), where the "OA" stands for *output analyzer*. As this article reports, as of a few years ago, the literature on transient length or *warm-up* analysis had presented as many as 44 different methods, many of which appear in the survey by Pasupathy and Schmeiser (2010). Exceedingly few of these methods have actually been adopted in simulation studies because of three compounding factors: they are too complex for the nonexpert, simulation tools do not provide ready to use implementations, and, worst of all, few of them have been proved to be general and effective. In the development of AutoSimOA, its authors have identified MSER-5 (White and Robinson 2010) as a robust and automatable method for warm-up analysis. Although the recommendation of MSER-5 for automation of transient length detection is a strong contribution in itself, the main take-away messages from this project are the architectural guidelines for the construction of data analysis and experiment control frameworks. The AutoSimOA's architecture allows for simulation experiments to be staged as one long run or as multiple replications. In both cases, the output data is funneled through a warm-up analyzer when it is known to have transients. Separate components work with steady-state data either to stage the appropriate number of replications or to determine the appropriate run-length. When single runs are used, the steady-state data is analyzed by a batch means calculator, which determines the confidence interval for the estimated metrics. This structure guarantees the use of the right output data analysis method for the chosen type of experiment (multiple replications or single run).

The experience with AutoSimOA indicates that there is a limit to how much automation there can be in simulation experiment control and data analysis. As knowledge of the model drives the decision of whether the experiment should use a single long run or multiple replications, the user is responsible for making this choice (as well as for the choice of length of the runs in the case of multiple replications). AutoSimOA makes *recommendations* to guide user choices, recognizing that there should be a user in the loop to make final decisions on run length and number of replications based on the graphical display on the evolution of the metrics of interest that is provided by the system. Although the lessons from AutoSimOA are transferrable to other simulation frameworks in a general sense, its implementation is proprietary. AutoSimOA has been implemented in the SIMUL8 (http://www.simul8.com) system and, therefore, reusable, modifiable, and redistributable source code is not available.

A third interesting framework for experiment execution and output data analysis can be found in the area of network simulation. The Statistically Rigorous Simulation (*STARS*) framework (Millman et al. 2011) implements the MRIP paradigm, as does Akaroa2, and works with OMNeT++ as the underlying simulator. To a great extent, the architecture of the framework is the same as those of Akaroa2 and AutoSimOA, which indicates that a common ground has been established for solutions to simulation experiment control and output data analysis. Arguably, the major point of departure between STARS and other frameworks is that it rekindles the search for methods for output data analysis and run-length control. The realization that simulations may include models driven by heavy-tailed, self-similar data distributions leads to the conclusion that most common methods for output data analysis are not universally applicable.

Experiment control and output data analysis have crucial importance in making simulation more accessible to nonexperts and less tedious and error-prone to experts. Other stages of the simulation workflow, however, can also benefit from automation. The Section "Adding in support for experimental design" focuses on systems that support the experimental design activity, which precedes the execution of simulation runs.

Adding in support for experimental design

The *Scripts for Organizing 'Spiriments* (SOS) (Griffin et al. 2002) had functionalities in common with Akaroa2, such as experiment execution management, the extraction of relevant data from the simulation output data stream, and the persistent storage of output data in a database. What SOS did differently was to augment the automated workflow with the generation of the DOE space. Starting from a file in which the user identified experimental factors and the levels they should be assigned in the experiment, SOS applied a combinatorial algorithm to generate the set of design points that constitute the DOE space. For each design point, the system launched a number of simulation runs, each of which driven by a different seed for random number generation.

Taking into account the lessons in DOE space management, execution control, and data analysis provided by SOS, SWAN Tools (Perrone et al. 2008) combined these functionalities into a web-based tool for network simulation. The SWAN Tools workflow was split into two stages: the first one to be carried out by experts and the second available to users of all levels of aptitude. In the first stage, the expert provided the system with a validated and verified model template. In the second stage, the user was guided to traverse a sequence of steps, using straightforward web-based interfaces. The SWAN Tools workflow consisted of experiment configuration, model configuration, experiment execution, and data analysis and visualization. These steps were supported by the additional functionality of creating a complete record of the experiment by archiving together the code used in the experiment, the experiment and model configuration data, and all simulation output data. With these features, SWAN Tools aimed to facilitate the *reproducibility* of experiments by making available on the web a persistent, complete record of the experiment that could be linked to a publication. The design of SWAN Tools systematically addressed several of the simulation methodology pitfalls identified by Camp et al. (2005), which are also discussed by Dalle (2012).

The contribution of the SWAN Tools project stood as a proof of concept in the automation of simulations. Although development and maintenance on its underlying simulator have ceased and SWAN Tools is no longer distributed, it left behind meaningful lessons. Several of the features of SWAN Tools can be found in SimProcTC (Dreibholz et al. 2009), another automation framework for network simulation. SimProcTC includes the generation of DOE, distributed experiment control, and data analysis, but it goes far in the support of data visualization. It supports the definition of templates that leverage the functionalities of the R Project for Statistical Computing for filtering and processing data samples, computing confidence intervals (when appropriate), and creating fairly elaborate static plots (in PDF format) that can be included directly into publications. SimProcTC, which works with the popular OMNeT++ network simulator, is actively maintained and has helped various published studies.

The experience with SWAN Tools guided the development of the *Simulation Automation Framework for Experiments* (SAFE) (Perrone et al. 2012), which is being constructed to support the *ns-3* network simulator (Riley and Henderson 2013). SAFE's feature set supplements and enhances the expertise of the systems engineer by providing guidance through in the simulation process, thereby avoiding known pitfalls. SAFE's design is guided by the following four design objectives:

1. It should be easy to handle for users with limited simulation skills. To this end, it should offer simple interfaces for model and experiment configuration, for simulation execution, and for output data visualization, and it should automate output data analysis. (The configuration interface should protect users from mistakes, validating their choices against best practices.)
2. It should support the needs of power users and allow them to bypass some of its safeguards, while allowing access to mechanisms that make it easy to stage experiments and visualize results.
3. It should support multiple users and provide protected compartments for each of them to store input configurations and output data.
4. It should implement the MRIP paradigm supporting networks of workstations and compute servers with multicore processors.

To meet these objectives, SAFE implements interfaces to support two distinct user stories, one for *power users* and another for *novice users*, described as follows. A power user is someone who prefers command-line interfaces, can write syntax experiment descriptions and model configurations, works comfortably with system tools for remote login and file transfer, and can build valid computational models for *ns-3* (which are often called *scripts*). A novice user is someone who works exclusively through web-based interfaces, does not have expertise to write configuration syntax, only runs experiments with previously written simulation scripts already stored in the system, and relies on automated output data processing. Power users are less interested in the niceties of a sheltering user interface than in having access to the core functionality of SAFE.

They work in a local *ns-3* installation, which they have downloaded from http://www.nsnam.org. They develop and/or customize *ns-3* models and write their own simulation scripts. The experimental factors in a simulation script are assigned levels through command-line arguments defined when execution is started.

The simulation script is ready for use in SAFE only after it has been compiled and debugged. The user creates a file containing details of the execution of the experiment, that is, defining the DOE space and the termination condition for the simulation. Next, the user invokes SAFE to launch the experiment. The system deploys the code in the collection of worker machines specified in its configuration and each one contributes to the computational effort of covering the DOE space by running simulations of design points. Each simulation run generates output data that is relayed to SAFE for persistent storage, and executes until the chosen termination condition is met. The simulation data is logged to a database, which can be queried by the user directly or via automated scripts that extract data and build plots.

The main functionality that SAFE provides for power users is support for the execution of experiments and for the safekeeping, analysis, and visualization of output data. The system records experiment scenarios in persistent storage together with output data, offers reliable mechanisms for the processing of results, and executes the experiment using the MRIP paradigm. As the usage constraints for this type of user are few, users have freedom to construct any scenario they envision. With that freedom comes the responsibility to verify the correctness of their *ns-3* simulation scripts and to debug them, as necessary.

For every simulator, there is a learning curve that users have to ascend to be able to make effective use of the tool's features. It takes significant time for one to develop proficiency with *ns-3* and the learning process can take longer than one may be able to afford. For this reason, it is highly desirable for tools to offer simplified interfaces that abstract away most complexities and provide plenty of guidance in staging experiments and analyzing their results. This has been the motivation behind the development of SAFE.

For the novice user, SAFE provides a web-based interface that only allows the execution of simulation scripts previously crafted by experienced programmers and installed in the system. The user points a web-browser to a given URL, presents credentials to authenticate with the system, and finds a collection of scripts that can be used in customizable experiments. These experiments are associated with configuration files that determine which factors are exposed through the web interface and how the user-supplied levels are validated. In addition, SAFE allows users to specify restrictions on the relationships between factors, which are used in the generation of the DOE space to yield only the design points of real interest. Once the design points are enumerated, SAFE starts executing simulation runs. The output data associated with each design point are logged to the database. Through the web interface, the user can explore the graphed data interactively, download portions of the data, and create custom static plots to be saved in the database and/or downloaded in different graphics formats (Main 2013).

The protections against common errors that SAFE offers to its users are multiple:

- Novices do not have to write their own *ns-3* simulation scripts. This does away with the need to learn to program in the simulator's environment. The user selects one instance among the *ns-3* simulation scripts that have been carefully constructed and debugged by experts. The user only has to define the DOE.
- As the user provides levels for the experimental factors previously defined, the system validates these values against model constraints, which guarantees that they are always within the permissible ranges for the simulation model. Alternatively, the user interface may constrain the user to select levels from a predefined list of discrete levels.
- Other protections include the organization of experiment output and the execution of experiments until enough samples are collected to yield the correct coverage for specified confidence levels, which uses a global analyzer in the spirit of Akaroa2's.

To guarantee experiment reproducibility, SAFE keeps records of the complete experimental scenario, which includes model and experiment configuration data, as well as the complete code base behind the experiment. This feature mirrors the growing trend in scientific communities to impose strict requirements of sharing to enable reproducibility, as exemplified by Gianni et al. (2010).

As identified by Dalle (2012), human error is one of the main deterrents to the reproducibility of published simulation experiments: lack of knowledge, insufficient details, and program/data manipulation errors often compromise the quality of a scientific study. Arguably, tools such as Akaroa2, AutoSimOA, STARS, SOS, SWAN Tools, and SAFE have been created in reaction to evidence that the complexity of rigorous simulation methodology requires more from experimenters than they are capable of handling without software support. The set of features in these and other simulation support tools signal a growing trend to provide comprehensive support for the simulation workflow. As much of the simulation workflow is relatively easy to automate, several of the authors cited in this chapter support this argument and indicate where contributions can make an impact in the near term.

Extracting, sharing, processing, storing, and visualizing output data

There are several mechanisms by which simulators produce output with the data they generate. They range from the simple logging of textual data via programming language constructs such as *printf* to sophisticated solutions that use persistent storage, enable the interoperability with other tools, and facilitate reporting for reproducibility.

The *ns-2* network simulator approaches data output generation according to an approach that is much more organized than the ad hoc insertion of *printf* calls throughout model code. The simulator defines a class for *traceable variables* that overloads the assignment operators in C++. Every time the value in a traceable variable changes, the new value is communicated to a *tracer object*, which in turn records it to a *trace file* following a well-specified syntax. A similar mechanism allows the simulator to also record network packets as they pass through points in the model that are designated as checkpoints. Several tools have been created to process *ns-2* trace files for statistical estimation and for animating the flow of packets in a network simulation. Although this approach has thrived in the *ns-2* community, it has significant drawbacks. First, the frequent invocation of input/output (I/O) operations slows down the simulation run. Second, as various simulation experiments execute and leave behind results in external artifacts (files), it becomes complicated to keep the output data organized as permanent records of results for posterior analysis.

To circumvent the first of these two problems, Akaroa2 provides a library that can be linked with C and C++, which offers a function through which the simulator transmits samples, or *observations*, to an analysis tool (Ewing et al. 1999). This approach has been used to integrate *ns-2*, and other simulators, with Akaroa2. Another alternative that provides similar functionality is *ns2measure* (Cicconetti et al. 2006), which is distributed under the GNU GPL, differently from Akaroa2. In later developments, ns2measure was combined with a tool for factorial experiment design (Cicconetti et al. 2009) and with *the Automated ns-2 Workflow Manager* (ANSWER) (Andreozzi et al. 2009), resulting in a comprehensive framework to support large-scale experiments with *ns-2*.

Similar to *ns-2*, the more recent *ns-3* network simulator also provides a mechanism for variable tracing, but allows the data generated to be passed on to *trace sinks*, which encapsulate data processing activities to be executed within the simulation run (Riley and Henderson 2013). Later on, Perrone et al. (2013) extended this mechanism with a collection of classes to incorporate a dataflow model of processing into the simulation and to allow more flexibility in marshaling data into various output formats to enhance interoperability with post-processing tools. The dataflow model allows for flexibility in filtering, transformation, and aggregation of data, which potentially reduces the cost of I/O when data is externalized. In retrospect, this model bears strong similarity to what was proposed by Luna (1991).

The approach of data collection via instrumentation of the simulation code, however, is not ideal. As Dalle and Mrabet (2007) argue that there are significant benefits in exploring the *separation of concerns* (SoC) design principle for the purpose of data collection. In the development of the component-based open simulation architecture (OSA), which is deeply rooted in the principles of SoC, the authors observe that instrumenting the code of the simulation model to produce data observations may compromise both its readability and its performance. The modeler may either underinstrument the code out of concern for excessive I/O and have to return later to incorporate support for additional observations, or overinstrument it in the attempt to capture all data that may be potentially useful. To circumvent these problems, the authors propose a solution based on *aspect-oriented programming* (AOP) that allows for observation code to be automatically inserted into model code. Using information kept separately in an instrumentation description file, an AOP compiler processes model code and embeds in it the right amount of instrumentation code to produce only the observations required at a given point in time. Within the context of data collection and processing framework, the instrumented model code is part of a lower-level sampling layer, which is complemented by a higher-level layer for statistical processing and I/O management.

Although the M&S tool development community has made noteworthy advances in solving problems related to data collection and processing, the issue of recording the *provenance* of simulation data has not been effectively addressed. After simulation experiments have completed and their data have been collected, processed, and stored in persistent media, it becomes hard to ascertain from where much of it originated and how it was transformed. As this information can help the experimenter understand relationships between model components and the correctness of results, it stands to reason that the development of this type of functionality needs to be investigated. Data provenance information currently resides mostly in the code of the simulation model and in the path of its traversal through online processing elements. How this information can be encoded for recording is a complex and open problem, which has received significant attention in the scientific workflows community (Crawl and Altintas 2008). Many of the lessons learned from that context are transferable to the M&S domain. Indeed, it seems likely that the notion of separation of concerns used in OSA to minimize code instrumentation may also be applicable in the recording of data provenance information in the simulation workflow. Recording simulation output data and its provenance may lie at the initial stage of what could be considered a workflow in itself, which would be focused on enabling one to learn much about the context of an experiment from its results. Yilmaz and Ören (2013) remark on the importance of this *traceability* property in their conclusions highlighting the need for tools that keep track of the processing path traversed by simulation data.

Once the simulator has produced data, importing it into tools for analysis, storage, and visualization is not a straightforward endeavor. There is a gap between the simulator and analysis tools that can be bridged by the use of strategies to enable the interoperability between these components. More often than not, it is necessary for one to use conversion programs to adapt the output format of a simulator to the input format of some analysis tool. To address this structural deficiency in the simulation process, the CostGlue framework (Saviç et al. 2008) proposes the architecture of a 3-layer reference model for painless interoperability shown in Figure 5.3. At the bottom, the *source* layer generates and records data, possibly combining streams produced by simulation and by real-world processes. The *processing* layer receives input from the source and applies post-processing functions, which might involve the computation of statistics or other forms of aggregation, and passes data upward to the *presentation* layer. The latter receives processed data and produces artifacts that can be consumed by the human analyst in the loop, which can include tables, graphs, animations, and textual reports. Being in the middle between the endpoints of this flow of data, the processing layer can work together with experiment control to determine the run-length and/or the number of replications of simulation runs.

The CostGlue model is implemented by four elements with well-defined roles:

- Core: receives calls from simulators to initiate operations related to data exchange and dispatches the operations to a designated processing element (or *plugin*)
- Database: stores simulation and real world data for efficient read and write access
- Multiple plugins: separate threads within the core, which work with the database for importing/exporting data, process and filter data in the database, and interconnect simulators with the framework
- API: an application programming interface that interconnects core and plugins

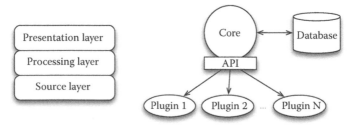

Figure 5.3 CostGlue architecture.

In addition, CostGlue defines an XML-schema for data exchange between components of its framework. The use of architectures similar to that of CostGlue can benefit the development of future simulation support tools, which will be better able to leverage the wealth of tools for data analysis and visualization such as Excel, MATLAB, Mathematica, GNU Octave, R, ROOT, and SciPy.

Whether conducted in academia or industry, many simulation studies culminate and terminate with some form of report published publicly in a scientific venue or internally in a business. Although such reports expose the products of data analysis and the conclusion of a simulation study, history has shown their frequent flaws compromise the reproducibility of the experiment and therefore retard scientific progress. A significant change in the status quo may come about if complete data sets and other simulation artifacts could be made widely available on the global Internet directly from the tool that generates them. (Particularly if all simulation *input* is disseminated with all its output.) This use case has been considered a priority in the design of SAFE. Main (2013) discusses the prototype of a web-based interface for SAFE that allows for interactive visualization, the generation of plots in various graphic formats, and the download of datasets. Built according to the best recommendations for data visualization, this module offers the user a simple interface through which the user can select the data series to include in the plot.

When the user clicks on one or more checkboxes to select a *highlighted metric*, SAFE's visualization module issues a query to a remote database, receives and processes the response, and plots the corresponding curve on the plot's template. This visualizer follows the micro/macro paradigm: it shows a scaled-down version of the larger curve (or *context*) over which it presents a *brush*, a small, sliding window that can be moved over the context allowing the user to select a fragment for detailed exploration in the plotting window. The fragment of curve displayed contains *hoverable points* that give the user numerical data values when pointed at by a mouse-controlled cursor. Figure 5.4 shows a screenshot of the SAFE time series visualization displaying data from a network simulation experiment.

When the pieces of the puzzle start coming together

Quite possibly, the clearest example that the lessons from simulation workflow automation are converging into a cohesive whole is the *Java-based Multipurpose Environment for Simulation II* (JAMES II) (Himmelspach et al. 2008; Ewald et al. 2010). This system provides users with a comprehensive solution that covers many of the functionalities discussed in this section in the areas of modeling, experiment control, experiment reporting and reproducibility, and data analysis and visualization. The architecture of the system follows practices that maximize flexibility and

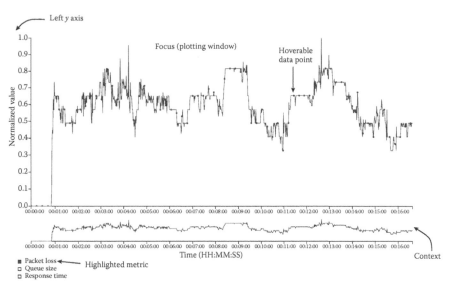

Figure 5.4 Screenshot of SAFE's interactive web-based visualization.

extensibility and can be viewed as embodying, if not defining, the state of the art in simulation tools. JAMES II is emblematic of the emergence of CAS systems and has pushed the boundaries of M&S in many ways. Although there have been numerous publications on the contributions created by the JAMES II project, this section focuses on its recent advances related to the development and the application of *domain-specific languages* (DSLs) in the simulation workflow.

Having acknowledged that the challenges in describing simulation experiments are many, the development team behind the JAMES II project went on to create the *Simulation Experiment Specification via a Scala Layer* (SESSL) (Ewald and Urhmacher 2014). SESSL was designed as an easy to learn language that is independent of the underlying simulator and yet can express a wealth of information on an experiment. Using the SESSL formalism, the experimenter can describe not only the assignments of levels to factors, but also specify the simulation algorithm for execution, number of replications, termination conditions, how output data are to be processed, and how results are to be reported. As shown in the forthcoming article, various case studies indicate that SESSL holds the promise of its design and is able to interoperate with different underlying simulators and systems.

SESSL reliance on the Scala programming language represents an approach to creating a DSL for simulation experiments, which is substantially different from mark XML-based languages. Still, the motivation behind SESSL is not dissimilar to that of the *Simulation Experiment*

Description Markup Language as it is able to convey the "specification of a conceptual model, its simulator, and the experimental frame, so that scientists can replicate the simulation experiment, possibly in a new context and platform" (Yilmaz and Ören 2013). With this, it becomes possible for experimenters to perform cross validation of models, which is essential for broad reproducibility.

Conclusions

As this chapter shows, the development of CAS systems has already begun and is advancing rapidly. Although comprehensive support systems that cover the combined M&S workflows do not seem to exist yet, two encouraging facts remain. First, the evidence suggests that the methods and the technologies exist to implement support for most, if not all, of the individual stages of the M&S workflows. Second, in recent years, the community has been making advances in integrating tools that support different activities in M&S to build more ambitious and powerful systems. When the following functional requirements are incorporated into one system, the objectives of CAS will have been realized:

[R1] Persistent storage of real-world data collected from physical processes to drive the construction of models and to aid dissemination.

[R2] Online access to repositories of modeling data, to analysis methods suitable for assisting model construction, and to the models created.

[R3] Platform-independent model description formalism that allows for the expression of relationships between subcomponents (composition, dependence, inheritance, etc.) and the automatic generation of code from them.

[R4] Automated tests for the verification of computational models against their specifications of constraints and goals.

[R5] Flexible design and execution of experiments driven by a formalism that allows for the expression of the DOE space, type of simulation, and termination condition.

[R6] Instrumentation of simulation code for output data collection, processing, and aggregation, as well as the recording of provenance (traceability) information.

[R7] Facilities to promote easy data exchange between the simulation platform and various solutions for data analysis and visualization.

[R8] Automated transient data deletion and steady-state detection.

[R9] Persistent storage of output data and automated recording with online access.

[R10] Generation of reports with processed simulation results and corresponding artifacts (graphics, files, etc.) and archival in persistent storage.

[R11] Aggregation of complete experimental scenario with all input data, model and simulator code, and output data.

[R12] User interfaces designed to provide assistance to experimenters with different levels of expertize including monitoring and control of simulation runs and automated construction of artifacts for data visualization (interactive and noninteractive).

Although integrating the components that implement all these functionalities into a single tool might seems like a complex problem in software engineering, it is a problem that has drawn great attention in the scientific workflows community. To a great extent, the lessons learned are transferrable from that problem area to the development of simulation software and are certain to help propel advances in the construction of CAS systems. For this reason, it seems reasonable to expect that fast advances toward CAS will ensue in the near future. Although the most immediately obvious benefit to the widespread adoption of CAS will be the enhanced usability of simulation platforms, the more important consequences will be the increased procedural rigor and reproducibility behind the experimental process.

Acknowledgments

This work is supported by award CNS-0958142 of the U.S. National Science Foundation (NSF). Any opinions, findings, and conclusions or recommendations expressed in this material is the author's alone and do not reflect the views of the NSF. The author acknowledges the following collaborators, who have been instrumental in exploring the topics presented in this chapter: Andrew W. Hallagan, Bryan C. Ward, Christopher S. Main, Tiago G. Rodrigues, Vinícius D. Felizardo, William S. Stratton, Thomas R. Henderson, and Mitchell J. Watrous.

References

Andreozzi, M. M., G. Stea, and C. Vallati. 2009. A framework for large-scale simulations and output result analysis with ns-2. In *Proceedings of the 2nd International Conference on Simulation Tools and Techniques (SIMUTools 2009)*, Rome, Italy.

Benjamin, P., M. Patki, and R. Mayer. 2006. Using ontologies for simulation modeling. In *Proceedings of the 2006 Winter Simulation Conference*, Monterey, CA.

Brumbulli, M. and J. Fischer. 2012. Simulation visualization of distributed communication systems. In *Proceedings of the 2012 Winter Simulation Conference*, Berlin, Germany.

Camp, T., S. Kurkowski, and M. Colagrosso. 2005. MANET simulation studies: The incredibles. *SIGMOBILE Mobile Computing and Communication Review* 9(4): 50–61.

Caponetti, L. and A. M. Fanelli. 1993. Computer-aided simulation for bone surgery. *Computer Graphics and Applications, IEEE* 13(6): 86–92.

Cicconetti, C., E. Mingozzi, and G. Stea. 2006. An integrated framework for enabling effective data collection and statistical analysis with ns-2. In *Proceedings of the 1st International Conference on Performance Evaluation Methodologies and Tools* (VALUETOOLS 2006), Pisa, Italy.

Cicconetti, C., E. Mingozzi, and C. Vallati. 2009. A 2^k r factorial analysis tool for ns2measure. In *Proceedings of the 4th International Conference on Performance Evaluation Methodologies and Tools* (VALUETOOLS 2009), Pisa, Italy.

Crawl, D. and I. Altintas. 2008. A provenance-based fault tolerance mechanism for scientific workflows. In *Provenance and Annotation of Data and Processes*, J. Freire, D. Koop, and L. Moreau (eds.), 5272: 152–159. Lecture Notes in Computer Science. Springer-Verlag, Berlin/Heidelberg, Germany.

Dalle, O. 2012. On reproducibility and traceability of simulations. In *Proceedings of the 2012 Winter Simulation Conference*, Berlin, Germany.

Dalle, O. and C. Mrabet. 2007. An instrumentation framework for component-based simulations based on the separation of concerns paradigm. *Proceedings of the 6th EUROSIM Congress* (EUROSIM 2007), Ljubljana, Slovenia.

Dreibholz, T., E. P. Rathgeb, and X. Zhou. 2009. SimProcTC—the design and realization of a powerful tool-chain for OMNeT++ simulations. In *Proceedings of the 2009 OMNeT++ Workshop*, Rome, Italy.

Ewald, R., J. Himmelspach, M. Jeschke, S. Leye, and A. M. Uhrmacher. 2010. Flexible experimentation in the modeling and simulation framework JAMES II—Implications for computational systems biology. *Briefings in Bioinformatics* 11(3): 290–300.

Ewald, R. and A. M. Uhrmacher. 2014. SESSL: A domain-specific language for simulation experiments. *ACM Transactions on Modeling and Computer Simulation* 24(2): Article 11.

Ewing, G., K. Pawlikowski, and D. McNickle. 1999. Akaroa-2: Exploiting network computing by distributing stochastic simulation. In *Proceedings of the 13th European Simulation Multi-Conference* (ESM '99), Warsaw, Poland.

Gianni, D., S. McKeever, T. Yu, R. Britten, H. Delingette, A. Frangi, P. Hunter, and N. Smith. 2010. Sharing and reusing cardiovascular anatomical models over the web: A step towards the implementation of the virtual physiological human project. *Philosophical Transactions. Series A: Mathematical, Physical and Engineering Sciences* 368 (1921): 3039–3056.

Griffin, T. G, S. Petrovic, A. Poplawski, and B. J. Premore. 2002. SOS: Scripts for Organizing 'Speriments. Available at http://ssfnet.org/sos/index.html; accessed on February 27, 2014.

Harless, W. G., G. G. Drennon, J. J. Marxer, J. A. Root, and G. E. Miller. 1971. CASE: A computer aided simulation of the clinical encounter. *Journal of Medical Education* 46(5): 443–448.

Himmelspach, J., R. Ewald, and A. M. Uhrmacher. 2008. A flexible and scalable experimentation layer. In *Proceedings of the 2008 Winter Simulation Conference*, Austin, TX.

Hoad, K., S. Robinson, and R. Davies. 2010. AutoSimOA: A framework for the automated analysis of simulation output. *Journal of Simulation* 5(1): 9–24.

Law, A. M. 2007. *Simulation Modeling and Analysis*. 4th Ed. McGraw-Hill Higher Education, Columbus, OH.

Luna, J. 1991. Application of hierarchical modeling concepts to a multi-analysis environment. In *Proceedings of the 1991 Winter Simulation Conference*, Phoenix, AZ.

Luna, J. J. 1993. Towards a computer aided simulation model engineering (CASME) environment. In *Proceedings of the 1993 Winter Simulation Conference*, Los Angeles, CA.

Main, C. S. 2013. Visualization Techniques for the Analysis of Network Simulation Results. Honors Thesis. Department of Computer Science, Bucknell University, Lewisburg, PA.

McPhillips, T., S. Bowers, D. Zinn, and B. Ludäscher. 2009. Scientific workflow design for mere mortals. *Future Generation Computer Systems* 25(5): 541–551.

Millman, E., D. Arora, and S. W. Neville. 2011. STARS: A framework for statistically rigorous simulation-based network research. In *Proceedings of the IEEE International Conference on Advanced Information Networking and Applications* (WAINA), Biopolis, Singapore.

Murphy, S. B., P. K. Kiejewski, H. P. Chandler, P. P. Griffin, D. T. Reilly, B. L. Penenberg, and M. M. Landy. 1986. Computer-aided simulation, analysis, and design in orthopedic surgery. *Orthopedic Clinics of North America* 17(4): 637–649.

Olson, K. A. and C. M. Overstreet. 2011. Enhancing understanding of models through analysis. In *Proceedings of the 1st International Conference on Simulation and Modeling Methodologies, Technologies and Applications* (SIMULTECH 2011), Noordwijkerhout, the Netherlands, pp. 321–326.

Olson, K. A., C. M. Overstreet, and E. J. Derrick. 2007. Code analysis and CS-XML. In *Proceedings of the 2007 Winter Simulation Conference*, Washington, DC.

Pasupathy, R. and B. Schmeiser. 2010. The initial transient in steady-state point estimation: Contexts, a bibliography, the MSE criterion, and the MSER statistic. In *Proceedings of the 2010 Winter Simulation Conference*, Baltimore, MD.

Pawlikowski, K. 2003a. Do not trust all simulation studies of telecommunication networks. *Information Networks Lecture Notes in Computer Science* 2662: 899–908.

Pawlikowski, K. 2003b. Towards credible and fast quantitative stochastic simulation. In *Proceedings of the International SCS Conference on Design, Analysis and Simulation of Distributed Systems* (DASD '03), Orlando, FL.

Pawlikowski, K., H.-D. J. Jeong, and J.-S. R. Lee. 2002. On credibility of simulation studies of telecommunication networks. *IEEE Communications Magazine* 40(1): 132–139.

Perrone, L. F., C. Cicconetti, G. Stea, and B. C. Ward. 2009. On the automation of computer network simulators. In *Proceedings of the 2nd International Conference on Simulation Tools and Techniques* (SIMUTools 2009), Rome, Italy.

Perrone, L. F., T. R. Henderson, M. J. Watrous, and V. D. Felizardo. 2013. The design of an output data collection framework for ns-3. In *Proceedings of the 2013 Winter Simulation Conference*, Washington, DC.

Perrone, L. F., C. J. Kenna, and B. C. Ward. 2008. Enhancing the credibility of wireless network simulations with experiment automation. In *Proceedings of the 4th IEEE International Conference on Wireless and Mobile Computing, Network, and Communications* (WiMob '08), Avignon, France.

Perrone, L. F., C. S. Main, and B. C. Ward. 2012. SAFE: Simulation automation framework for experiments. In *Proceedings of the 2012 Winter Simulation Conference*, Berlin, Germany.

Riley, G. and T. R. Henderson. 2013. The ns-3 network simulator. In *Modeling and Tools for Network Simulation*, K. Wehrle, M. Günes, and J. Gross (Eds.), pp. 15–34. Springer-Verlag, Berlin/Heidelberg, Germany.

Röhl, M. and A. M. Uhrmacher. 2006. Composing simulations from XML-specified model components. In *Proceedings of the 2006 Winter Simulation Conference*, Monterey, CA.

Rybacki, S., J. Himmelspach, E. Seib, and A. M. Uhrmacher. 2010. Using workflows in M&S software. In *Proceedings of the 2010 Winter Simulation Conference*, Baltimore, MD.

Sanchez, S. M. 2007. Work smarter, not harder: Guidelines for designing simulation experiments. In *Proceedings of the 2007 Winter Simulation Conference*, Washington, DC.

Savić, D, J. Bester, M. Pustišek, S. Tomazic, F. Potorti, and F. Furfari. 2008. CostGlue: Simulation data exchange in telecommunications. *Simulation* 84(4): 157–168.

Takano, H., S. M. Ulhaq, and M. Nakaoka. 1997. Computer-aided simulation technique of digitally controlled switched-mode power conversion circuits and systems using state variable matrices. In *Proceedings of the Power Conversion Conference* 1: 411–418.

Uhrmacher, A. M. 2012. Seven pitfalls in modeling and simulation research. In *Proceedings of the 2012 Winter Simulation Conference*, Berlin, Germany.

White, K. P. and S. Robinson. 2010. The problem of the initial transient (again) or why MSER works. *Journal of Simulation* 4: 268–272.

Xia, J., F. Qi, W. Yuan, D. Wang, W. Qiu, Y. Sun, Y. Huang, G. Shen, and H. Wu. 1995. Computer aided simulation system for orthognathic surgery. In *Proceedings of the 8th IEEE Symposium on Computer-Based Medical Systems*, Lubbock, TX.

Yilmaz, L. and T. Ören. 2013. Toward replicability-aware modeling and simulation: Changing The conduct of M&S in the information age. In *Ontology, Epistemology, and Teleology for Modeling and Simulation: Philosophical Foundations for Intelligent M&S Applications*, A. Tolk (ed.), pp. 207–226. Springer-Verlag, Berlin/Heidelberg, Germany.

chapter six

Model-driven method to enable simulation-based analysis of complex systems

Paolo Bocciarelli and Andrea D'Ambrogio

Contents

Introduction

Traditionally, the design of a complex system relies on a systems engineering process that makes use of text documents and engineering data in multiple formats.

The inherent limitations of the document-based manual approach have been targeted by the *model-based systems engineering* (MBSE) approach, promoted by the International Council on Systems Engineering (INCOSE), which defines MBSE as "the formalized application of modeling to support system requirements, design, analysis, verification, and validation activities beginning in the conceptual design phase and continuing throughout development and later life cycle" (INCOSE 2007).

In this respect, Systems Modeling Language (SysML) is the Unified Modeling Language (UML)–based language that provides the *modeling* capability required in the systems engineering domain (OMG 2010). SysML is now considered the standard modeling notation adopted in the MBSE context.

The advantages obtained by the MBSE approach, in terms of enhanced communications, reduced development risks, improved quality, increased productivity, and enhanced knowledge transfer, can be further scaled up by innovative approaches that treat models as the primary artifacts of development, by increasing the level of automation throughout the system lifecycle. Such approaches, denoted as *model-driven systems engineering* (MDSE) approaches, represent a radical shift from a merely contemplative use of models to a productive and more effective use. MDSE applies metamodeling techniques and automated model transformations, introduced in the more general *model-driven engineering* (MDE) context (Atkinson and Kuhne 2003; Schmidt 2006), to the systems engineering domain, thus boosting the aforementioned advantages of the MBSE approach.

This chapter focuses on the use of MDSE to enable the simulation-based analysis of modern complex systems, that is, large-scale heterogeneous systems, which are usually composed of several subsystems.

The intrinsic complexity of such systems requires the adoption of quantitative analysis techniques to allow an early evaluation of the system behavior, to assess, before the implementation phase begins, whether or not the *to-be* system will satisfy the stakeholder's requirements and constraints.

In this context, simulation-based techniques may be effectively introduced to enact a design-time evaluation of several behavioral characteristics of the system under study. Thus, the adoption of simulation techniques constitutes a valuable strategy both to cut the cost of developing experimental prototypes and to mitigate the risk of time/cost overrun due to redesign and reengineering activities.

To face the complexity of the systems addressed in this work and to take into account their distributed topology, the use of *distributed simulation* (DS) techniques has proven to be a natural and viable choice.

Unfortunately, the concrete use of DS-based approaches is often limited in practice by the nonnegligible effort and the significant skills that are required to make use of DS frameworks and environments, such as the high-level architecture (HLA) framework and its related implementation technologies (IEEE 1516-2010; IEEE 1516.1-2010; IEEE 1516.2-2010).

To overcome such limitations, this chapter proposes a method to support the automated generation of HLA-based DSs, starting from system descriptions specified using *SysML models.**

The proposed method exploits principles and standards introduced in the MDSE field and more specifically within the model-driven architecture (MDA), the Object Management Group's (OMG) incarnation of MDE (OMG 2003). The method introduces appropriate model-to-model and model-to-text transformations that a system engineer can execute to derive the HLA-based simulation code of a system specified in SysML, without being required to own specific skills of DS standards.

The proposed method consists of a two-phase transformation. Starting from the SysML model of the system under study, it first derives the UML model of the relevant HLA-based DS and then generates the corresponding executable code.

To this purpose, two UML profiles, namely *SysML4HLA* and *HLAProfile*, are used to annotate the models with the details required to drive the model-to-model and model-to-text transformations, respectively.

The rest of this work is organized as follows: a literature review of existing SysML and HLA work is first discussed. Then, background concepts that constitute the basis of this work are briefly recalled. The proposed model-driven method is then illustrated, along with the profiles and the model transformations that are the bases of the proposed method. Finally, an example application is given.

Related work

This section reviews the existing literature dealing with both the use of SysML in the modeling and simulation (M&S) domain and the modeling/development of HLA-based DS systems.

As regards the use of SysML in the M&S context, to the best of our knowledge, there are no contributions that specifically address the generation of Java/HLA code from SysML specifications, with the exception of our previous contribution (Bocciarelli et al. 2012) that this work extends and improves. Specifically, the proposed model-driven method has been revised and redesigned according to the HLA Federation Development

---

* It should be noted that SysML is formally specified as an UML profile. In this respect, according to a rigorous terminology, "SysML models" should be more correctly referred as "UML models annotated with SysML profile." Nevertheless, the term "SysML models" has been used here for the sake of conciseness.

and Execution Process (FEDEP) (IEEE 1516.3-2003). Moreover, the model transformations at the core of the proposed method have been adapted to exploit the enhancements promoted by the recently published HLA-Evolved standard (IEEE 1516-2010).

More generally, several contributions are available that propose the use of SysML as a notation suitable not only for defining systems specification but also for supporting the system simulation, such as Weyprecht and Rose (2011), Peak et al. (2007), and Paredis and Johnson (2008).

In Weyprecht and Rose (2011), a simulation core, implemented using fUML, has been proposed. This paper advocates SysML as a standardized simulation language, and model-driven techniques are introduced to generate the code of a simulation software, starting from SysML behavioral models, such as activity diagrams. In such a paper, the adoption of a model-driven paradigm is limited to the use of the Eclipse Modeling Framework (EMF), specifically the Java Emitter Templates (JETs), to generate a basic source code skeleton for each needed class.

Moreover, from the implementation point of view, the paper only describes a prototypal implementation of the simulation core. The implementation of a complete simulation solution is planned as a future work.

In Peak et al. (2007), SysML is used as a notation to support the simulation-based design of systems. By presenting several examples, the paper shows how SysML is able to capture the engineering knowledge needed to derive executable parametric models.

In Paredis and Johnson (2008), the use of SysML is proposed to support the system simulation. More specifically, such a paper introduces the use of a graph transformation approach to accomplish an automated transformation between SysML and domain-specific languages.

Differently from the aforementioned contributions, this chapter describes a model-driven method to generate a Java/HLA-based implementation of a DS software, starting from a SysML specification.

Regarding the representation of DS systems, a large effort has been spent in defining UML profiles for modeling HLA federations, such as in Topçu et al. (2008), Topçu et al. (2003), and Zhu et al. (2008). Similarly to such contributions, this chapter proposes a UML profile to model a HLA-based simulation system that has been partially based on the HLA Object Model Template (OMT) (IEEE 1516.2-2010). Nevertheless, this chapter goes far beyond and also takes into consideration the HLA metamodel proposed in Topçu et al. (2008) to improve the expressiveness of the proposed profile. Moreover, it should be underlined that this chapter contribution is not limited to a UML extension for modeling an HLA federation. This work also proposes a model-driven method to generate the implementation of a DS software, starting from the SysML specification of the system under study.

Regarding the issue of implementing (or supporting the implementation of) simulation systems, several contributions can be found in literature that apply a model-driven paradigm in the modeling and simulation domain, such as D'Ambrogio et al. (2011), Tolk and Muguira (2004), Jimenez et al. (2006), and Haouzi (2006).

In D'Ambrogio et al. (2011) and D'Ambrogio et al. (2010), methods to generate a Java/HLA-based and a DEVS/HLA-based implementation of a DS software from a UML representation of the system have been proposed, respectively. On the one hand, such contributions constitute a starting point for this work, which shares the same objective (i.e., the automated generation of the DS implementation) and the adopted approach (i.e., the model-driven paradigm). On the other hand, this work extends the previous contributions in several ways. The proposed model-driven method is applied to the systems engineering domain and thus, in this chapter, the model transformations are driven by the SysML representation of the system under study. Moreover, this chapter gives a more extensive description of the adopted UML profiles and of the model-to-model and model-to-text transformations at the core of the method implementation.

In Tolk and Muguira (2004), a model-driven approach is proposed in the M&S application domain. Specifically, such a contribution proposes the creation of a specific domain for M&S within MDA. A novel aspect of our approach is that we adopt MDA techniques to the production of simulation systems treated as general-purpose software systems. This means that, for a given software system, the same approach can be adopted to eventually generate both the operational system and the simulation system from the same model specification. Moreover, the implementation of the proposed method is not complete in terms of both MDA compliance and software. Differently, this chapter approach implements a MDA-compliant process by introducing two UML profiles and a set of model-to-model and model-to-text transformations for generating the simulation code from a SysML model specification.

In Jimenez et al. (2006), a MDA-based development of HLA simulation systems is also proposed. Such a contribution is limited to the definition of an initial UML profile for HLA and, differently from the contribution proposed in this chapter, does not take into consideration the application of the profile for the implementation of the simulation system.

Finally, in Haouzi (2006), the main concepts behind the application of MDA techniques to the development of HLA systems are outlined. Such a contribution is limited to a theoretical discussion about the use of MDA-based techniques in HLA domain. Differently, this work proposes a model-driven approach to reduce the gap between the model specification and the distributed system implementation. An example application is also discussed to show how the application of the proposed method allows one to reduce the DS development effort by automating the production of the Java/HLA code from an initial SysML-based system specification.

Background

This section gives an outline of the most relevant background concepts, technologies, and standards at the basis of this work. The section starts with an overview on MDA. Then, a summary of the different approaches for the execution of a simulation model is given. Finally, the HLA-Evolved standard is briefly illustrated.

Model-driven architecture

MDE is an approach to software design and implementation, which addresses the raising complexity of execution platforms by focusing on the use of formal models (Atkinson and Kuhne 2003; Schmidt 2006). According to this paradigm, a software system is initially specified by use of high-level models. Such models are then used to generate other models at a lower level of abstraction, which are used in turn to generate other models until stepwise refined models can be made executable.

One of the most important initiatives driven by MDE principles is the MDA (OMG 2003), the OMG's incarnation of MDE.

MDA-based software development is founded on the principle that a system can be built by specifying a set of model transformations, which allow to obtain models at lower abstraction levels starting from the model at higher abstraction levels.

To achieve such an objective, MDA has introduced the following standards:

- *Meta object facility* (MOF): for specifying technology neutral metamodels (i.e., models used to describe other models) (OMG 2004)
- *XML metadata interchange* (XMI): for serializing MOF metamodels/models into XML-based schemas/documents (OMG 2007)
- *Query/view/transformation* (QVT): for specifying model transformations (OMG 2008)

The relationship among MDA standards is summarized by the scenario depicted in Figure 6.1. Model M_A and model M_B are instances of their respective metamodels, namely metamodel MM_A and metamodel MM_B, which in turn are expressed in terms of MOF constructs. A model transformation, specified at metamodel level using QVT, allows one to generate model M_B from model M_A. Both model M_A and model M_B can be serialized using XMI rules to obtain the corresponding XMI documents, that is, the XML-based representation of such models. XMI rules can also be used at metamodel layer to serialize metamodels and obtain XMI schemas for XMI document validation.

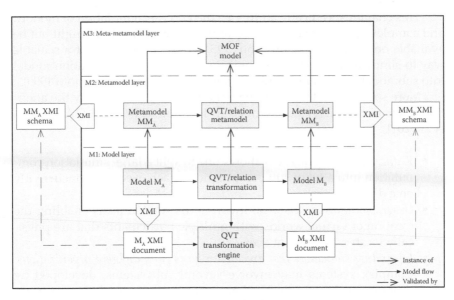

Figure 6.1 Relationship among MDA standards.

In this chapter, MDA has been used to design and develop the method to enact the DS-based analysis of systems specified using SysML, according to a MDSE paradigm.

Simulation execution paradigms

A given simulation model is implemented into a simulation program that can be executed according to three different paradigms, that is, local, parallel, and distributed.

In a *local simulation* the simulation program is deployed onto a single processor platform that completes its execution.

Differently, according to a parallel/distributed approach, the simulation program may be executed over multiple processors. In this chapter the following taxonomy is considered (Fujimoto 2001):

- *Parallel Simulation* (PS) is concerned with execution on multiprocessor computing platforms that contain multiple CPUs that interact frequently, for example, thousands of times per second.
- *Distributed Simulation* (DS) is concerned with the execution on loosely coupled systems where interactions take much more time, for example, milliseconds or more, and occur less often. The execution is carried out on a distributed system composed by a set of hosts interconnected through a LAN or a WAN (e.g., the Internet).

This chapter specifically addresses the DS case. Simulation of modern and complex systems requires computational resources that might not be available on a single host, and thus DS is often used to enact a scalable way to simulate a complex system by partitioning the simulation model into submodels, each simulated on an independent host (Fujimoto 1999).

More specifically, the following benefits can be obtained by the adoption of DS techniques, as underlined in Fujimoto (1999) and D'Ambrogio et al. (2011):

- *Reduced execution time*: DS allows one to split a large simulation computation into a set of sub-computations, each executed concurrently on a different host.
- *Geographical distribution*: DS involves distributed hosts enabling the creation of virtual worlds with multiple participants that are physically located at different sites.
- *Integrating simulators that execute on hosts from different organizations*: Complex systems may involve several subsystems, developed by different manufacturers. Rather than porting a simulation program for each submodel to a single host, it may be more cost-effective to integrate the existing simulators, each responsible to simulate a specific submodel and each executing on a different host, to create a new virtual environment.
- *Fault tolerance*: DS increases tolerance to failures. If one host goes down, it may be possible for other hosts to pick up the work of the failed machine, allowing the simulation computation to proceed despite the failure.

High-level architecture evolved

HLA is an IEEE standard (IEEE 1516-2010; IEEE 1516.1-2010; IEEE 1516.2-2010) providing a general architecture for the implementation of DS systems. The standard, which promotes interoperability and reusability of simulation components in different contexts, is based on the following concepts (Kuhl 1999):

- *Federate*: A simulation program that represents the unit of reuse in HLA.
- *Federation*: A DS execution composed of a set of federates.
- *Run-time infrastructure* (RTI): The simulation-oriented middleware for managing federates interaction. The RTI consists of the following:
  - *Local components*, which reside on the federate sites
  - *RTI executive component*, which is deployed on a central server

The major benefits deriving by the use of HLA is that the provided application programming interface (API) relieves developers from all concerns related to the communication and the synchronization among the several components (i.e., federates) constituting the DS system, obtaining considerable effort savings in the development process. Despite this improvement, HLA still suffers from the complexity of its API that limits its concrete use in practice. In this respect, this work addresses such drawbacks and proposes a model-driven method that supports the automated generation of executable HLA code from SysML specifications.

HLA was released in the early 1990s and, as DS is a constantly evolving domain, it was continuously improved to face novel challenges and to meet the practitioners evolving requirement. In 2010, the *HLA-Evolved* standard was released (IEEE 1516-2010). Such a new standard includes several enhancements, as illustrated in Möller et al. (2008). Beside the technological improvements to ease and make less error-prone the development of federations, HLA-Evolved provides a revised API, based on Web Service and Web Services Description Language, which makes possible to provide an HLA-based simulation over the Internet, according to a *Simulation-as-a-Service* (SIMaaS) paradigm. This work specifically adopts the HLA-Evolved standard.

High-level architecture federate development and execution process

The development of a HLA federation is a complex and challenging issue that requires a tough expertise to be conducted in a cost- and time-effective manner. In this respect, the HLA standard proposes the adoption of the so-called FEDEP (IEEE 1516.3-2003), which defines best practices for developing and executing federate and federations. The FEDEP rationale is depicted in Figure 6.2. The process consists of the following steps:

1. *Define federation objectives.* The first task of the FEDEP deals with the definition and understanding of the problem and a description of the federation objectives.
2. *Perform conceptual analysis.* In the second step, a conceptual analysis is carried out to develop a representation of the real-world domain. Moreover, federation objectives are refined into a set of federation requirements that will be used to drive the remaining activities of the FEDEP.

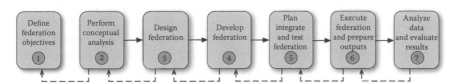

Figure 6.2 Federation development and execution process.

3. *Design federation.* HLA promotes the reuse of simulation components, so that the federation design activity includes both the design of new federates and the discovery of existing federates that are suitable for reuse.

4. *Develop federation.* The federation development activity includes the definition of the Federation Object Model (FOM), the implementation of new federates, and the adaption of those existing federates that need to be modified to be compliant with federation requirements and constraints.

5. *Plan, integrate, and test federation.* In this step the federation execution is planned and the integration activities required to interconnect federates are performed.

6. *Execute federation and prepare outputs.* In this step the federation is executed.

7. *Analyze data and evaluate results.* In the last step of the FEDEP the simulation outcomes are evaluated and results are reported back to interested stakeholders.

The method for enacting the HLA-based systems analysis proposed in this work has been inspired by the FEDEP.

Model-driven method for simulation-based design-time analysis

This section outlines the proposed MDSE method for carrying out a simulation-based systems analysis. The method, which has been inspired by the HLA FEDEP, is founded on model-driven principles and standards and exploits two UML profiles, namely SysML4HLA and HLAProfile, to support both the generation of the model and the implementation of the HLA-based DS.

According to the context outlined in the introduction, the system development process is concerned with two different domains. On the one hand, it is related to the system design and implementation. On the other hand, it also addresses the simulation development domain, where simulation engineers are engaged in supporting the system development process by introducing simulation techniques to allow an early evaluation of the system under study. In this respect, the proposed method aims at supporting both system and simulation engineers, as depicted in Figure 6.3.

At the beginning, the system under study (and that is going to be simulated) is initially specified in terms of a SysML model (e.g., block definition diagrams, sequence diagrams). According to the model-driven terminology (OMG 2003), such a model constitutes the platform independent

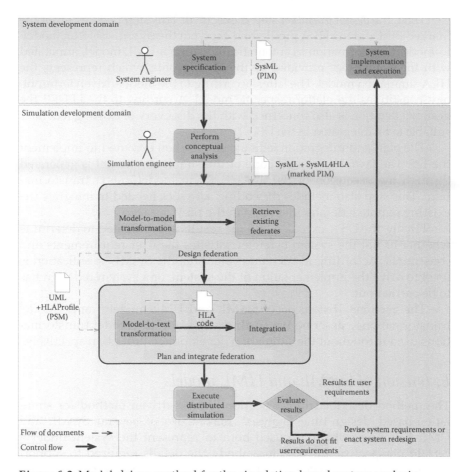

Figure 6.3 Model-driven method for the simulation-based system analysis.

model (PIM) of the system. At the system development level, the system engineer in charge of producing the system model is not concerned with any details regarding the simulation model, but she is strictly focused on the specification of a SysML-based system design model, starting from the system requirements.

The SysML model constitutes the input of the subprocess that is related to the development of the DS. In this respect, simulation engineers carryout a *conceptual analysis* of the required simulation and use the SysML4HLAProfile to annotate the PIM to enrich such a model with information needed to derive the HLA simulation model. Specifically, the HLA profile allows one to specify both how the system has to be partitioned in terms of federation/federates and how system model elements have to be mapped to HLA model elements such as object classes and interaction classes.

Then, the *design federation* step is executed. This step takes as input the marked PIM and the HLA profile, and carryout the SysML-to-HLA model-to-model transformation to automatically obtain a UML model, annotated with the stereotypes provided by the HLA profile, which represents the HLA simulation model. The latter, according to the model-driven terminology, constitutes the platform-specific model. According to the FEDEP, the design federation is also concerned with the discovery of existing federates suitable to be integrated in the DS.

The plan and integration federate step is then executed to implement the DS. The Java/HLA-based code of the simulation model is generated through the execution of the HLA-to-Code model-to-text transformation.* This step also includes the coding activities needed to integrate the existing remote federates identified in the previous step.

Finally, the DS is executed and the results are evaluated to determine whether or not the system behavior satisfies the user requirements and constraints. According to such an evaluation, the SysML specification is used to drive the implementation of the system, or a system redesign has to be carried out.

The Sections "Extension of SysML and UML models" and "Model transformations" describe the UML profiles and the model transformations used throughout the simulation development process, respectively.

Extension of SysML and UML models

The method described in the Section "Model-driven method for simulation-based design-time analysis" makes use of several SysML and UML diagrams, which are used both to represent the system under a certain perspective and also to embody the details required to support the model transformations. To this purpose, such diagrams are annotated using stereotypes provided by the SysML4HLA and HLAProfile UML profiles (Bocciarelli et al. 2012). The SysML4HLA profile is used to annotate a SysML-based system specification to support the automated generation of the HLA-based DS model. Differently, the HLAProfile, is used to annotate the UML HLA-based DS model to represent HLA-based implementation details. For the readers' convenience, such profiles are briefly recalled in the Subsections "SysML4HLA profile" and "High-level

---

* This step requires the choice of a specific HLA implementation (e.g., Pitch, Portico) that provides the HLA services in a given programming language (e.g., Java, C++). Specifically, the implementation of the HLA-to-Code transformation provided in this work makes use of Portico and Java. Nevertheless, the model-driven approach at the basis of the proposed method allows to use different HLA implementations or programming languages. All such cases can be easily dealt with by revising the specification of the HLA-to-Code model-to-text transformation.

architecture profile," respectively. A detailed description of SysML4HLA and HLAProfile can be found in Bocciarelli et al. (2012).

SysML4HLA profile

One of the main objectives of this work is to exploit model-driven techniques to develop a HLA-based DS of systems, starting from SysML specifications. Unfortunately, SysML does not natively provide any information to support such development process.

In this respect, the proposed method makes use of the SysML4HLA profile, a UML profile specifically introduced in Bocciarelli et al. (2012). SysML4HLA provides stereotypes that extend the block element of SysML, which in turn is an extension of the UML class metaclass. These stereotypes have been introduced to represent the basic elements of an HLA simulation: federates, ObjectClasses, and InteractionClasses. The stereotypes provided by the SysML4HLA profile are summarized in Table 6.1.

High-level architecture profile

As described in Figure 6.3, the marked PIM (i.e., a SysML model extended with the SysML4HLA profile) is taken as input by the model-to-model transformation that yields as output the HLA-based model of the system under study. Such model is then annotated using the HLAProfile, which specifies concepts, domain elements, and relationships required to represent a HLA-based simulation under an implementation-oriented perspective. The HLAProfile, which has been inspired by the HLA metamodel proposed in Topçu et al. (2008), consists of several stereotypes and has been structured in several packages, as shown in Figure 6.4.

Table 6.1 Summary of the SysML4HLA Profile

Stereotype	Extension	Description
Federation	Block (UML::Class)	Defines an element that represents the whole federation
Federate	Block (UML::Class)	Defines an element that acts as federate within the federation
ObjectClass	Block (UML::Class)	Defines an element, managed by federates, treated as an object class
InteractionClass	Block (UML::Class)	Defines an element, exchanged among federates, treated as an interaction class

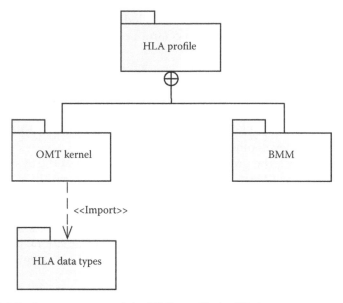

Figure 6.4 Package structure of the UML profile for HLA.

High-level architecture datatypes package

This package includes the datatypes of the several attributes used to specify the stereotypes included in the OMTKernel package. Its structure is shown in Figure 6.5. This package contains the following enumerated types:

- *PSKind* specifies the publish/subscribe capabilities of a federate. It assumes one of the following values: *Publish, Subscribe, PublishSubscribe,* or *Neither.*
- *UpdateKind* specifies the policy for updating an instance of a class attribute. It assumes one of the following values: *Static, Periodic,* or *Conditional.*
- *DAKind* specifies whether the ownership of an instance of a class attribute can be released or acquired during the simulation execution. It assumes one of the following values: *Divest, Acquire, DivestAcquire,* or *NoTransfer.*
- *TransportationKind* specifies the policy adopted by the RTI for transporting the messages exchanged by federates. It assumes one of the following values: *HLAreliable* or *HLAbesteffort.*
- *OrderKind* specifies the delivery order of attribute instance and interactions among federates during the federation execution. It assumes one of the following values: *Receive* or *TimeStamp.*

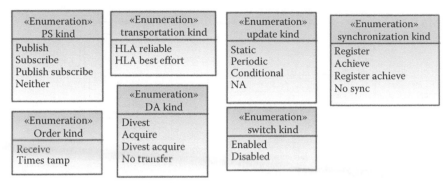

Figure 6.5 HLA datatype package.

- *SwitchKind* defines an enumeration used to specify the initial setting of RTI parameters. It assumes one of the following values: *Enabled* or *Disabled*.
- *SynchronizationKind* specifies the capability of federates with respect to the synchronization points that they attempt to reach during the federation execution. It assumes one of the following values: *Register, Achieve, RegisterAchieve,* or *NoSync*.

OMTKernel package

This package includes the stereotypes defined in the HLA OMT specification (IEEE 1516.2-2010). It provides the following stereotypes and the related associations, as shown in Figure 6.6:

- Federation and Federate: The UML elements representing the whole federation and associated federates, respectively. Both stereotypes extend the UML Class metaclass.
- HLADimension: A specific dimension for an attribute of an Object-Class or an InteractionClass. Dimensions are used to define a multidimensional coordinate system through which federates either express an interest in receiving data or declare their intention to send data. It extends the UML Class metaclass and is specified by the following tagged values:
    - *DataType*: The datatype for the federate view of the dimension.
    - *UpperBound*: The upper bound for the dimension that meets the federation's requirement for dimension range resolution.
    - *UnspecifiedValue*: A default range for the dimension that the RTI uses if the specified dimension is related to an available dimension

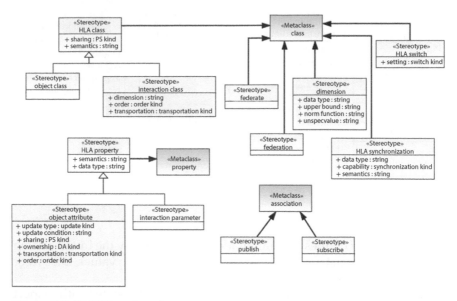

Figure 6.6 OMTKernel package.

of an attribute or interaction that has been left unspecified when a federate creates a region for specifying update/subscription interests.

- HLASwitch: The initial value for the RTI configuration parameters. Extends the UML Class metaclass and it is specified by the following tagged value:
  - *Setting:* The setting value.
- HLASynchronization: The synchronization point that federates attempt to reach during the federation execution. It extends the UML Class metaclass and is specified by the following tagged values:
  - *Datatype:* The datatype of the user-supplied tag for the synchronization point.
  - *Capability:* The level of interaction that a federate is capable of accomplishing.
  - *Semantics:* A textual description that specifies the element semantics (for documentation purposes).
- HLAClass: The common properties shared by the ObjectClass and the InteractionClass stereotypes. It extends the UML Class metaclass and is specified by the following tagged values:
  - *Sharing:* The publish/subscribe capabilities of a federate with respect to the related ObjectClass or Interaction instance.

- *Semantics*: A textual description that specifies the element semantics (for documentation purposes).
- ObjectClass: An Object Class. It extends the UML Class metaclass.
- InteractionClass: An interaction. It extends the UML Component metaclass and is specified by the following tagged values:
  - *Dimension*: The available dimension. It is used when federates adopt the data distribution management (DDM) services to reduce the volume of data delivered to federates.
  - *Transportation*: The transportation type to be used.
  - *Order*: The order of delivery.
- HLAProperty: The common properties shared by ObjectAttribute and InteractionParameter stereotypes. It extends the UML Property metaclass and is specified by the following tagged value:
  - *DataType*: Datatype of attributes or parameters.
  - *Semantics*: A textual description that specifies the element semantics (for documentation purposes).
- ObjectAttribute: An object class attribute. It extends the UML Property metaclass and is specified by the following tagged values:
  - *UpdateType*: The policy for updating an attribute.
  - *UpdateCondition*: Conditions for the update.
  - *Ownership*: Indicates whether ownership of an instance of a class attribute can be released or acquired.
  - *Sharing*: The publish/subscribe capabilities of a federate with respect to the stereotyped attribute.
  - *Dimension*: The available dimension. It is used when federates adopt the DDM services to reduce the volume of data delivered to federates.
  - *Transportation*: The type of transportation.
  - *Order*: The order of delivery.
- InteractionParameter: An interaction class attribute. It extends the UML Property metaclass.
- Publish: The association between a Federate and a published element. It extends the UML Association metaclass.
- Subscribe: The association between a Federate and a subscribed element. It extends the UML Association metaclass.

High-level architecture behavioral metamodel

The behavioral metamodel (BMM) package includes the stereotypes derived by the HLA BMM (Topçu 2008), which provides the UML extensions required to model the observable behavior of the federation, as summarized in Table 6.2.

Table 6.2 BMM Stereotypes

Stereotype	Extension	Description
Initialization	UML::Message	Identifies messages exchanged to setup the distributed simulation infrastructure
Action	UML::Message	Identifies a local action executed by a federate that does not require any interaction with other federates
Message	UML::Message	Identifies a communication between two federates

Model transformations

This section outlines the rationale of the model transformations that form the basis of the method depicted in Figure 6.3, specifically the following:

1. The `SysML-to-HLA` model-to-model transformation, which takes as input the SysML-based system specification and yields as output the HLA-based DS, model.
2. The `HLA-to-Code` model-to-text transformation, which takes as input the simulation model and yields as output the code that implements the HLA-based DS.

`SysML-to-HLA` *model-to-model transformation*

The `SysML-to-HLA` model transformation takes as input a SysML model representing the system that is going to be simulated and yields as output a UML model that specifies the HLA-based DS.

The input model is annotated with stereotypes provided by the *SysML4HLA profile*, whereas the output model makes use of the *HLA profile*. Specifically, the source model consists of the following SysML diagrams:

- A *Block Definition Diagram*, that defines the static structure of the system to be developed in terms of its physical components.
- A set of *Sequence Diagrams* (SDs), which specify the interactions among the system components.

The target model is specified by the following UML diagrams:

- *Structural model*, constituted by a UML Class Diagram, that shows the partition of the simulation model in terms of federates. The diagram also shows how federates publish or subscribe HLA resources (i.e., object classes and interaction classes).
- *Behavioral model*, constituted by a set of SDs, showing the interactions among federates and RTI.

The model transformation has been implemented using the QVT language (OMG 2008), which is provided by the OMG as the standard language for specifying model transformations that can be executed by available transformation engines (Eclipse Foundation 2014; Medini QVT 2011).

The Subsections "Structural model" and "Behavioral model" specify in natural language the mapping rules defined to generate the structural model and the behavioral model, respectively.[*]

Structural model

- Rule 1: The main `block` element in the block definition diagram stereotyped as <<federation>> represents the HLA federation and is thus mapped to a UML `class` element, stereotyped as <<federation>>. Other `block` elements in the binary decision diagram (BDD) stereotyped as <<federate>> represent the HLA federates that participate to the federation and are thus mapped to UML `class` elements stereotyped as <<Federate>>. Moreover, in the target model, such `class` elements representing federation and federates are connected by aggregation associations.
- Rule 2: A `block` element, stereotyped as <<ObjectClass>> and connected by a composition relationship with a `block` element representing a federate, is mapped to a UML class element, stereotyped as <<ObjectClass>>. The related block `properties` are mapped to class `attributes` in the target model, stereotyped as <<objectattribute>>.
- Rule 3: A `signal` element, stereotyped as <<Interaction-Class>>, is mapped to a UML `class` element, stereotyped as <<InteractionClass>>. The related signal `properties` are mapped to class attributes in the target model, stereotyped as <<interactionparameters>>.
- Rule 4: Each composition association between a block A stereotyped as <<federate>> and block B stereotyped as <<ObjectClass>> is mapped to an association between the corresponding class elements C and D in the target model, stereotyped as <<publish>>. It is assumed `Class` elements C and D are generated from A and B by applying Rules 1 and 2.

---

[*] It should be noted that some stereotypes provided by the HLA profile (e.g., dimension, synchronization, and switch) are used to represent FOM elements that cannot be automatically derived by the SysML model and, as such, are not considered by the transformation rules. The HLA profile has been designed to be as complete as possible, and suitable to fully describe a HLA federation, regardless the capability of the current prototypal implementation of the SysML-to-HLA transformation. In this respect, the HLA model could be manually refined.

- Rule 5: Each association between a block A stereotyped as <<federate>> and a block B stereotyped as <<ObjectClass>>, where there exists a composition association between block B and a block C ≠ A stereotyped as <<federate>>, is mapped to an association between the corresponding elements D and E in the target model, stereotyped as <<subscribe>>. It is assumed Class elements D and E have been generated from A and B by applying Rules 1 and 2.
- Rule 6: Each association between blocks A and B, both stereotyped as <<ObjectClass>>, where (1) C and D are two blocks stereotyped as <<federate>> and (2) there exist composition associations between A and C, and A and B, respectively, is mapped to an association between the corresponding elements E and F in the target model, stereotyped as <<subscribe>>. It is assumed Class elements E and F are generated from C and B by applying Rules 1 and 2.
- Rule 7: Each association that connects flowports belonging to blocks A and B, being C the exchanged signal, is mapped to two associations in the target model. The first one between the class element D and F, where D and F correspond to A and C according to Rules 1 through 3. The second one between the class element E and F, where E corresponds to B, according to Rule 1 and Rule 2. Such associations are stereotyped according to the flowport direction:
  - <<subscribe>>, if the port direction is out
  - <<publish>>, if the port direction is out
  - both <<publish>> and <<subscribe>>, if the port direction is inout

Table 6.3 details the mapping from SysML to UML, with regards to Rules 1 through 3. The rationale of mapping rules from 4 to 7, which are involved in the generation of the publish/subscribe relationships, is instead depicted in Figure 6.7.

Table 6.3 SysML to UML Mapping

SysML		HLA	
Element (stereotype)	Diagram	Element (stereotype)	Diagram
Block (federation)	BDD	Class (federation)	Class diagram
Block (federate)	BDD	Class (federate)	Class diagram
Block (object class)	BDD	Class (object class)	Class diagram
Signal (interaction class)	BDD	Class (interaction class)	Class diagram
Block attribute	BDD	Attribute (object parameter)	Class diagram
Signal property	BDD	Attribute (interaction parameter)	Class diagram

Rule 4

Rule 5

Rule 6

Rule 7

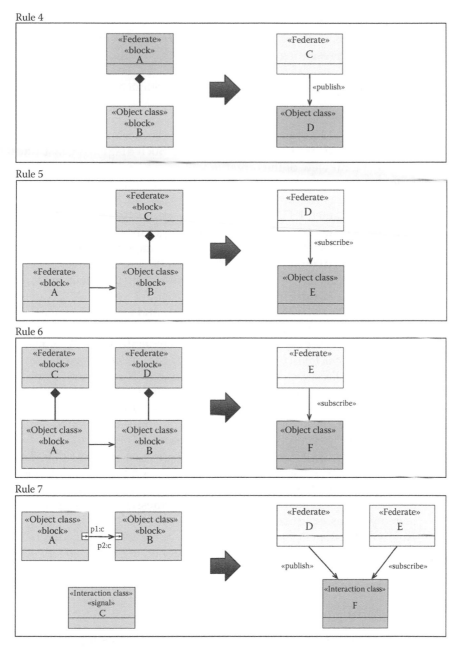

Figure 6.7 Publish/subscribe relationship generation rules.

Behavioral model

Sequence diagrams in the source model specify the interactions among model elements representing federates (i.e., block elements stereotyped as <<federate>>).

The behavioral view of the target model is obtained according to the following rules:

- Rule 1: Each UML sequence diagram in the source model is mapped to a sequence diagram in the target model. Such diagrams constitute the behavioral view of the target model.
- Rule 2: To represent interactions between federates and the HLA run-time infrastructure, each sequence diagram contains a UML component named RTI.
- Rule 3: Each sequence diagram in the target model represents the behavior of a federate interacting with its component and/or other federates. The diagram must contain the following messages:
  - A self-message named *createRTIAmbassador()*, stereotyped as <<initialization>>
  - A message exchanged between federate and RTI, named *joinFederation()*, stereotyped as <<initialization>>
  - A set of messages exchanged between federate and RTI, namely *publishInteractions()*, *subscribeInteractions()*, *publishObjects()*, and *subscribeObject()*, stereotyped as <<initialization>>, according to the publish/subscribe capability of each federate
  - Main flow according to subsequent Rules 4 and 5
  - A message exchanged between federate and RTI, named *leaveFederation*, stereotyped as <<message>>
- Rule 4: Messages included in the source sequence diagram exchanged between a federate and one of its component (i.e., an element stereotyped as <<federate>> and an element stereotyped as <<ObjectClass>>) are mapped to self-messages to the federate, stereotyped as <<action>>
- Rule 5: Messages included in the source sequence diagram exchanged between two federates are mapped to messages between federates and RTI (and vice versa). Such messages are stereotyped as <<messages>>

HLA-to-Code *model-to-text transformation*

The HLA-to-Code model-to-text transformation takes as input a UML model representing the HLA-based simulation system and yields as output the corresponding implementation code.

The code generation makes use of the Pitch pRTI (2012) RTI implementation and Java as the language for implementing federates and ambassadors. The model-to-text transformation has been implemented using the template-based *Acceleo* language (Eclipse Foundation 2014), provided as a model-driven Eclipse plugin.

The implementation of the proposed transformation includes the following templates:

- *generateFederate* for each element in the HLA model stereotyped as <<federate>> generates a Java class that implements the corresponding federate.
- *generateObjIntClass* for each UML class in the HLA model stereotyped as <<ObjectClass>> or <<InteractionClass>> generates a tag in the XML file in which classes representing *ObjectClass* and *InteractionClass* are serialized. In other words, such file gives the representation of the FOM. Moreover, this template creates, in each class generated by the *generateFederate* template, the methods for publishing and subscribing resources, according to the *Publish/Subscribe relationships* in the structural HLA model.
- *generateAmbassador* generates a set of java classes constituting the implementation of the required federate ambassadors.
- *completeFOM* generates the FOM sections related to switch, synchronization points, and dimension, whereas the corresponding elements are specified by the HLA model.

Example application

This section presents an example application that shows how the proposed model-driven method effectively exploits the UML profiles to support the simulation of a system by generating the HLA/Java code starting from the initial SysML specification.

The proposed example application deals with the development of an aircraft.

System specification

The first step of the method includes the definition of the system model using the SysML notation. For the sake of brevity, this example only takes into account the diagrams needed to derive the HLA simulation model and, consequently, the code of the simulation program. Specifically, the SysML aircraft model includes the following diagrams:

- A block definition diagram, to specify the structural view of the system. The diagram shows the aircraft components (e.g., the structural parts such as ailerons, elevator, autopilot, and powerplant) and their relationships.

- A set of sequence diagrams, to specify the behavioral view of the system. Such diagrams describe the interactions between system components.

Conceptual analysis: The specification of the aircraft federation

At the second step, the SysML model is refined and annotated with stereotypes provided by the HLAProfile. This step marks the PIM by adding to the SysML model the information needed to map each SysML domain element to the corresponding HLA domain element and to make the PIM model ready to be automatically processed by the model-driven transformations included in the next steps. As an example, Figures 6.8 and 6.9 show the BDD and a SD that specify the interactions between the Autopilot (composed itself by ControlLogic, GPS, and Gyroscope components) and the Structure (composed itself by Aileron, Elevator, and PositionSensor components) component, respectively.

Designing the federation

The UML model annotated with SysML and HLA profiles is given as input to the SysML-to-HLA model-to-model transformation, to generate the HLA-based simulation model. Such model consists of a UML class diagram (i.e., the HLA *structural model*) and a set of UML sequence diagrams (i.e., the HLA *behavioral model*).

As an example, Figures 6.10 and 6.11 depict the structural view (UML class diagram) and the behavioral view (UML sequence diagram) corresponding to the block definition diagram and the sequence diagram shown in Figures 6.8 and 6.9, respectively.

Once the federation structure has been specified, the federation design step also includes the optional execution of a discovery that aims to retrieve existing federates suitable to be integrated in the federation. In this example it is supposed that the federation is fully implemented from scratch.

Integrating the federation

Once the federation model has been specified, the HLA-to-Code model-to-text transformation is carried out. Such a transformation takes as input the HLA model produced in the previous step and yields as output the Java/HLA code that implements the HLA-based DS. It should be noted that the proposed method does not generate the *complete* code that implements the HLA-based DS. The model-driven method should be considered as a supporting tool. More specifically, the HLA-to-Code transformation allows to generate a template of the Java classes that contains the class

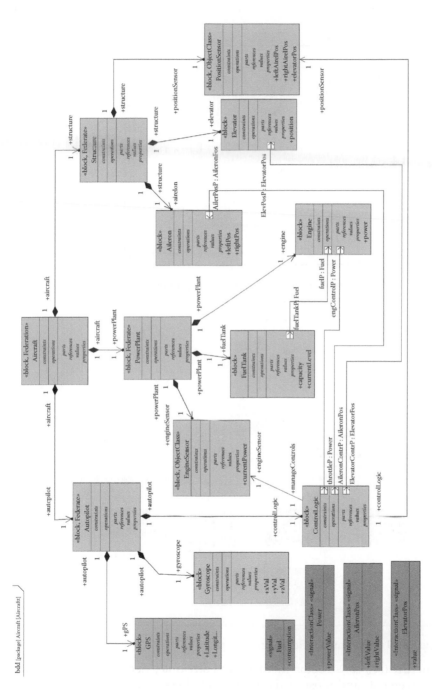

Figure 6.8 Source model: Block definition diagram.

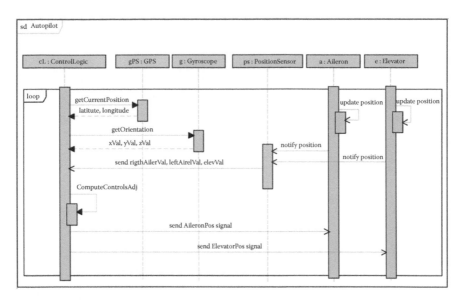

Figure 6.9 Source model: Sequence diagram (interaction between `Autopilot` and `Structure` components).

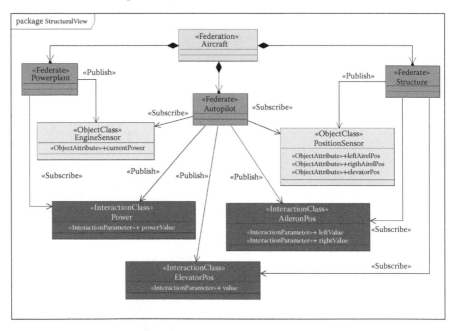

Figure 6.10 Target model: Structural view.

structure (i.e., constructor, methods and attributes declarations, exception management) and most of the HLA-related code (i.e., data types definition, RTI interaction methods), but the code implementing the federate simulation logic has to be added manually.

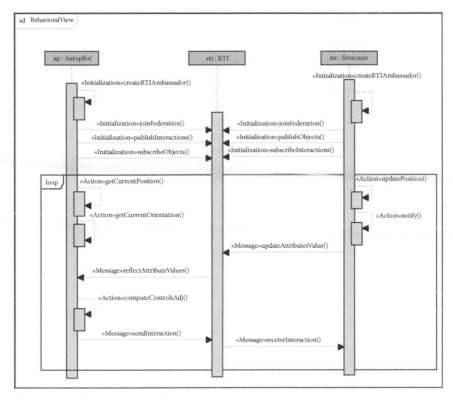

Figure 6.11 Target model: Behavioral view (interaction among `Autopilot` and `Structure` federates and the RTI).

As an example, a portion of the code that implements the Autopilot federate is shown in Listing 6.1. A comment acts as placeholder to indicate where the code that implements the federate simulation logic has to be placed.

The final step of the method consists of the simulation execution and the results evaluation, which are not here discussed for the sake of conciseness, being the main focus of this chapter centered on the model-driven method to support the generation of both the HLA model and the HLA code.

Conclusions

This chapter has introduced a method to enable the simulation-based analysis of complex systems through a MDSE approach. Specifically, the proposed method has shown how the UML model and the Java code of a HLA-based DS can be derived from a SysML specification of the system under study. The approach has introduced two automated model

```
package it.uniroma2.sel.simlab.aircraftsim.autopilot;

import hla.rti1516e.*;
import hla.rti1516e.exceptions.*;
...
public class Autopilot extends NullFederateAmbassador{
...
//create ambassador
RtiFactory rtiFactory = RtiFactoryFactory.getRtiFactory();
_ambassador = rtiFactory.getRtiAmbassador();
try {
_ambassador.connect(this, CallbackModel. HLA_IMMEDIATE,
localSettingsDesignator);
}catch (AlreadyConnected ignored)
...
//join federation
_ambassador.joinFederationExecution(federateName +
federateNameSuffix, "MapViewer", federationName, new URL[]
{url});
...
//defines handlers and publish/subscribe capabilitiestry{
getHandles();
subscribeObjects();
publishInteractions();
} catch (FederateNotExecutionMember e){
throw new RTIinternalError ("HlaInterfaceFailure", e);
...
//start method must provide the federate simulation logic
start();
...
}
```

Listing 6.1 Portion of the code generated for the Autopilot federate.

transformations and two UML profiles. The transformations automatically map the source SysML model into a HLA-specific model and eventually into the Java/HLA source code, whereas the profiles are used to annotate the SysML and HLA-specific models to drive the relevant transformations.

An example application to the development of a HLA-based DS from the SysML model of an aircraft has been presented to show how the proposed approach can be effectively used by systems engineers who are not familiar with DS standards and technologies (HLA in this chapter case).

The method allows to automatically obtain a significant portion of the final HLA-based code by limiting the manual activity to the implementation of the federate simulation logic.

References

Atkinson, C. and Kuhne, T. 2003. Model-driven development: A metamodeling foundation. *IEEE Software* 20(5): 36–41.

Bocciarelli, P., D'Ambrogio, A., and Fabiani, G. 2012. A Model-driven Approach to Build HLA-based Distributed Simulations from SysML Models. SIMULTECH 2012, pp. 49–60, Rome, Italy.

D'Ambrogio, A., Gianni, D., Risco-Martín, J. L., and Pieroni, A. 2010. A MDA-based approach for the development of DEVS/SOA simulations, In *Proceedings of the Spring Simulation Multiconference 2010* (SpringSim'10), Orlando, FL.

D'Ambrogio, A., Iazeolla, G., and Gianni, D. 2011. A software architecture to ease the development of distributed simulation systems. *Simulation*, 87(9): 813–836.

D'Ambrogio, A., Iazeolla, G., Pieroni, A., and Gianni, D. 2011 A model transformation approach for the development of HLA-based distributed simulation systems. In *Proceedings of 1st International Conference on Simulation and Modeling Methodologies, Technologies and Applications* (Simultech 2011), pp. 155–160, Noordwijkerhout, the Netherlands.

Eclipse Foundation. 2014. Acceleo. http://www.acceleo.org/pages/home/en.

Eclipse Foundation. 2014. QVT Transformation Engine. http://www.eclipse.org/m2m.

Fujimoto, R. M. 1999. *Parallel and Distribution Simulation Systems*, 1st Ed. Wiley, New York.

Fujimoto, R. M. 2001. Parallel simulation: parallel and distributed simulation systems. In *Proceedings of the 33nd conference on Winter simulation* (WSC '01), pp. 147–157. IEEE Computer Society, Washington, DC.

Haouzi, H. E. 2006. Models simulation and interoperability using MDA and HLA. In *Proceedings of the IFAC/IFIP International conference on Interoperability for Enterprise Applications and Software* (I-ESA'2006), Bordeaux, France.

IEEE 1516-2010. Standard for Modeling and Simulation High Level Architecture: Framework and Rules.

IEEE 1516.1-2010. Standard for Modeling and Simulation High Level Architecture: Federate Interface Specification.

IEEE 1516.2-2010. Standard for Modeling and Simulation High Level Architecture – Object Model Template (OMT) Specification.

IEEE 1516.3-2003. Recommended Practice for High Level Architecture Federation Development and Execution Process (FEDEP).

INCOSE. 2007. Systems Engineering Vision 2020 v.2.03. Available at: http://www.incose.org/ProductsPubs/pdf/SEVision2020_20071003_v2_03.pdf.

Jimenez, P., Galan, S., and Gariia, D. 2006. Spanish HLA abstraction layer: towards a higher interoperability model for national. In *Proceedings of the European Simulation Interoperability Workshop*, Stockholm, Sweden.

Kuhl, F., Weatherly, R., and Dahmann, J. 1999. *Creating Computer Simulation Systems: An Introduction to the High Level Architecture*. Prentice Hall PTR, Upper Saddle River, NJ.

Medini QVT. 2011. IKV++ Technologies Ag. http://projects.ikv.de/qvt.

Möller, B., K. Morse, M. Lightner, R. Little, and R. Lutz. 2008. HLA Evolved—A summary of major technical improvements. In *Proceedings of 2008 Euro Simulation Interoperability Workshop*, No. 08F-SIW-064, Edinburgh, Scotland.

OMG. 2003. MDA Guide, version 1.0.1.

OMG. 2004. Meta Object Facility (MOF) Specification, version 2.0.

OMG. 2007. XML Metadata Interchange (XMI) Specification, version 2.1.1.

OMG. 2008. Meta Object Facility (MOF) 2.0 Query/View/Transformation, version 1.0.

OMG. 2010. System Modeling Language, v.1.2.

Paredis, C. J. J. and Johnson, T. 2008. Using OMG's SysML to support simulation. In *Proceedings of the 40th Conference on Winter Simulation* (WSC '08), pp. 2350–2352, Miami, FL.

Peak, R. S., Burkhart, R. M., Friedenthal, S. A., Wilson, M. W., Bajaj, M., and Kim, I. 2007. Simulation-based design using SysML. Part 1: A parametrics primer. In *Proceedings of the INCOSE International Symposium*, San Diego, CA.

Pitch pRTI. 2012. http://www.pitch.se/.

Schmidt, D. C. 2006. Model-driven Engineering. *IEEE Computer* 39(2): 25–31.

Tolk, A. and Muguira, J. A. 2004. M&S within the Model Driven Architecture. In *Proceedings of the Interservice/Industry Training, Simulation, and Education* (I/ITSEC), Orlando, FL.

Topçu, O., Adak, M., and Oğuztüzün, H. 2008. A metamodel for federation architectures. *ACM Transactions on Modeling and Computer Simulation* 18: 1–10.

Topçu, O., Oğuztüzün, H., and Gerald, H. M. 2003. Towards a UML profile for HLA federation design, part ii. In *Proceedings of the Summer Computer Simulation Conference* (SCSC'03), pp. 874–879, Montreal, Canada.

Weyprecht, P. and Rose, O. 2011. Model-driven development of simulation solution based on SysML starting with the simulation core. In *Proceedings of the 2011 Symposium on Theory of Modeling & Simulation: DEVS Integrative M&S Symposium* (TMS-DEVS '11), pp. 189–192, San Diego, CA.

Zhu, H., Li, G., and Zheng, L. 2008. A UML profile for HLA-based simulation system modeling. In *Proceedings of the 6th IEEE International Conference on Industrial Informatics* (INDIN 2008), Daejeon, Korea.

chapter seven

Collaborative modeling and simulation in spacecraft design

Volker Schaus, Daniel Lüdtke,
Philipp M. Fischer, and Andreas Gerndt

Contents

Introduction

Collaboration is a key factor for successful development of spacecraft. It is a broad term and has many meanings because exchange of information is needed all along the development life cycle. A successful design directly depends on sharing necessary information with others and, on the other hand, comprehending and using relevant information as input for your own work. Modeling and simulation is extensively used in spacecraft design and appears in many different varieties throughout the whole development life cycle. It covers a wide range, from simple orbit simulations based on analytical equations to high-fidelity and multidomain simulations using distributed high-performance computers. In today's large-scale, complex projects, traditional engineering approaches reach their limits. The amount of information (e.g., models, reports, specifica-tions, design parameters, and simulation results) is vast and cannot be handled effectively. In addition, it is hard to trace design changes and predict possible implications of a modification. The design and analy-sis methods have changed, too. Today, they are mainly computer based and involve simulations. There is a scattered landscape of models and tools, which contains inconsistencies and makes it hard to exchange data. Traditional engineering tries to tackle this issue with reports and review meetings. Yet, this document-centric approach does not overcome the shortcomings. For a long time, modeling and simulation was a task done by an individual on a specific topic on a local resource. Current projects, however, demand for collaborative model development covering multiple fields of expertise and use distributed simulation environments.

The ongoing research and implementations in the field of model-based systems engineering (MBSE) show a way out of this document-centric approach dilemma. The main idea is the application of formal modeling techniques in combination with a central system model containing all the design information. Formal representations have the major advantage that they are machine readable. Such models can be interpreted by computers and thus, exploit their full power. This enables automation in many ways. Needed artifacts for the design, such as models for domain-specific analy-sis, simulation setups, and reports can be generated. Model transformations bridge the gap between disciplines and allow for easy data exchange. This procedure is similar to the model-driven approach in software engineering.

MBSE has great potential to change the way of designing. It provides a consistent data basis for accompanying the whole design process from early phases to operations. It can support working at the right level of detail and triggering the desired analysis method. It connects disciplines and allows multidomain analysis. Especially considering the system level, the central accessible data model offers new capabilities to run special analysis methods such as sensitivity analysis or domain structure matrix on the whole system. This immediately increases the knowledge of the system under development and has a positive impact on informed decision making. In addition, MBSE is a key factor to support the collaboration of engineers of different domains to work concurrently on a design. It allows distributing changes of the design without any delay and, thus, helps to create a common understanding of the current state of the design throughout the development process.

Yet considered as an enabling technology, MBSE is still emerging and faces many challenges. The chapter covers the following aspects in this regard:

- Modeling languages: There are already many existing modeling languages, but do they cover all the special demands of a certain domain, for example, spacecraft design? A lot of ongoing work is related to language development and metamodeling. It is important to provide an appropriate level of abstraction for the design activity. Accordingly, this involves an in-depth analysis of the design process and the extraction of the necessary core elements that drive the activity.

- Tool support: The rise and fall of MBSE heavily depends on tool support. This part of the chapter mentions practical examples of tool development and thus describes the current state of MBSE implementations.
- Model development and reuse: This part presents a process and guidelines for sustainable model development. Because models contain a lot of knowledge about the system, it is of great importance to access and reuse this knowledge in the future.
- Team collaboration and distributed tool interaction: A big challenge toward setting up a system simulation is to overcome the segmentation of information in various design and analysis tools. Tackling this issue, this section discusses model transformations. It also gives ideas to be considered when developing a central data model. It presents an example from the aircraft design community, which shows the adoption of a special middleware to set up a transparent workflow across many involved domains. This allows for efficient data exchange and links several remote simulations.

The rest of this chapter is organized as follows: the next two subsections of the introduction, Subsections "Characteristics of space system design" and "Way of designing," give background information on special challenges in spacecraft design and describe the structured design process that is usually followed in space mission projects. Those two subsections are on an introductory level and may be skipped, if the reader is already familiar with spacecraft engineering and systems engineering in general.

The Section "Model-based systems engineering" deals with MBSE in theory and practice. It first pins down the expectations of MBSE. It includes comments on modeling languages. Then, it gives examples from the current state of implementations and deals with future challenges.

The Section "Collaborative modeling and simulation throughout the whole development life cycle" focuses on different aspects of collaborative modeling and simulation along the development life cycle of spacecraft. This section starts with the early phases where the concurrent engineering (CE) work methodology is now widely used. It continues with the Subsecton "Model development process and reuse." Here, the reader learns about a collaborative model development process that was successfully used at the German Aerospace Center (DLR) to develop a spacecraft system simulator covering several subsystems, the space environment, and the orbit dynamics.

The Subsection "Collaborative development of operational simulators" ends with an overview of the current practice of collaboratively developing simulators for the operational phase of a spacecraft. This is a special case of modeling and simulation because these simulators are developed when the spacecraft has already been built or is at least fully specified. This is significantly different from the modeling and simulation activities during the development phases of the spacecraft itself where the specification is subject to change and the full system specification is one main result of the development activity. The work here bears great analogy with current practices in software engineering.

The chapter ends with the Section "Conclusion."

Characteristics of space system design

Spacecraft are complex systems and designing them is a challenging task. In some ways, it can be compared to other large-scale projects from other domains like ship building or architecture and construction. In many other aspects, it is set apart and forms a niche with special characteristics. Examining these characteristics and challenges throughout the life cycle helps to understand the demands and to derive the requirements for the implementation of MBSE methodologies in the future.

Looking back in time, it was only a couple of decades ago since mankind entered the space age in 1957 with the launch of *Sputnik 1*. So, the space domain is still relatively young compared to other engineering fields. In addition, in the beginning, space programs were prestigious national projects with strong political motivations. Since then, the space domain has changed.

Some mission scenarios, such as telecommunication, have been successfully commercialized. The mission requirements have been very well understood and are mapped to engineering solutions. Today, private companies offer high quality products for these purposes. Standard solutions have evolved and design principles have reached a high level of maturity. The customer can select from a range of product families, and products are almost commercial-off-the-shelf like in other engineering fields. Altogether, the competition across agencies, governments, and industry has increased, putting more pressure on today's quality demands and prices.

Other areas of spacecraft design, such as exploration, observation, and the implementation of new technologies, are usually one-of-a-kind mission scenarios. They can be seen as prototypes that are typically built only once. As result, there is no series production. Navigation systems such as the European initiative Galileo are an exception, where a total of 30 satellites are needed to ensure coverage and reliability. However, in this particular case, the whole system was not ordered at once but split into several competitive calls for tender leading to different suppliers. So in the end, the satellites will be similar, but not identical in terms of serial production.

Yet, also in the area of exploration and research-oriented missions, the trend is to reuse existing concepts as much as possible. This can be seen, for instance, when looking at the structural design of spacecraft. Many space telescopes use a long cylinder as main structure. The trapezoidal, doghouse-like structural concept can be recognized on several earth observation missions like Champ, GRACE, and CryoSat. The idea is not only to take over individual concepts from one mission to the other, but to establish satellite families that cover a broader range of mission scenarios. Another example is the satellite infrastructure of the Bi-Spectral Infrared Detection mission, which was just recently reused with modifications for the on-orbit verification TET-1 mission. Comparing the schematic drawings of the two spacecraft depicted in Figure 7.1, it is easy to recognize similarities in the structure, for example, the cubic shape, the arrangement of subsystems or components, and the deployment concept of the solar arrays. Reuse was also successful with the on-board attitude and control software (Maibaum et al. 2011). In general, the reuse principle applies to all the artifacts that are part of the development process including modeling and simulation. The benefit of reuse is only possible with a clear understanding of the system and its internal interfaces.

Space projects are conducted by international and intercultural teams in various different organizations, agencies, and companies. The members of the design team are distributed around the world in different time zones. Collaboration is one key factor for a successful design over the full life cycle. As in the reuse case earlier, it is essential to define the interfaces between the various involved parties and to make sure they are understood and respected.

Projects have a very long duration. It can take 10–20 years or even more from the first mission concept to reach an operational state, and even longer when considering end-of-life disposal scenarios. The missions are expensive. The largest cost contribution of missions with prototypic characteristics is the nonrecurring cost of research, development, test, and evaluation of a spacecraft (Wertz and Larson 1999; U.S. Congress 1990).

Leaving mission- and project-management-specific aspects aside and focusing more on the actual engineering, other unique implications must be considered. Spacecraft are exposed to various environments that influence the design in many ways. Some of them are very different from Earth, for example, as follows:

- Extreme temperature jumps between sunlight and eclipse
- Only residual atmosphere in low Earth orbit or vacuum, meaning there will be no convective flow of heat
- Microgravity environment or weightlessness, affecting the behavior of fluids and causing design implications for pressure tanks
- Radiation, which can interfere with the experiments on board and cause failures in the electronics

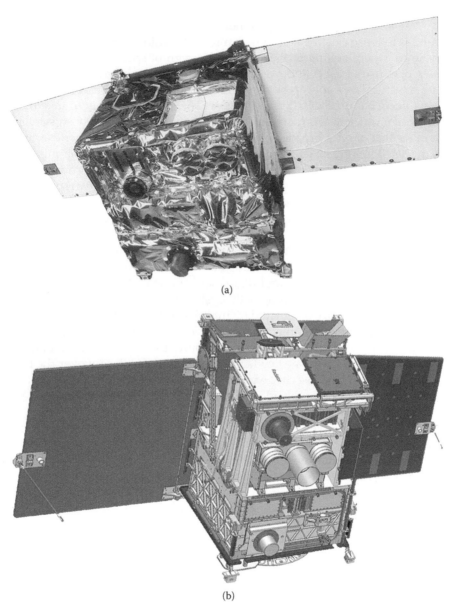

(a)

(b)

Figure 7.1 Schematics of (a) the Bi-Spectral Infrared Detection satellite and (b) the next in the family, on-orbit verification mission TET-1.

For the designers, it is difficult to develop a feeling for the effects because they can hardly be experienced here on Earth. Tests are possible, but very expensive. Usually, only one effect can be reproduced in a lab at a time. So if, for example, the engineers want to test a deploy mechanism in microgravity, this can be done in free fall towers, parabolic flights, or with special support structures. As an example, the project Gossamer uses deployable booms as support structures of solar sails. The verification of the deployment mechanism of these booms took place during a parabolic flight campaign (Block et al. 2011). However, it is difficult and expensive to combine the test setup for microgravity with extreme temperatures or vacuum at the same time (Lichodziejewski et al. 2004). Because space conditions are hard to reproduce in labs on Earth, it is important to have accurate models to predict the behavior of the spacecraft and its subsystems and components in its operational environment (Wormnes et al. 2013).

Maintenance is another challenge in spacecraft design. After a successful launch, the satellite is up in space, so it is hardly possible to bring the hardware physically back in the workshop on the test bench and take it apart. Maintenance of the hardware in orbit is difficult and expensive, often enough not possible at all.

Like in other fields, the trend for space systems is to implement more and more functionality in software, instead of hardware. This reduces complexity in the hardware design and therefore shifts the focus on development and quality assurance with respect to software. Bugs in the software can potentially be fixed and maintained when the satellite is already operational in orbit. In this case, the ground team can develop a software patch. Once available, the new version of the software can be radioed up to the spacecraft similar to receiving an update over the Internet on a standard home computer. This is a huge advantage compared to nonfixable hardware. However, this approach needs a considerable amount of knowledge about the system on the ground to be able to develop a proper software fix. Extensive testing and quality assurance using various models have to be performed before such a software patch is uploaded to the running system.

Finally, long space projects today suffer from change of personnel because of fixed term contracts and the many partners that are involved. Nevertheless, it is the many involved people who are driving the design and their ability to communicate and collaborate efficiently is a key factor for a successful space mission.

Way of designing

Generally, designing is a creative task, which cannot be tackled with a fixed process that assures a working solution in the end. Changes occur frequently and at any time during the process. At the beginning, it is usually easier to accommodate a change and propagate the effects of the change to other disciplines. Later, design changes are significantly more expensive.

Complex design tasks are carried out by teams. Collaboration plays an important role. At the beginning, a design team is relatively small and can fit in one room. At later stages, the number of involved people is much larger and projects are carried out by many international partners distributed around the world.

Space projects follow a structured approach to allow both creativity and flexibility on one hand, and account for project management, planning, and quality on the other hand. The process is basically the systems engineering V-model as described in the design methodology for mechatronic systems (Association of German Engineers 2004). This process consists of the six well-known steps: requirements definition, design, development, integration, verification, and validation. These steps are organized along a V, as shown in Figure 7.2. The first three steps go down the left leg; the remaining three steps are going up the right leg of the V. The basic idea behind the V-model is that testing becomes an integral part of the design activity. The objective is to reduce the risk in the development of a new product and to improve the communication between all stakeholders. Each definition step on the left leg has a matching counterpart on the right leg, the testing leg. This means in practice that hardware components are tested against a specification, that the integrated system is verified against the design requirements, and that the system in the end meets the needs of the customer and all other stakeholders.

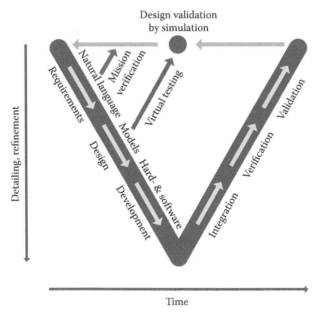

Figure 7.2 V-model for systems engineering highlighting the benefit of modeling and simulation through virtual testing and mission verification by model checking.

For Europe's space projects, this process is made more distinct by the Space Project Management Standard (European Cooperation for Space Standardization 2009). Regarding the development process, the standard has the following features:

- Defines six design phases that cover the full life cycle of a spacecraft from concept to operation and disposal
- Introduces review meetings (preliminary design review, critical design review, etc.) as quality gates and decision points
- Integrates a customer–supplier relation in the process

Nowadays, with powerful computers and IT solutions, modeling and simulation becomes an integrated part of the design process. Looking again at Figure 7.2, the use of modeling and simulation provides a short-cut or small V in the process. One is labeled virtual testing and means that as soon as the design can be modeled, it can be validated in a simulation. The idea is to do all the steps that are later done with the real hardware in the testing leg already in a virtual environment as early as possible in the design. The objective is again to reduce the risk: the risk of spending a lot of money on the development and in the end realizing that the product does not match the requirements. According to INCOSE's *Systems Engineering Handbook,* up to 85% of the overall cost of a project is defined during concept and design phase (Haskins 2006). So there is a strong interest to do as much frontloading of design information and validation as possible, to know very early, if the current approach is an appropriate solution for the design problem.

Since the beginning, design organizations use a document-centric approach. Information is gathered in reports that pass a signature loop for release; presentations are used for communicating the results and function as basis for discussion. On the project management side, the systems engineer tries to keep track of all changes in large action item tables, and writes minutes of meetings to document the outcome of discussions. Many domains have to work in accordance with certain procedures, guidelines, and standards like the DO178B for aeronautical software development or their counterparts in the space domain, the ECSS-E-ST-40 and ECSS-Q-ST-80. Consider designing the software of a reentry vehicle like Spaceship One. Which standard is to be used? Can they be both applied at the same time or are some aspects mutually exclusive? One solution in this case is to condense all the applicable regulations in one book specific to the project like it was done in the Galileo project (Eickhoff 2009). Then, at least the hope is, there is one source of information that is correct.

Altogether, the common document-centric approach creates a vast information and document flood in complex projects. And it seems to be impossible for one person to oversee it. Many organizations have introduced product life cycle management systems. They are, for sure, one step in the right direction, moving from paper-based documents to a digital database. Some offer version control, a search engine, authorization mechanisms, and support templates and other minor features. However, there are still these many sources of information that lead to inconsistencies or discrepancies in the design, and if overlooked can cause major design flaws that are not recognized.

Model-based systems engineering

Theory and vision

The *INCOSE Systems Engineering Vision 2020* gives the following definition of MBSE (2007):

> Model-based systems engineering (MBSE) is the formalized application of modeling to support system requirements, design, analysis, verification and validation activities beginning in the conceptual design phase and continuing throughout development and later life cycle phases. MBSE is part of a long-term trend toward model-centric approaches adopted by other engineering disciplines, including mechanical, electrical and software. In particular, MBSE is expected to replace the document-centric approach that has been practiced by systems engineers in the past and to influence the future practice of systems engineering by being fully integrated into the definition of systems engineering processes.

In general, MBSE is an enabling technology for innovative, interdisciplinary product design. It is used in various domains. Spacecraft development is one of them and for this field the following aspects are especially important:

- Overcoming the document-centric approach: The amount of documentation is huge. It takes a lot of time to write it; it is very hard to maintain an overview and the consistency between all the various documents. The model-based approach improves this situation. Documents can be generated automatically to a large extent. The consistency of the data is assured by a central source of information, the data model.
- Integration of all stakeholders' needs: Because of the complexity and the many interdependencies in spacecraft design, it is important to integrate all domains and create a system definition. This is not limited to the engineering disciplines; it also covers, for example, cost and risk.

- Bridging the information and communication gaps between the development phases and the stakeholders: The central model holds all information and is used as reference for all involved parties. The work results of suppliers have to be integrated in the model. The integration step can be used as review and interface check.
- Improving the traceability: Complex design tasks require many iterations and decisions in the process. It is extremely important to track changes in order to question decisions and to be able to go back to a previous state in case a design flaw is recognized. This means recording who did what, when, and why. A second aspect in this context is traceability of requirements. This means to know which requirement is fulfilled by which components or function of the system. The mapping is usually not so easy because generally, the combination and interplay of certain components will fulfill a requirement. Requirements engineering is a research domain of its own.

The key element of the model-based approach is, of course, the system model that holds all relevant design information. Usually, the model is saved in a repository on a central server and managed by a version control system. All stakeholders access and work with that system model, so it acts as a single point of truth for the design team. Such models are also advantageous in terms of data exchange between the engineering tools. The result of the growing complexity is the necessity of an all-to-all communication between the involved tools. In the worst case, when all the domains are dependent on each other and need to exchange data, the number of required interfaces will be $n \times (n-1)/2$; here n is the number of tools. In comparison, the number of interfaces for the data exchange via a central data model is just equivalent to the amount of domains, meaning one interface for each tool to the central data model. The interfaces should be bidirectional to realize an integrated approach. Looking at data exchange between tools from an IT point of view, it is always a model transformation. The direct transformation of a data model from one tool to another is sometimes not possible because the two models require different parameters or data for the conversion. This is known as the problem of different dimensionalities of the models. The additional information needed can be put in the model transformation tool directly, but the more appropriate way is to enrich a central data model that interfaces to both applications (Gross and Rudolph 2012).

This leads to the question: What kind of information should be present in the central model? As designing is a creative task, we cannot foresee all eventualities of what might be needed. So first of all, the data model should be easily extensible and customizable. Besides the core information that describes the system, we need auxiliary information. This can be information to drive external tools, for example, the settings for a numeric solver defining the algorithm and step size that is supposed to be used or setup procedures to configure the tool environment for an analysis. Sometimes it is not so easy to make the decision. Geometry modeling needs the size, shape, and relative positions of all parts and components to display a virtual assembly. But to really work with the virtual model, we need to assign colors to easily find and recognize subassemblies, specific parts, or materials. We also need to set a transparency value to look inside our product. The color and the transparency is clearly auxiliary information. They are parameters of the geometry model introduced by the computer-aided design tool.

Now, there are different ways to handle these parameters. The first one is to include them in the central model, which makes it more complex and it will be more difficult to understand and to work with the model. There are ways to handle complexity in software, for example, the use of certain views on the data. In this case, there could be a specific visualization view that only shows parameters related to visual appearance of the parts. The inclusion is generally useful, if more than one domain uses these parameters. Many engineering domains base their analysis on a simplified geometric model like the thermal analysis or the structural analysis and also use color and transparency settings. So, it makes sense to define the parameters in the central model. It leads to unified appearance of the design in different engineering tools. The design team or the organization could define standards that make it easier and faster to compare and understand different designs.

The second solution is to hide the parameters in the interface of the model. The color and transparency values can then be defined by setting the preferences of the interface. This approach is appropriate, for example, if the color has a different meaning in different domains. There are reasons for both solutions. The decision, if a parameter should be included into the central data model, needs to be made case by case.

In terms of collaboration, the idea of everyone working on the same data set at the same time is a synonym for chaos and a nontrivial challenge from the computer science point of view. A role management and authorization mechanism is one reasonable way to organize the access to the central resource. The desired situation is that conflicts are not possible and that only one person can modify a part of the model at a time. In reverse, this means that the responsibilities of tasks are clearly defined and communicated among the design team. In practice, this clarification goes down to single parameters if necessary. A typical example is the question of which domain will add the weight of solar panels to the system. Will the power engineer supply the value or will it come from the structure engineer? It is also important to assure that the data model holds only one parameter with a specific meaning and it is not introduced several times at different locations. Redundant information leads to confusion and inconsistency and has to be avoided.

A major part of the ongoing work in the MBSE field is formalization of the design. A formal representation is the foundation for any computer-based interpretation, analysis, or automation. The way we describe systems will define the way engineers will work in the future. One trend surely is the development of modeling languages. The most prominent language is SysML™. It is a standardized language that was especially developed to describe complex systems. It is based on the Unified Modeling Language (UML) and uses a similar graphical notation. The diagram types allow for defining static properties of the system such as interfaces, relations, hierarchies, and decomposition. In addition, it is possible to describe the system's behavior with several diagram types like the activity diagram, the sequence diagram, or the state machine diagram. There is also a special diagram for the requirements definition.

SysML is widely used today, and there are many good examples and books available (Gesellschaft für Systems Engineering e. V. 2012; Friedenthal et al. 2011). The tool support has advanced a lot in recent years. It has reached a state where it is possible to collaboratively work on executable models and generate the necessary artifacts for large distributed simulations fully automatically (Bocciarelli et al. 2012; Chapter 6).

Nowadays, modeling languages spring up like mushrooms. Many different communities come up with formal languages addressing their special needs in their field. Some examples for the so-called domain-specific languages are the following:

- Architecture Analysis and Design Language (AADL): AADL is managed by the Society of Automotive Engineers. A dialect of the AADL language was used in the European Space Agency's (ESA) activities on formal system validation (Chapter 14).
- MARTE: An UML profile focusing on modeling and analysis of real-time and embedded systems.
- ModelicaML: An extended subset of UML allowing graphical modeling and the generation of executable Modelica simulation code.

Many of these languages follow the common pattern that they are derived from a parent language (in many cases UML). UML offers a convenient mechanism to define profiles that extend and tailor a language to specific needs. Such profiles can be loaded in model editors that directly provide a specialized environment. This makes it possible to come up with, for example, a spacecraft language. Ideally, such a language offers all necessary vocabulary and syntax to define spacecraft. Whether a specified spacecraft semantically makes sense is, of course, out of question. Domain-specific languages give a lot of freedom in terms of model development. However, a limiting factor is, if one comes across a design problem that goes beyond the scope of the language.

Other approaches are based on more individual data models with their own specific methodology and framework to support the design activity. An example is ESA's Virtual Spacecraft Design project. The data model, which is the basis for the Virtual Spacecraft Design, is released as Technical Memorandum by European Cooperation for Space Standardization (ECSS). A complete tool suite was built around the data model providing a design editor, a reference database, and a visualization component (European Cooperation for Space Standardization 2011; European Space Agency 2012).

Model-based systems engineering in practice

The MBSE community rejoices in an active development of technologies and emerging tools. The DLR initiatives in this area are grouped under the framework activity Virtual Satellite. The project is a research environment for model-based spacecraft development in general, but also supports the productive use case in the early design phase during CE activities.

The backbone of the software is a data model. It is shared via a central repository that is managed by a version control system. The model offers basic functionality such as the following:

- Levels of decomposition
- Design options
- Hierarchical breakdown
- Role management and authorization
- Parameter definition, sharing, and exchange
- Calculations
- Constraints
- Unit checking and automatic conversion using the conceptual model for Quantities, Units, Dimensions, Values (QUDV)*

The collaboration and interaction with the data model is done via a rich client application. It hides a lot of the model-based concepts by providing a modern and easy-to-use graphical user interface, so that the end users work in a model-based fashion without realizing it. This makes it easier to introduce the new methodology. The screenshot of the Virtual Satellite software in Figure 7.3 gives an overview of the application. A general challenge of tools for working on complex tasks is breaking down the information in manageable and understandable parts or processes.

---

* The conceptual model for QUDV is part of the OMG SysML definition, version 1.2, Annex C5.

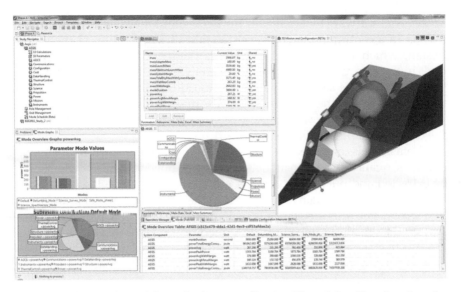

Figure 7.3 Screenshot of the rich-client application Virtual Satellite showing the hierarchical decomposition of the system on the left hand side and various views of the design parameters organized in tables, charts, and a 3D visualization of the configuration on the right hand side.

There are typical ways to handle complexity, for example, filtering (selections and views) or guided processes (wizards). One example for a specific view on the data is the mass summary view of a spacecraft. Such a view typically focuses on a single aspect, in this case the mass, and presents the content of the data model in a structured and organized fashion. On the basis of these representations, the design team will discuss next actions and make decisions. The mass summary is a simple example for an evaluation view on the data. The automatic creation of the view will work only if the correct parameters are provided by each contributor, so it requires a certain formalism and convention of how the information is present in the data model. Within Virtual Satellite, we defined templates with sets of standard parameters respecting a naming convention on different hierarchical levels of the system. We also incorporated certain equations (mass budget) and basic relations that link certain parameters. The users can start a new design using these templates. In our experience, the users of the software found it very useful to begin with a template although this increases the learning effort. The users have to not only understand the tool, but also the template. It is important to find the right balance.

In essence, the formal representation of design information in a data model is the basis for any more sophisticated computer-based analysis. The term analysis in this context has a broad meaning. It can mean displaying simple graphs, charts, and 3D plots. It can also mean triggering an external software tool that runs numeric simulations. Having all information in a central model, however, enables the evaluation of the whole design, which is important for systems engineering. Therefore, MBSE tools in the future will provide ways to analyze the design on system level such as automatic sensitivity analysis, multidisciplinary simulations, or the integration of the domain structure matrix methods (Schaus et al. 2012). It also paves the way for new techniques like formal verification of the design (Chapter 14; Fischer et al. 2013).

Data models can be seen as the abstract representation of domain knowledge. The quality of the data models is related to how well the knowledge can be extracted and put into an abstract model description. So the first step is to understand the domain and to develop the theoretical model foundation. In a second step, this model needs to be filled with life in MBSE tools. Here, various data models can be linked. A good example for this is a subpart of the SysML standard related to the definition of physical units and quantities called QUDV (Object Management Group 2012). This standard defines how physical units are defined and related to each other. It covers all aspects of unit conversion and quantity checks. The data model was integrated in the Virtual Satellite software. All parameters in the data model of Virtual Satellite can be equipped with a physical unit. At first glance, this looks like an extra modeling effort, because everyone working on the design has to assign units to the design parameters. However, the benefit is that the MBSE software can use this additional information to automatically verify consistency. One advantage for collaboration is that the engineers can simply work in their preferred system of units, for example, SI units or imperial units. They can use the system that they are familiar with and the software in the background does all the conversion automatically. This removes the workload and frees time for actually solving the complex design problem. In combination with calculations and a solver engine, the unit and quantity model also allows for sanity checks in the calculations. In addition, the software can check whether all summands of the equation are having the same quantity kind, for example, if they are all of the quantity mass. If not, the software raises an error and highlights this to the user. So the additional modeling effort of assigning units is quickly compensated by automatic conversion and sanity checks that help to improve the quality of the design (Schaus et al. 2013).

Collaborative modeling and simulation throughout the whole development life cycle

Concurrent engineering

CE is a work methodology that has been successfully applied to space systems development in recent years, in particular in the early conceptual design phase. Soon after the first studies have been performed, the benefits of CE became already apparent. The studies delivered consistent and complete mission designs and led to a better understanding and awareness of the design at system level. The time needed for conceptual designs has dropped from months to weeks only. Because CE parallelizes tasks, many people are involved at the same time. Still the cost for the mission concept studies has been reduced by 50% (Fortescue et al. 2011).

Concurrent design teams and facilities have been established in many organizations. ESA's Concurrent Design Facility started in 1998. National Aeronautics and Space Administration/Jet Propulsion Laboratory works with Team X in the Project Design Center and evolved CE more toward mission architecture teams exploring a broader range of the trade space (Moeller et al. 2011; Heneghan et al. 2012). Rather than going too much into depth with one single mission and doing a so-called point design, these activities aim for a broad understanding of the possibilities to achieve critical mission goals. Consequently, such studies take place before a specific scenario for the mission is selected. The DLR inaugurated a Concurrent Engineering Facility at the Institute for Space Systems in Bremen in 2008.

The development processes have evolved over time. At the beginning, projects were following a sequential process, also known as waterfall process. Space systems engineering is traditionally organized by domains of knowledge. In the sequential process, this means that after defining the requirements, each domain is doing their design work one after the other (payload, mission analysis, attitude and orbit control, thermal, etc.). When the last domain is finished, the whole design is considered to be finished. In practice, of course, the waterfall model is not practical because in reality there are design iterations. There are interdependencies between the domains. Design iterations are necessary so that the design converges to a proper solution for the problem at the end. Allowing iterations in a sequential process is leading to a spiral development process. However, the spiral process is not efficient because all the sequential steps have to be performed one after the other until a new iteration is possible.

Actually, what is needed is a methodology that allows frequent iterations. This approach was used in the so-called centralized design process. Here, the systems engineer acts as the central manager between all the domains. This means that all iterations are managed by the systems engineer. It puts an immense workload on the key person because all changes have to be communicated among the design team. Finally, looking at the CE process, iterations are allowed between each individual domain directly. A team leader acts as moderator and guides the study, where the interdisciplinary domain experts and the customer are working in close collaboration in the same room at the same time.

The most associations with CE are linked to the actual study phase where the whole design team is regularly coming together for joint design sessions. Certainly, this is the core element of the process as shown in

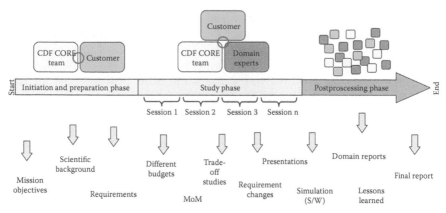

Figure 7.4 A time sequence of the three phases of concurrent engineering showing the participation of the main actors, the tasks, and results of process. (From D. Schubert et al, "A New Knowledge Management System for Concurrent Engineering Facilities," *Proceedings of the 4th International Workshop on System & Concurrent Engineering for Space Applications (SECESA)*, Lausanne, Switzerland, 2010. Used with permission.)

Figure 7.4 (Schubert et al. 2010). Here, the baseline design is actively and openly discussed. Fast iterations are changing the design rapidly. In between the joint sessions, there is time for individual work, for example, checking one specific aspect of the design with a detailed analysis. There is also time to look into possible design alternatives in side meetings. The study phase is the center element of CE, but to achieve a thorough well-defined mission concept, this phase is preceded with a preparation phase and followed by a postprocessing phase. During the preparation phase, the concurrent design team works together with the customer to fully understand his or her needs. A typical task during preparation is to do background research on previous missions or similar experiments, which have been already performed. It also includes planning the study phase, checking for availability and inviting the experts, preparing the tools, and setting up the infrastructure. The crucial part of the preparation is to turn the mission objective in a set of initial mission requirements, and pose this design question to the interdisciplinary design team in the actual study phase.

The postprocessing phase is following the study phase and gives extra time to do more detailed analysis with simulations. Nowadays, it still needs time to set up complex simulations that are not feasible during the agile and fast pace of the study phase. It also allows for reflecting the proposed design in quiet or discussing it with other colleagues to get a second or third opinion. The domain experts write up reports for their subsystem design and all the information is collected in a final report.

The final report is the main outcome of the conceptual phase and is used as basis to decide whether a project enters the next programmatic phase.

As mentioned in the last paragraph, the time for model development and running complex simulation is very limited during the hectic study phase of CE. The experts here mainly work with predefined, individual models. This can be simple parametric models that are available as spreadsheets. During the study, the experts receive critical parameters for their domains from other domains, derive them from the requirements, or make initial assumptions. These parameters are used as inputs and the calculation model takes the information of the cells and calculates output parameters. Usually, the predefined models are not modified during the design.

Especially at the beginning, the design studies are subject to a lot of uncertainties. Large safety margins are used to compensate the uncertainties, but nevertheless any complex design task requires some trade-off analysis between different possible implementations, performances, functionalities, or solutions. Typically several design trade spaces are explored simultaneously. The design team tries to pick the configuration that best meets the requirements. Applying safety margins is just one practical approach to compensate for uncertainty in the design. In recent times, there is also a new methodology called evidence-based robust design optimization. Rather than using margins, the new approach introduces uncertainties as additional parameters in the design. These parameters are then used as part of a multiobjective optimization problem; over project time, these uncertainties of course become smaller. The methodology introduces quantitative measures to describe the credibility of the design budgets. Having these measures at hand helped to improve the design of space systems (Vasile et al. 2011; Alicino and Vasile 2013).

CE is well established in the early feasibility studies of space missions. A part of current research is focusing on extending the use of CE to the later, more detailed design phases. The nature of the more detailed design phase is that more people are involved and the analysis that are carried out are more complex and therefore need more time to be set up, run, and interpreted. Still first attempts showed that CE sessions are also beneficial here and increase the robustness and the performance. It became apparent that we need to overcome the information gap between the phases. The underlying data model should be used seamless and support all the phases (Findlay et al. 2011).

The remaining part of the Subsection "Concurrent engineering" presents two examples of new opportunities that are arising from consequently working in a model-based way. The first example deals with the configuration of a spacecraft. In CE, we found that there are hurdles to verbally discuss positions and orientations of different parts relative to each other or to a certain reference coordinate system. Our approach uses shared interactive visualization for the configuration task, which enables every subject matter expert to individually place parts and to alter them in case of discrepancies. The already existing system model was extended to hold necessary parameters for visualization. This includes the basic shape (geometric primitives), size, position, orientation color, and transparency. This information allows the on-the-fly generation of a visual representation and enables intuitive manipulation. Figure 7.5 explains this process in more detail. To visualize the data model, it is transformed into the special visualization data model. The result is a scene that can either be locally displayed in an integrated view in Virtual Satellite (as shown in Figure 7.3 on the right hand side) or transmitted to other visualization systems, for example, a projection screen or a power wall. The connection to the visualization model is bidirectional, meaning that changes within either one of the models are always synchronized across all instances (Fischer et al. 2012).

Another possibility of MBSE is the idea to have a continuous verification process in CE, which checks the design against certain mission requirements. The starting point for this approach is again the central data model that is fed by all the participants in a CE study. The model is enriched with additional information, which turns it into an executable state machine representation of the spacecraft and its mission. The created mission scenario allows for verification against formalized requirements. If desired, the simulation runs after each iteration or design change. Once the results are interpreted, they enable immediate feedback to the design team. The verification framework is part of the Virtual Satellite and uses techniques from the field of formal methods, model checking, and way finding algorithms. It focuses on the very early

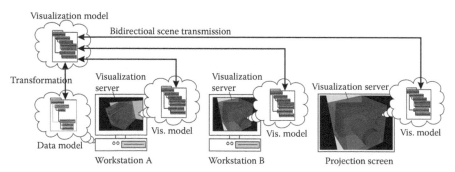

Figure 7.5 The graphic is showing the interactive visualization setup with one local and two remote visualizations. Directly interpreting the visualization parameters and providing an interactive view of the spacecraft configuration help to make individual expert knowledge (of a computer-aided design expert in this case) accessible for the whole team.

design stages (Schaus et al. 2013). For the application of formal verification in later phases, refer to Chapter 14.

Model development process and reuse

For many organizations, models are a very important part of their business. A lot of time and effort is spent on model development and maintenance. The models contain an immense knowledge of the system or the product, which is being developed. The goal is clearly to reuse as much information as possible in future projects. In consequence for the models, this means to provide solutions to make this knowledge accessible. This situation is especially true for space agencies or research institutes, which are involved throughout the whole life cycle of a space mission. They are typically responsible for the conceptual design. Later, the agencies or institutes are the contracting authorities for many suppliers. There is a constant need for reviewing, validating, and verifying design information, and manage the interfaces to assure overall system performance and quality.

The reuse of models for future research and development projects belongs to the much wider field of knowledge management (Kotnour et al. 1997). General purpose knowledge management strategies and tools address the collection, cataloging, and making available of knowledge that lies in documents, presentations, and so on. However, complex models need special treatment compared to these human-readable artifacts. A lot of the knowledge lies implicitly in the design, the modeling assumptions, and the program code of the models.

Modeling and simulation is an integral part of spacecraft development. Throughout the life cycle, many models are created by different domains or entities. They exist in various different levels of detail, granularity, and fidelity. Usually, because the next project deadline is pushing for fast results, models are directly developed in the platform-specific target language (C, C++, Java, or other programming languages) and are sometimes poorly documented.

Especially for collaboration between partners and reusing models in future projects, this situation is unsatisfactory. The model development process should be harmonized and include collaborative aspects, as it is suggested in Figure 7.6 (Schaus et al. 2011). This sequential process consists of five steps in total. Each step generates a documentation artifact. The four steps, requirements analysis, design, implementation, and validation, are similar to the waterfall model in software engineering. The second step in the process, the model analysis, is introduced because of the special nature of modeling in this context. A model is generally some kind of approximation of an ideal (or real) behavior. For example, the purpose can be to model a physical effect such as Earth's gravitational

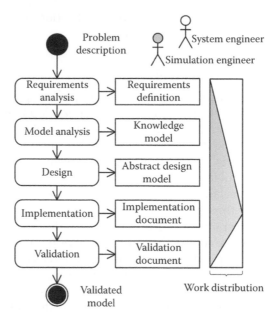

Figure 7.6 Suggested model development process to be used throughout the whole development life cycle. (V. Schaus et al., "Collaborative Development of a Space System," *20th IEEE International Workshops on Enabling Technologies: Infrastructure for Collaborative Enterprises (WETICE)* © 2011 IEEE.)

field. The idea of the model analysis step is to capture the essential background information of the underlying modeling problem in a so-called knowledge model. This knowledge model can, for example, contain the following:

- Analytical equation (for the physical effect to be modeled)
- Range of validity
- Level of accuracy
- Used algorithm
- Links to references

The modeling process is entered by different actors together and followed in close collaboration throughout the stages. The work contribution may vary over time, as in this example the workload is shifted from the system engineer to the simulation engineer. Reviews are important for harmonization and to assure high quality. The review meetings can be done after each step when anyhow a piece of documentation is postulated in the modeling process.

The prospect of reuse is simple—reducing the design time! A consequent reuse strategy postulates ways to categorize, document, find, understand, and derive variants. An abstract representation is needed to understand the model quickly and decide whether it can be applied in the specific reuse case. The interfaces should be stated clearly for integration. A powerful, yet simple-to-use search engine is needed to find and compare potential model reuse candidates. Modularity and small granularity are needed to support specialization and to enable the creation of model variants. The documentation should be brief, comprehensible, and not contain repetitions. To organize and provide access to the models, a central library or database, which can be accessed through a graphical user interface, for example, a website or special client application (Lüdtke et al. 2012), has been proven to be useful.

When such a model development process is realized in an organization, it is advisable to define a set of mandatory attributes as meta-data for the models, establish modeling guidelines, and a review process. The whole process should be supported by IT tools as much as possible.

As design time is precious, the documentation is usually the first thing that is omitted. Therefore, the additional effort for documentation during development should be minimal for the engineer and sharing the model in a central library should not be allowed, when the documentation is missing. Also, when deadlines are pushing for fast results, the tendency is to do a quick and dirty modeling and not respect the mandated modeling guidelines. Here, the modeling tool should run model checks on-the-fly against the set of predefined guidelines. A model that does not pass the checks cannot be used in the project. More and more tools integrate such syntax checking engines in their modeling products. They can check for required settings or parameters, data type checks, port mappings, and analyze the model dependencies. Examples are the Model Advisor in MATLAB/Simulink® or ModelicaML, an extended subset of the UML, which can be added as profile to the graphical modeling environment for Modelica simulation models (Schamai 2009; MathWorks 2014). Confidence in the model plays an important role in reuse, and many of the aforementioned aspects are helping to increase confidence.

When the model development process is respected and all the artifacts are created properly, reuse is possible at every step. One can take the implemented model directly given that the target programming language is the same. Or one can start a step higher and use the abstract design as basis for a new implementation. It is even possible and useful to just reuse the knowledge model or the requirements, or just parts of the artifacts. There will always be a benefit compared to starting all over.

The described modeling process in this section was applied during the development of the TET satellite. In this project, the small enterprise Astro- und Feinwerktechnik developed a closed loop dynamic simulation of the satellite in close collaboration with the DLR to analyze the behavior. The simulation covers the four main categories: the mission, the space environment, the dynamics (equations of motion), and the satellite model. The latter satellite model itself consists of four subsystems, namely power supply and distribution, a four node thermal model, the on-board data handling, and the attitude control system. In total, over one hundred models of different complexity and granularity have been created and are now available in a library for future projects.*

Team collaboration and distributed tool interaction

The vision is that a team collaborates via a central data model. The current situation is that the involved domains of knowledge have developed complex software suites to perform the necessary analysis. Examples are Satellite Toolkit for orbit analysis and optimization, the finite element software NASTRAN for structural analysis, or CATIA for configuration and the geometric model. The list of examples can be continued by looking at all the individual domains. In the end, we find that they are all working with quasi-standard tools.

One issue in this approach is that there are interdependencies in the design, which are omitted by this segmentation. Let us consider the following example: the solar array of a spacecraft is generating power. The power is consumed by the on-board computer and electronics. They are releasing heat. In the first iteration, we foresee a radiator to dissipate the heat against the space background. The result of the thermal analysis now shows that the on-board electronics are overheating. We change the design and add Peltier elements at the critical locations to cool the electronics. In order that the Peltier elements do as intended and cool the critical locations, they consume extra power. So the power engineer has to check the dimensioning again for the solar arrays. So in this case, the power and thermal domain are coupled. The analysis is done in separate tools. The question now is how we overcome the limitations of the domain specific modeling?

One possible solution is model transformation. The model of one domain is transformed to the other and then both can be run at the same time. We found that automatic model transformation works in simple cases. A research project, where the DLR was involved, developed a transformation workbench, which demonstrated that it is generally possible to automatically convert an executable model that exists in one source language to another target language. The workbench was able to parse a MATLAB/Simulink model and create an abstract semantic model from it.

* The research activities were funded by the German Federal Ministry of Education and Research (BMBF) within the Framework Concept "KMU innovativ" (grant number 01IS08015x).

This abstract representation was then used to generate the model of the selected target language, for example, Modelica (Voß 2013).

Even a simple transformation cannot guarantee the same performance of the model in another language. The simulation environments and numerical core libraries are different and therefore cause the models to behave differently as well. This requires test cases for comparing the performance of the two environments. The results can later be used to compensate for errors that come from the simulation runtime environment.

The bottleneck of the transformation is the semantic model. The transformation is only as powerful as the content of the semantic model. Different simulation platforms require additional information that must be present in the transformation tool. They require constant maintenance because often old models will not automatically run in a newer version of the simulation platform. It is also very often the case that a model that exists in one language does not have a matching representation in another. This is typically the case for models that are closely connected to the simulation environment, for example, the Scope block in MATLAB/Simulink. As soon as such a model is present, the model might not be convertible because there is no equivalent model in the target simulation environment. Nevertheless, for clearly defined use cases, automatic model transformation is feasible, like ESA's MOSAIC (Model-Oriented Software Automatic Interface Converter). MOSAIC provides automatic transformation of MATLAB/Simulink models to ESA's Simulation Model Portability (SMP) standard (Lammen et al. 2011).

One way to handle this is to define a subset of models that are used, for example, in a specific project. Checking whether a model is convertible is again a tedious procedure when one has to do it manually. Even if it is automatized by a transformation test suite, it requires constant maintenance. Defining reduced subsets of models is common practice and usually part of modeling guidelines. Such guidelines exist in many organizations, very often as a document that a designated simulation engineer has to respect in daily work. The presence of a document, however, will not prevent violations against these guidelines. What is really desired in this context is smart integration of modeling guideline checks in the model development tools. The tool run checks automatically in the background as models are developed and raise warnings directly to the user.

Interdependencies in this case can be visualized as workflows. Figure 7.7 shows an example of such a workflow as it is used in parametric aircraft design with the DLR design tools Remote Component Environment and VAMPzero.* The advantage of such a graphical

* Both tools, the Remote Component Environment and VAMPzero are open source projects of the DLR. They are available through the public software distribution portal of the DLR: http://software.dlr.de/.

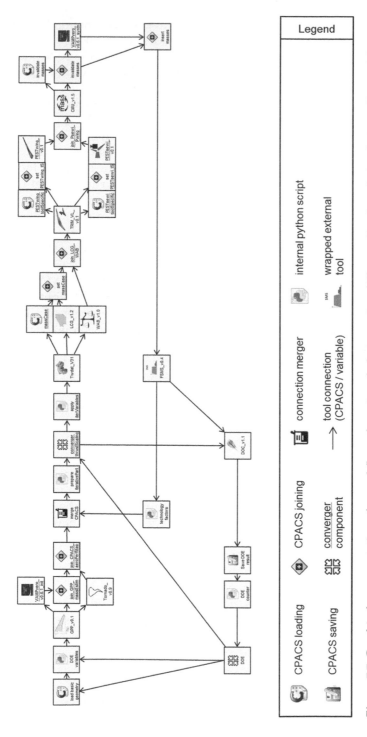

Figure 7.7 Graphical representation of a workflow of an aircraft design using different distributed simulations. (Derived from E. Moerland et al. "Collaborative Understanding of Disciplinary Correlations Using a Low-Fidelity Physics Based Aerospace Toolkit," *Proceedings of the 4th CEAS Air & Space Conference, Linköping, Sweden,* 2013. Without annotations.)

representation for collaboration is that the data flow is transparent and can support the design team. It can be used to identify parallel or sequential tasks, for instance.

Design loops are present in many design tasks. Also the aforementioned example with the power and thermal subsystem has such a loop. A change in one discipline affects the other and vice-versa. These loops can be directly seen in a graphical workflow representation. Usually, the experts check the results of the analysis at least once during a design iteration and decide whether the next iteration can start or not (human-in-the-loop). Step by step, the design converges toward a solution that fulfills the requirement. Given that the execution of the workflow is fully automatic, these design loops are an optimization problem that can be solved numerically. The requirements are transformed into goals or cost functions for an optimization algorithm.

Workflows showing the connections between the different disciplines and tools are the starting point for cross-domain modeling. The continuous links between the partial models enables efficient management of the development process and allows traceability of change propagation. Modeling the dependencies should be done right from the beginning of the design process, when the system model is still relatively simple, for example, a set of analytic equations. Later, the analysis of some partial models might require high-performance computers and cluster resources that are distributed or only available at dedicated time slots. The whole design process is managed by the workflow. The design team collaboratively edits and modifies the workflow. For example, in a refinement case, one engineer replaces the initial spreadsheet-based calculation with a detailed analysis using modeling and simulation software. In this case, he would remove the spreadsheet node from the workflow and add a new node that is interfacing to the more sophisticated tool. Of course, the interface has to be configured correctly, so that the workflow can trigger the run of the complex modeling and simulation represented by the new node. In case the more detailed analysis requires additional parameters from other disciplines, new relations can be added to the diagram.

The approach of workflows and cross-domain modeling increases the consistency of the overall design. To use this in everyday practice, it is necessary to standardize the way of modeling so that collaboration and data exchange is easily possible between the domains. It is also very important to show only a reduced amount of information that is relevant or demanded with respect to the current context (Stark et al. 2010).

In this final paragraph, we pick up the example of the parametric aircraft design, to give the reader a better understanding of the workflow presented in Figure 7.7. First of all, it is important to know that the aircraft design community has defined the Common Parametric Aircraft Configuration Schema, a data definition that can be used to exchange data between various tools: in other words, a common language for aircraft design. In brief, the main parts of the workflow depicted in Figure 7.7 are the following:

- An initialization part for adjusting the baseline geometry of the aircraft and getting first rough estimations for the mass and the aerodynamic coefficients.
- A part that is incorporating an engine performance module and estimating the structural wing mass. The more detailed information on the masses are directly used to update the overall mass budget.
- A part determining the fuel requirements for a given flight profile, the emissions, and the engine scaling factors (which introduce subsequent requirements).
- A part that is running design iterations until the required engine for the investigated configuration and profile is obtained.
- A part that is evaluating the direct operating cost of the aircraft.
- A part that is using the design of experiments approach to vary the aircraft geometry parameters.

Regarding the distribution of the workflow, the loading and adjustment of the geometry, as well as the postprocessing of the results, is performed locally on the client computer. The mass and aerodynamic coefficients estimation and also the flight mission simulation tools are hosted on a dedicated aircraft predesign server of the institute of air transportation systems. The engine performance data is obtained using a tool hosted on a server at the institute of propulsion technology (Moerland et al. 2013).

In summary, the distributed simulation example from the aircraft design shows that design iterations can be automated to a large extent. The prerequisites are the common data definition that allows data exchange and a proper description of the interfaces. The example involves local resources as well as distributed servers depending on the required performance or the geographical location of project partners. The invocation of remote components also respects the concerns of project partners regarding the protection of their intellectual property and copyright. Often enough, there are hesitations and restrictions to give away developed models and integrate them in a larger simulation. The clear interfaces provide the necessary separation of responsibilities; the possibility to include remote components provides the required privacy. Both aspects are enabling factors for collaboration.

Collaborative development of operational simulators

Operational spacecraft simulators are developed for the operation control centers that oversee the whole mission on the ground. They are used to train the operational engineers, so that they are capable of handling the system later during their shift in the control room. Another use for such simulations is the validation of command sequences before they are actually sent to the spacecraft. They are also used to generate the predicted reference scenarios of what is expected to happen and also to develop emergency procedures in case a failure occurs.

It is important to mention that these simulators are usually developed when the design of the spacecraft is finished. They are developed against the fixed specification that is a big advantage: the goal is clear and there is no need for iterations. In addition, the resulting simulator runs purely as software. There are no requirements for real-time systems or hardware-in-the-loop capabilities.

The ESA consolidates their efforts in this field to increase the development productivity. The current stage is that the existing products and tools are integrated in one Simulator Software Development Environment (Ellsiepen et al. 2012).

In terms of collaboration, the task to create an operational simulator can be very much compared to the current standard practice in software engineering. There is a distributed team that has access to a central repository with version control. The central repository holds all relevant information of the project. The work is organized in team meetings, face to face, or remote via video conference. Requirements and tasks are assigned to members of the team with an issue tracker. Changes in the repository are always linked to specific issues, allowing end-to-end traceability. The current models of the repository are automatically integrated to a fully functional simulator, tested, and deployed to the customer with a continuous integration system.

Besides the interlocking tools, mechanisms, and automation, the other methodology that has been adopted from the software domain is model-driven architecture (MDA). The three primary goals of MDA are portability, interoperability, and reusability and all of them are also desired in the simulator development. The following list gives an overview of technologies and tools used in the simulator development and states the analogy in software development.

- SMP 2.0—The SMP standard defines a modeling language for simulation models. In a so-called meta-model, it describes all the necessary concepts of simulation models like, for instance, inheritance, interfaces, relations, and scheduling. An analogy would be the UML (European Space Agency 2005).
- Profile concept—With the SMP standards, the engineers can define their models in an abstract way, for example, with a graphical editor. In practice, this graphical representation makes it easy to discuss the design topology and exchange ideas. The SMP standard can be

added as profile to graphical editors, similar to the SysML profile. With such profiles in the background, the validity of the diagram can be automatically checked.

- Spacecraft Simulator Reference Architecture—the reference architecture is the top-level design guide for operational simulators. It can be compared to the design patterns in software development (Walsh et al. 2010).

- Generic Models—a set of SMP models. They contain plenty of domain knowledge and are used as a starting point to derive new models (by inheritance). The collection covers three major aspects of space simulators: Telecommand and Telemetry (SIMPACK), Spacecraft position and environment (PEM), and the electrical network (SENSE). The analogy to software development is a library of which classes are directly used or inherited and overwritten.

- Universal Modeling Framework (UMF 2.0)—UMF is an Eclipse-based integrated development environment for simulators. It integrates all the necessary tools for SMP development in one suite and therefore is a productive work environment for simulation engineers. In this case, the analogy is simple, because UMF builds on Eclipse.

- Library of Models (LoM)—The LoM is a server-based repository for simulation models. It is used to save and share models across projects. It provides a web-based search client and integration in UMF. Dependencies between simulation models are managed directly by Maven, a tool also used for software project management. The analogy in this case is, Maven repositories on the Internet providing Open Source software libraries (Parson et al. 2012).

The Section "Collaborative development of operational simulators" of spacecraft shows that collaborative modeling and simulation is quite similar to software engineering.

Conclusion

This chapter gave an overview of modeling and simulation in collaborative environments as it is used in spacecraft development. One major trend is, of course, MBSE. Unified data models and modeling languages are the backbone for many activities.

Probably the biggest advantage is the automatic generation of various artifacts needed for design and analysis. Formal and executable models enable, for example, automatic checks, dependency analysis, performance analysis, and report generation. This reduces the time spent on repetitive and tedious work in the design process like updating design parameters or writing boilerplate code for simulations. It helps to free resources and boosts productivity, thus, the human can focus on the creative part of activity.

Many of the driving ideas in MBSE come from software development. In consequence, the methodologies are similar. With modern IT resources and tool support, it is possible to achieve fast turnaround times and informed decision making during the design process. This suits very well to today's agile projects. MBSE as an enabling technology has great potential to meet future challenges in design.

References

Alicino, S. and Vasile, M. "Surrogate-based Maximization of Belief Function for Robust Design Optimization," in *AIAA Space Conference*, San Diego, CA, 2013.

Association of German Engineers (VDI), "Design methodology for mechatronic systems (VDI 2206)," 2004.

Block, J., Straubel, M., and Wiedemann, M. "Ultralight deployable booms for solar sails and other large gossamer structures in space," *Acta Astronautica*, 68(7–8), 984–992, 2011.

Bocciarelli, P., Gianni, D., Pieroni, A., and D'Ambrogio, A. "A Model-driven Method for Building Distributed Simulation Systems from Business Process Models," in *Proceeding of the 2012 Winter Simulation Conference*, Berlin, Germany, 2012.

Eickhoff, J. *Simulating Spacecraft Systems*, Springer, Berlin/Heidelberg, Germany, 2009.

Ellsiepen, P., Fritzen, P., Reggestad, V., and Walsh, A. "UMF—A Productive SMP2 Modeling and Development Tool Chain," in *Simulation and EGSE facilities for Space Programmes (SESP)*, Noordwijk, the Netherlands, 2012.

European Cooperation for Space Standardization (ECSS), "Space Engineering—Space System Data Repository," ESA-ESTEC Requirements and Standards Division, Noordwijk, the Netherlands, 2011.

European Cooperation for Space Standardization (ECSS), "Space Project Management—Project Planning and Implementation," ESA-ESTEC Requirements and Standards Division, Noordwijk, the Netherlands, 2009.

European Space Agency, *Simulation Model Portability 2.0 Handbook*, 2005.

European Space Agency, "Virtual Spacecraft Design Homepage," 2012. Available at http://www.vsd-project.org/; accessed on February 26, 2014.

Findlay, R., Braukhane, A., Schubert, D., Pedersen, J., Müller, H., and Essmann, O. "Implementation of concurrent engineering to Phase B space system design," *CEAS Space Journal*, 2(1–4), 51–58, 2011.

Fischer, P. M., Lüdtke, D., Schaus, V., and Gerndt, A. "A Formal Method for Early Spacecraft Design Verification," in *IEEE Aerospace Conference*, Big Sky, MT, 2013.

Fischer, P. M., Wolff, R., and Gerndt, A. "Collaborative Satellite Configuration Supported by Interactive Visualization," in *IEEE Aerospace Conference*, Big Sky, MT, 2012.

Fortescue, P., Swinerd, G., and Stark, J. *Spacecraft Systems Engineering*, 4th ed., John Wiley & Sons Ltd, West Sussex, United Kingdom, 2011.

Friedenthal, S., Moore, A., and Steiner, R. *A Practical Guide to SysML: The Systems Modeling Language*, 2nd ed., The MK/OMG Press, Waltham, MA, 2011.

Gesellschaft für Systems Engineering e. V., German Chapter of INCOSE, "FireSAT: Model vs Documents Alone," June 03, 2012. Available at http://mbse.gfse.de/documents/45.html; accessed on February 21, 2014.

Gross, J. and Rudolph, S. "Generating Simulation Models from UML - A FireSat Example," in *2nd International Workshop on Model-Driven Approaches for Simulation Engineering (Mod4Sim)*, Orlando, FL, 2012.

Haskins, C. (Editor), *Systems Engineering Handbook*, International Council on Systems Engineering, San Diego, CA, 2006.

Heneghan, C., Warfield, K., and Hihn, J. "The Concurrent Engineering Working Group: Learning to Work Together," in *Proceedings of the 5th International Workshop on System & Concurrent Engineering for Space Applications (SECESA)*, Lisbon, Portugal, 2012.

International Council on Systems Engineering, *INCOSE Systems Engineering Vision 2020*, International Council on Systems Engineering, San Diego, CA, September 2007.

Kotnour, T., Orr, C., Spaulding, J., and Guidi, J. "Determining the benefit of knowledge management activities," in *IEEE International Conference of Systems, Man, and Cybernetics, Computational Cybernetics and Simulation*, Orlando, FL, 1997.

Lammen, W. F., Moelands, J. M., and Wijnands, Q., "Push and pull of models compliant to the Simulation Model Portability standard," National Aerospace Laboratory NLR, Amsterdam, the Netherlands, 2011.

Lichodziejewski, D., Derbès, B., Belvin, K., Slade, K., and Mann, T. "Development and Ground Testing of a Compactly Stowed Scalable Inflatably Deployed Solar Sail," in *45th AIAA/ASME/ASCE/AHS/ASC Structures, Structural Dynamics & Materials Conference*, Palm Springs, CA, 2004.

Lüdtke, D., Ardeans, J., Deshmukh, M., Lopez, R. P., Braukhane, A., Pelivan, I., Theil, S., and Gerndt, A. "Collaborative Development and Cataloging of Simulation and Calculation Models for Space Systems," in *3rd IEEE Track on Collaborative Modeling and Simulation (CoMetS)*, Toulouse, France, 2012.

Maibaum, O., Terzibaschian, T., Raschke, C., and Gerndt, A. "Software Reuse of the BIRD ACS for the TET Satellite Bus," in *8th IAA Symposium on Small Satellites for Earth Observation*, Berlin, Germany, 2011.

MathWorks, "Simulink Documentation—Modeling—Verify Model Syntax," Available at http://www.mathworks.de/de/help/simulink/verify-model-syntax.html; accessed on February 21, 2014.

Moeller, R., Borden, C., Spilker, T., Smythe, W., and Lock, R. "Space missions trade space generation and assessment using the JPL Rapid Mission Architecture (RMA) team approach," in *IEEE Aerospace Conference*, Big Sky, MT, 2011.

Moerland, E., Becker, R. G., and Nagel, B. "Collaborative Understanding of Disciplinary Correlations Using a Low-Fidelity Physics Based Aerospace Toolkit," in *Proceedings of the 4th CEAS Air & Space Conference*, Linköping, Sweden, 2013.

Object Management Group, "OMG Systems Modeling Language (OMG SysML™)," Version 1.3, 2012.

Parson, P., Estades, P., Rodriguez, M., and Ellsiepen, P. "Simulator Development Consolidation—LoM (Library of Models)," in Simulation and EGSE Facilities for Space Programmes (SESP), Noordwijk, the Netherlands, 2012.

Schamai, W. *Modelica Modeling Language (ModelicaML): A UML Profile for Modelica*, Linköping University Electronic Press, Linköping, Sweden, 2009.

Schaus, V., Fischer, P. M., and Gerndt, A. "Taking Advantage of the Model: Application of the Quantity, Units, Dimension, and Values Standard in Concurrent Spacecraft Engineering," in *23rd Annual INCOSE International Symposium*, Philadelphia, PA, 2013.

Schaus, V., Fischer, P. M., Quantius, D., and Gerndt, A. "Automated Sensitivity Analysis in Early Space Mission Design," in Proceedings of the 5th International Workshop on System & Concurrent Engineering for Space Applications (SECESA), Lisbon, Portugal, 2012.

Schaus, V., Grossekatthöfer, K., Lüdtke, D., and Gerndt, A. "Collaborative Development of a Space System," in *20th IEEE International Workshops on Enabling Technologies: Infrastructure for Collaborative Enterprises (WETICE)*, Paris, France, 2011.

Schaus, V., Tiede, M., Fischer, P. M., Lüdtke, D., and Gerndt, A. "A Continuous Verification Process in Concurrent Engineering," in *AIAA Space Conference*, San Diego, CA, 2013.

Schubert, D., Romberg, O., Kurowski, S., Gurtuna, O., Prévot, A., and Savedra-Criado, G. "A New Knowledge Management System for Concurrent Engineering Facilities," in *Proceedings of the 4th International Workshop on System & Concurrent Engineering for Space Applications (SECESA)*, Lausanne, Switzerland, 2010.

Stark, R., Beier, G., Wöhler, T., and Figge, A. "Cross-Domain Dependency Modelling—How to achieve consistent System Models with Tool Support," in *European Systems Engineering Conference (EuSEC)*, Stockholm, Sweden, 2010.

U.S. Congress, Office of Technology Assessment, *Affordable Spacecraft: Design and Launch*, Washington, DC: U.S. Government Printing Office, 1990.

Vasile, M., Minisci, E., Zuiani, F., Komninou, E., and Wijnands, Q. "Fast Evidence-based Space System Engineering," in *Proceedings of the 62nd International Astronautical Congress*, Cape Town, South Africa, 2011.

Voß, V. "Re-usable simulation models for multi-domain and interdisciplinary product development," *PhD Thesis*, University of Stuttgart, in German Language, 2013.

Walsh, A., Wijnands, Q., Lindman, N., Ellsiepen, P., Segneri, D., Eisenmann, H., and Steinle, T. "The Spacecraft Simulator Reference Architecture," in *11th International Workshop on Simulation & EGSE Facilities for Space Programmes (SESP)*, Noordwijk, the Netherlands, 2010.

Wertz, J. R. and Larson, W. J. *Space Mission Analysis and Design*, 3rd ed., McGraw-Hill, New York, 1999.

Wormnes, K., de Jong, J. H., Krag, H., and Visentin, G. "Throw-Nets and Tethers for Robust Space Debris Capture," in *64th International Astronautical Congress*, Beijing, China, 2013.

chapter eight

Performance engineering of distributed simulation programs

Giuseppe Iazeolla and Alessandra Pieroni

Contents

Introduction

Modeling and simulation (M&S) consists of modeling a real system Σ and developing its simulation program (or simulator of Σ). Assume a local simulator (LS) of a given system Σ is available and we wish to turn it into a distributed simulator (DS). In the DS case, the LS is partitioned into segments called federates, each federate being run by a separate host. Before implementing the DS (i.e., at design time) we wonder: will the DS execution time be shorter than LS one? (In some cases the DS may run slower than the equivalent LS.) To answer this question, we argue that the performance engineering (PE) process should be applied to various development phases of the DS.

183

Tools are available that may automatically transform an LS into a DS (Gianni et al. 2011). Such tools, however, only operate at implementation time. At that time, it is generally too late and very expensive to discover that the DS execution time is worse than the LS one, and also too late to reengineer the DS. PE operates, instead, at design time, a time at which the work of removing the causes of performance degradations is easier and less expensive.

PE is based on the concept of *life cycle* software performance validation (PV). PV is the process of evaluating the software ability to satisfy the user performance goals. In this case, the performance goals to seek will be the DS execution time, which should be lower than the execution time of the equivalent LS.

As we shall see in the Section "Production of the performance model of the distributed simulator," the M&S approach, which supports the development of a k-federate DS of a given system Σ, requires the development of a PV program (called PM [DS(Σ)]), which itselfconsists of k-performance submodels that may, in principle, be run on a distributed platform.

In the Sections "Performance engineering process" and "Performance validation process," we will first outline the PE and the PV concepts and then we shall apply such concepts to the production of DSs.

Performance engineering process

Research in PE (Leveson 1996; Lyu 1996; Smith 1992) has emphasized the difference between two approaches and to software *validation* (Iazeolla et al. 2000):

1. The conventional approach, called end-cycle validation (Figure 8.1), or the process of evaluating (at the end of the software product development) the product ability to satisfy the user performance goals.
2. The less conventional approach called life-cycle validation (Figure 8.2), or the process of predicting (at the early phases of the development *life-cycle*) the ability of the to-be product to satisfy the user performance goals, and finally evaluating the ability of the as-is product at the end.

The PE process implements the *life-cycle* approach and can be carried on according to the scheme described in Figure 8.3 that includes PV as a part of the process. The scheme applies to each phase of the *life-cycle* to determine and optimize performance. It applies to *life-cycle* artifacts. The term artifact is conventionally used to mean either the final software program or an intermediate version of it. The requirement document, the analysis document, the design documents, and so forth are examples of artifacts. When applied to the (i)-th phase, the scheme assumes that the

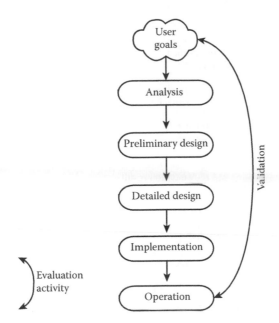

Figure 8.1 End-cycle validation approach. (Courtesy of CRC Press. Iazeolla, G. et al., System Performance Evaluation, CRC Press, Boca Raton, FL, 2000.)

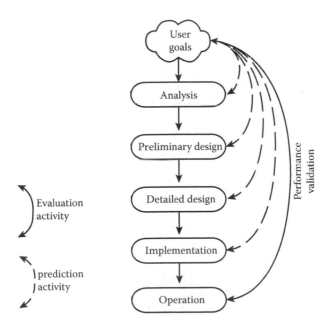

Figure 8.2 Life-cycle validation approach. (Courtesy of CRC Press. Iazeolla, G. et al., *System Performance Evaluation*, CRC Press, Boca Raton, FL, 2000.)

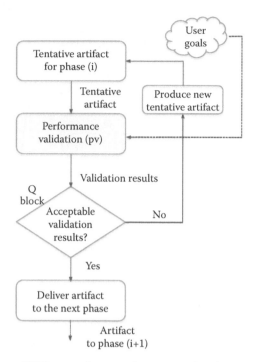

Figure 8.3 Software PE Process (seen at the generic development phase [i]).

previously validated (i–1)-th phase artifact is received in input. On its basis, a tentative phase (i) artifact is produced, and validated in the PV block by comparing the predicted performance (the DS execution time) of the to-be product with the user performance goals.

In case of unacceptability, a new tentative artifact is produced for better performance. Otherwise, the artifact is approved and delivered to the next phase.

Performance validation process

In the following we first provide background information on the PV process and then we detail the PV process for the DS case (Section "Performance validation process for the distributed simulator").

The PV process is performed by the PV block of the PE scheme shown in Figure 8.3. Such a process consists of three main steps.

- Step 1 (performance model [PM] production) consists of developing a PM of the software system.
- Step 2 (PM evaluation) uses the PM model to provide predictions of the software performance. To this scope any available performance evaluation tool can be used.

- Step 3 yields the validation results to be used for the subsequent decisions (Q block) shown in Figure 8.3.

The illustration in Figure 8.4 assumes that the phase (i) artifact in Figure 8.3 to develop is the software preliminary design. In this case, a tentative preliminary design is given in input to the PV block in Figure 8.3.

This is illustrated in more detail in Figure 8.4. Such a figure expands the PV block of Figure 8.3 and refers to a generic PV process, that is, not relative to the DS case. Figure 8.4 also shows that the PM production step is divided into two further steps:

- "Basic PM" production
- "Parameterized PM" production

These details for the DS case are given in the Section "Performance validation process for the distributed simulator."

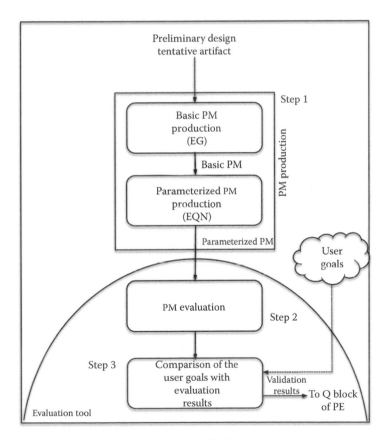

Figure 8.4 Generic PV process (seen at preliminary design phase).

Performance validation process for the distributed simulator

On the basis of the Section "Performance validation process" (Figure 8.4), the PM of the DS is now developed to be used in the PV process to predict the DS capability to run faster than the equivalent LS. We consider that such capability depends on three interacting factors: (1) the partitioning of the LS into federates, (2) the delays in the network N_S the federates use to exchange synchronization messages, and (3) the delays in the network N_D the federates use to exchange data messages. The combination of such factors makes a nontrivial task predicting the benefits of the LS-to-DS transformation.

Assume an LS of a given system Σ is available, and that we wish to turn it into a DS. In the DS case, the LS is partitioned into segments called federates, each federate being run by a separate host. Figure 8.5 shows the two federate cases, with N_S denoting the network for the exchange of synch messages and N_D the one for data messages (Pieroni and Iazeolla 2012).

Before implementing the DS (i.e., at design-time) we wonder: will the DS execution time be shorter than LS one? In some cases the DS may run slower than the equivalent LS. To answer this question the PM of a *k*-federate system is derived to be used at step 1 of PV process (Figure 8.4), for distributed simulation software.

As mentioned in the introduction, we see (Section "Production of the distributed simulator basic performance model") that the PM will itself consists of *k* PM submodels one for each federates of the DS.

The following terminologies are used:

- Σ = system to be simulated
- $LS(\Sigma)$ = local simulator of Σ
- T_{LS} = LS execution time
- $DS(\Sigma)$ = distributed simulator of Σ
- T_{DS} = DS execution time
- $PM(DS[\Sigma])$ = Performance model of $DS(\Sigma)$ to predict the execution time T_{DS}.

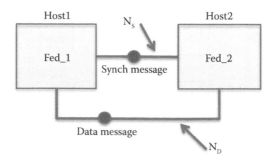

Figure 8.5 DS system with two federates.

On the basis of such terminology, the generic PV process shown in Figure 8.4 turns into the PV process specific for DS case (Figure 8.6). The only difference with Figure 8.4 is the more detailed view of step 3. In Figure 8.6, indeed, the fact that the user goal is an acceptable execution time (T_{DS}) of the DS to develop is now detailed.

Production of the performance model of the distributed simulator

This section describes step 1 in Figure 8.6. According to the PV process in Figure 8.6, the PM production consists of the production of a basic PM(DS[Σ]) and of a parameterized PM(DS[Σ]), which are discussed in the Sections "Production of the distributed simulator basic performance

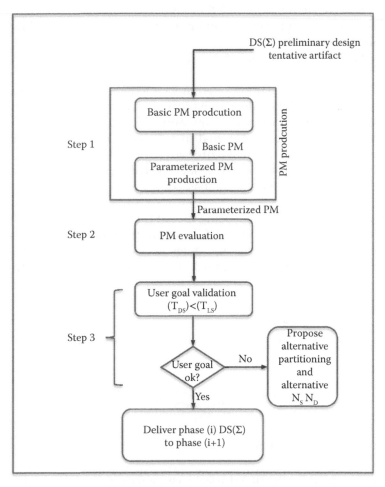

Figure 8.6 The PV process specific for the DS (seen at preliminary design phase).

model" and "Production of the distributed simulator parameterized performance model," respectively.

Production of the distributed simulator basic performance model
The DS basic PM will model both the federates and the primitives of the high-
level architecture (HLA) DS middleware that coordinates the federates.[*]

It is assumed the reader is familiar with the structure of an HLA
federation (IEEE Std. 1516 2010), based on the so-called run-time infrastructure (RTI) (Kuhl et al. 2000). The RTI is the software that allows the
federates to execute together. In Figure 8.7, the interface between the RTI
and the federates is illustrated (Kuhl et al. 2000). The federates do not talk
to each other directly. They are instead connected to the RTI and communicate with each other using services provided by the RTI. The RTI offers
to each federate an interface called RTI ambassador.

Each federate on the other hand presents an interface, called federate
ambassador, to the RTI.

In the following, we shall denote by

- LEX, the local execution internal to a federate, in other words, the
  standard simulation operations such as event processing, event routines, scheduling of local events, and so forth.
- HLAR, the execution internal to a federate of an RTI service, for
  example, an invocation of a time advance request.
- HLAF-Ex, the execution internal to a federate of a service request
  coming from the federate ambassador.

Figure 8.8, shows our queueing model (Pieroni and Iazeolla 2012) of
a k-federate DS. Such a model includes both the conventional queueing
nodes and several AND/OR logic gates necessary to bring in the logic
of the HLA standard. In Figure 8.8, the details of the PM of only one

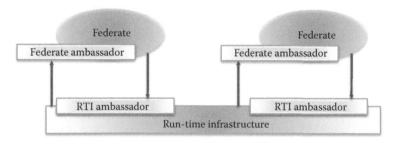

Figure 8.7 HLA federation structure.

---

[*] It is important to notice that HLA standard is not the only DS standard, as illustrated in
Tolk (2012).

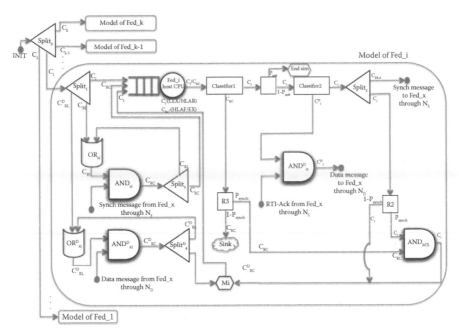

Figure 8.8 View of the federation PM(DS[Σ]) with details of the i-th federate model.

federate (Fed_i) are illustrated, all remaining federates being identical. All such federate models (or PM submodels) can, in principle, be run on a distributed platform (see the Section "PV comparison results: The OMNet++ version of the PM[DS(Σ)])."

In the Fed_i model shown in Figure 8.8, the interactions are shown between Fed_i itself and all remaining federates (in the publish/subscribe RTI assumption [Kuhl et al. 2000]). In Fed_i, the set of all remaining federates is denoted by using the bold **Fed_x** notation. Therefore, the xi (or ix) notation is used in the subscript of various components in Figure 8.8. For example, gate ANDxi relates to the synch messages exchanged between **Fed_x** and Fed_i. Consequently, we have to figure out the existence of number $k-1$ AND gates. The same can be said for all other AND and OR gates with a bold **x** in the subscript.

As can be read from Figure 8.8, each Fed_i submodel sends (receives) messages to (from) **Fed_x** through the N_S and N_D networks. The entire federation PM will thus consist of a set of Fed_i submodels (i = 1, ..., k, as in Figure 8.8) that interact between themselves through the N_S and N_D nodes as in Figure 8.9, which shows how messages from various federates are enqueued in front of N_S or N_D to be served, that is, forwarded to the destination federates.

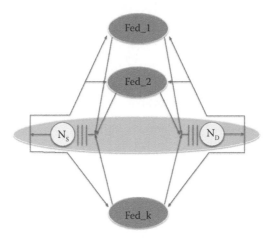

Figure 8.9 View of the federation PM including the communication networks.

The Fed_i model in Figure 8.8 includes the following:

- A time-consuming node (Fed_i Host CPU) that synthetically represents the host that runs the federate. Such node service–time parameters vary with the serviced job class (C_i or C_{RC}).
- A set of non-time-consuming nodes, namely
  - AND nodes that perform AND-logic operations.
  - OR nodes that perform OR-logic operations.
  - SPLIT nodes that split an incoming job class into two or more outgoing classes.
  - Classifier nodes that, based on the class of the input job, forward the job in one or the other direction.
  - Router nodes that perform probabilistic routing of the incoming jobs.
  - A Merge node that merges two job classes.

The computation performed by the federation starts by launching the RTI interface and by initializing the HLA components local to each federate. Such initial computations are performed only once and do not substantially affect the federation execution time, and thus are omitted from the modeling. They are synthetically represented by the *INIT* job on the top in Figure 8.8.

The *INIT* job enters the *Split0* node and yields a main thread for each federate belonging to the federation. The main thread C_i for Fed_i and its follow-on is detailed in Figure 8.8.

The entire federation model is illustrated in Figure 8.9 that includes the following $k + 2$ time-consuming nodes:

- One central processing unit (CPU) node for each of the k federates
- One node for the N_S
- One node for the N_D

As already said, the CPU parameters will be specified in the Section "Production of the distributed simulator parameterized performance model," where we will also specify the N_S and N_D parameters.

It is assumed that the conservative time management HLA option is used (Kuhl et al. 2000); in other words, no federate advances logical time except when it can be guaranteed not to receive any events in its past. The zero-lookahead option is also assumed (Kuhl et al. 2000) (actually HLA does not accept a null value for lookahead and, thus, a very small value is given to the lookahead). In this way, there is a guarantee that federates do not receive events in the past and, thus, that they are fully synchronized.

For such choices, the federates will not process events in parallel and parallelism will only be found when federates include intrinsically parallel portions of LS. If intrinsically parallel portions exist in LS, a positive speedup will be obtained when transforming the LS into its DS version.

The computation performed by Fed_i is carried out by jobs of various classes that circulate in its PM, namely

- Class C_i jobs
- Class C^D_i jobs
- Class C_{HLA} jobs
- Class C_{RL} jobs
- Class C^D_{RL} jobs
- Class C_{RC} jobs
- Class C^D_{RC} jobs

The only jobs that consume CPU execution time are C_i and C_{RC}. The class C_i job simulates the so-called federate main thread (Kuhl et al. 2000), performing LEX and HLAR computations. The class C_{RC} job simulates the so-called federate RTI callback (Kuhl et al. 2000), performing HLAR-Ex computations. The class C^D_i job is a job derived from C_i and holding the data payload to be forwarded to **Fed_x** through network N_D, when the RTI-Ack arrives from **Fed_x** (see the AND^{D}ix node). A class C_{HLA} job is a job derived from C_i and holding the synch message to be forwarded to **Fed_x** through network N_S. A class C_{RL} job represents the so-called federate request listener thread, waiting for synch messages from **Fed_x** (see the ANDxi node). A class C^D_{RL} job is the federate request listener thread, waiting for data messages from **Fed_x** HLAF-Ex computations. A class C^D_{RC} job is the federate request callback thread holding the data payload coming from **Fed_x** and to be used by the C_i job class.

In summary, synchronization and data messages that Fed_i exchanges with other federates **Fed_x** are enqueued in front of the Fed_i Host CPU to be processed. Considered that in the publish/subscribe assumption Fed_i interacts with all remaining $k-1$ federates, the message flow arriving into the queue of the Fed_i Host CPU scales-up with the dimension k of the federation.

Another element that may increase the message flow into the CPU queue is the use of lookahead. Indeed, the frequency of the synchronization messages exchanged between federates per wall-clock-time unit may be affected by the value of the lookahead parameter set by the user.

Let us conclude this section by pointing out that in building the Fed_i model we did not make any mention of the simulated system Σ. This is because the Figure 8.8 model is independent from Σ and thus it is of general use, in other words, it is valid for any Σ. So the Figure 8.8 model can be used for any HLA-based simulation. Only its parameters (the Fed_i Host CPU parameters and the routing probabilities) may depend on Σ, as better seen in the Section "Production of the distributed simulator parameterized performance model."

Production of the distributed simulator parameterized performance model

According to what was stated in the Section "Production of the distributed simulator basic performance model," the only time-consuming nodes are the Fed_i Host CPU, for $i = 1, ..., k$; the N_S node; and the N_D node.

The service parameters to introduce for models in Figures 8.8 and 8.9 are as follows:

1. Service time distribution and moments for the CPUs
2. Service time distribution and moments for N_D, N_S nodes
3. Routing probabilities p_{QUIT} and p_{SYNC} (Figure 8.8)

As stated at the end of the Section "Production of the distributed simulator basic performance model," such parameters depend on the system Σ to simulate. For this chapter's scope, Table 8.1 gives such parameters. They relate to a Σ example case illustrated in Gianni et al. (2010).

Knowledge of such specific Σ is out of the scope of this chaper. Indeed, we only discuss the use of the model (Figure 8.8) that is independent of Σ, as said at the end of the Section "Production of the distributed simulator basic performance model."

Table 8.1 refers to a $k = 2$ federates case (Fed_1 and Fed_2) communicating through the N_S and N_D nodes (Figure 8.9).

Table 8.1 Model Parameters for a 2-Federate Distributed Simulator

	Distribution	Parameters
Fed_i Host CPU service time t_{CPU} (i = 1, 2)	Positive truncated normal	$E(t_{CPU}) = 10$ minutes (Scenario A) $E(t_{CPU}) = 500$ minutes (Scenario B) $\sigma^2(t_{CPU}) = 1$
N_S, N_D service time t_S	k-Pareto, $k = 4$	$E(t_S) = 21$ minutes
Routing probabilities	p_{QUIT}	0.001 (Fed_1); 0.001(Fed_2)
	p_{SYNC}	0.82 (Fed_1); 0.74 (Fed_2)

The Fed_i Host CPU service time parameters vary with the job class (C_i or C_{RC}) and can be derived from the CPU capacity and the Fed_i software run by the CPU. Indeed, the service time parameters for each job class (distribution, mean $E(t_{CPU})$, and variance) can be obtained (according to methods illustrated in D'Ambrogio and Iazeolla [2003]), based on the software model run by the Fed_i CPU, and on the CPU capacity. For our study case, a common positive truncated normal distribution with mean $E(t_{CPU})$ of 10 microseconds or 500 minutes, (for Scenarios A and B respectively, see the Section "PV comparison results: The OMNet++ version of the PM[DS(Σ)]") and $\sigma2(t_{CPU}) = 1$ variance was derived for both classes, by use of the methods in D'Ambrogio and Iazeolla (2003).

The parameters for the N_D and N_S networks can be similarly derived from the software run by the network components and their capacity. By use of D'Ambrogio and Iazeolla (2003), a common 4-Pareto distribution with $E(t_s) = 21$ microseconds was derived for both N_S and N_D.

The routing parameters p_{QUIT} and p_{SYNC}, finally, can be derived from measurements on LS(Σ), in particular, by counting the number of events $n_{intEvents}, n_{disEvents}, n_{disToIntEvents}$ which respectively denote the number of local events (internal events), the number of events sent from a potential Fed_1 to a potential Fed_2, and the number of events received from a potential Fed_2. Such counting can be easily performed by collecting the number of LS events in a simulation experiment for a given hypothetical LS partitioning into two federates. Under the conservative time management assumption, the following was proved in Gianni et al. (2010):

$$p_{QUIT} = \frac{1}{n_{cycles}}$$

where n_{cycles} is the number of local-HLA processing cycles. Value n_{cycles} can be estimated by the number of events locally processed within the model partition.

More specifically,

$$n_{cycles} = n_{IntEvents} + n_{disToIntEvents}$$

Similarly, under the same assumption, the following was proved in Gianni et al. (2010):

$$p_{SYNC} = \frac{n_{IntEvents}}{(n_{IntEvents} + n_{disEvents})}$$

In summary, one may partition the LS code into the *k*-portions of the DS code in various ways. The effect of the partitioning choice is reflected in the PM(DS[Σ]) model in the values taken by parameters p_{SYNC} and p_{QUIT}.

Evaluation of PM(DS[Σ]) and DS validation

This section describes step 2 in Figure 8.6 and step 3 in Figures 8.6 and 8.10.

The PM(DS[Σ]) evaluation is intended to answer the question: when does DS(Σ) run faster than LS(Σ)? In other words, when does $T_{DS} < T_{LS}$ hold?

As stated at the beginning of the Section "Performance validation process for the distributed simulator," the speedup depends on three interacting factors:

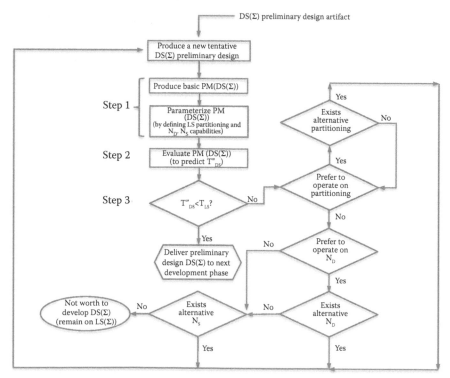

Figure 8.10 Detail of the PV process specific for the DS.

- LS partitioning: By partitioning the LS into federates spread across many hosts, which operate in parallel, one may obtain: $T_{DS} < T_{LS}$, in other words a run-time gain, or a positive speedup S, the speedup being defined by $S = T_{LS}/T_{DS}$, a positive speedup meaning $S > 1$. Let us call S the *no-delay* speedup, for reasons that will be soon clear.
- Synch communication overhead: All DS simulations must incorporate techniques to coordinate the execution of federates across the many hosts by synchronization messages. Such messages travel along a synch network N_S (which may be a LAN, a MAN or a WAN, or a composition thereof) whose delay ΔN_S may yield a $T'_{DS} > T_{DS}$ thus reducing the no-delay speedup S to a *synch-delay* speedup $S' < S$ with $S' = T_{LS}/T'_{DS}$.
- Data communication overhead: The Federates also need to exchange data packets by way of data messages. Such messages travel along a data network N_D (that may or may not coincide with N_S), whose delay ΔN_D may yield a $T''_{DS} > T'_{DS}$, thus reducing the synch-delay speedup S' to a *synch&data-delay* speedup $S'' < S'$ with $S'' = T_{LS}/T''_{DS}$.

The question above then becomes: When does T''_{DS} turn out to be lower than T_{LS} ($T''_{DS} < T_{LS}$), thus still yielding a positive speedup $S'' > 1$? This is the question stated in Figure 8.6 step 3 and detailed in step 3 of Figure 8.10.

In other words, which is the appropriate LS partitioning that yields no-delay speedup able to win over the synchronization and data communication overheads? The Section "Speedup/communication overhead trade-off" tries to answer such a question.

Speedup/communication overhead trade-off

As with most parallel computations, to obtain a positive speedup the portion of LS that can be parallelized must be large relative to the portion that is inherently serial. Let us denote by S(N) the maximum speedup that can be achieved using N processors and by Q the fraction of computation that is inherently serial. According to Amdahl's law (Fujimoto 1999; Park 2008) even with an arbitrarily large number of processors ($N \to \infty$), S(N) can be no larger than the inverse of the inherently serial portion Q of LS.

$$S(N) = \frac{1}{Q + \frac{1-Q}{N}} \tag{8.1}$$

Thus, one requirement for the DS code to achieve positive speedups is that the fraction Q should be small. An appropriate partitioning of LS into a set of federates should then be found at design-time that improves S while maintaining the synch and data overheads low. In other words, a partitioning that yields a high computation-to-communication ratio (i.e., a large amount of computation between communications). On this basis, the PV procedure of Figure 8.10 can be used in the PV block of Figure 8.3 to decide whether to remain on the LS version of the simulation system or carry out the implementation of its DS version. In other words, assume an $LS(\Sigma)$ has been developed and that its T_{LS} is not satisfactory. A search for an appropriate partitioning of $LS(\Sigma)$ into federates and for an appropriate choice of the N_S and N_D networks has to be performed by the iterative use of the $PM(DS[\Sigma])$, to obtain a $T''_{DS} < T_{LS}$.

At each iteration, if the T''_{DS} predicted by the PM is sufficiently lower than T_{LS}, the decision to implement the $DS(\Sigma)$ can be taken. Otherwise, one may either try a new tentative partitioning or try alternative networks N_S and N_D of improved capabilities. In case neither partitioning nor network improvements can be found, one may decide not to implement the $DS(\Sigma)$.

An example use of the PM in the LS/DS decision procedure is illustrated in the Section "PV comparison results: The OMNet++ version of the $PM(DS[\Sigma])$." The PM cannot be evaluated by analytic methods and thus its evaluation is simulation based. The coding of the PM is done in the OMNet++ simulation language (OMNet++ 2011) and an example coding is provided in the Section "PV comparison results: The OMNet++ version of the $PM(DS[\Sigma])$."

PV comparison results: The OMNet++ version of the $PM(DS[\Sigma])$

This section describes step 3 in Figure 8.6. To perform an example prediction of the $DS(\Sigma)$ execution time (T_{DS}) to be used in the Figure 8.10 decision procedure, we developed the OMNet++ simulation version (Figure 8.11) of the Figure 8.8 model for a $k = 2$ federates case (Fed_1 and Fed_2).* Only the Fed_1 part (Figure 8.8) and the N_S and N_D nodes are shown in Figure 8.11 (Iazeolla and Pieroni 2012). As mentioned above, the model structure is valid for any system Σ and only its parameters, illustrated in Table 8.1 (i.e., the CPU service time, the N_D and N_S service times, the p_{QUIT} and p_{SYNC} routing probabilities) may change with Σ.

On the basis of Fed_i Host CPU parameters, the N_S and N_D parameters and the routing parameters, the OMNet++ code simulation model has been run to obtain the T''_{DS} predictions shown in Table 8.2. This was carried out (Iazeolla et al. 2010; Iazeolla and Piccari 2010) in two scenarios A and B: Scenario A being one in which the fraction Q of inherently serial computation was high and Scenario B in which Q was low.

---

* As said in the Section "Production of the distributed simulator basic performance model," the $PM(DS[\Sigma])$ model can in principle be run on a distributed platform. In this case a platform consisting of two hosts. One host for the Fed_1 submodel and the other for the Fed_2 submodel. In our case, however, the $PM(DS[\Sigma])$ simulator will be a local one, as OMNet++ cannot be run in a distributed way.

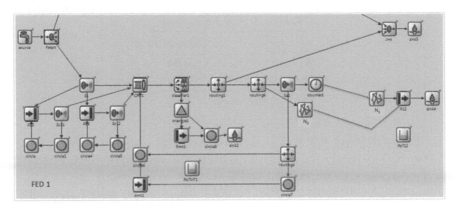

Figure 8.11 OMNet++ simulation version of the PM(DS[Σ]) Fed_i.

Table 8.2 Distributed Simulator Execution Time Results

	T_{LS}	PM results (OMNet++ predictions)	PM validation (real DS measurements)
A (high Q)	0.7 seconds	$T''_{DS} = 8.3$ seconds	$T''DS = 8.2$ seconds
B (low Q)	33 minutes	$T''_{DS} = 12.5$ minutes	$T''DS = 12$ minutes

PM, performance model.

The second column in Table 8.2 reports the LS execution time T_{LS}. The third column reports the DS execution time T''_{DS} predicted by OMNet++ simulator of the PM and the fourth column the times of the real DS (that was implemented in Java+HLA). Such a column thus provides a validation of the PM results and shows how the predicted results adequately match the real ones. Note that in Scenario B the execution times are in minutes whereas in Scenario A they are in seconds. This is because Scenario B is built in a way to yield a high computation-to-communication ratio, in other words, a large amount of computation between communications.

Table 8.2 also shows how in Scenario B the DS outperforms the local one. Indeed, in such a Scenario the DS execution time (T''_{DS}) is much lower than the LS time (T_{LS}).

Finally, by using the expression $S'' = T_{LS}/T''_{DS}$, the results in Table 8.2 were used to obtain the speedup results shown in Table 8.3.

Table 8.3 shows that a quite good speedup ($S'' = 2.64$) is obtained in Scenario B. In other words, in this case the run-time gain obtained by the parallel execution on two hosts compensates for the data and synch communication overheads. In Scenario A, instead, the parallelism does not yield a sufficient run-time gain to compensate for the overheads, and the resulting speedup (S = 0.08) is practically irrelevant.

Table 8.3 Speedup Results

	PM results
A: High Q	S = 0.08
B: low Q	S = 2.64

PM, performance model.

Tables 8.2 and 8.3 results are used by the decision procedure of Figure 8.10 to decide at design-time whether to remain on the LS version of the simulator or implement its DS version. In case the T''_{DS} execution time are not considered *ok* (see Figure 8.10), one may either try a new tentative partitioning (to modify the p_{SYNC} parameters, see the Section "Production of the distributed simulator parameterized performance model") or try alternative networks N_S and N_D of improved capabilities (to modify the $E(t_S)$ parameters, see the Section "Production of the distributed simulator parameterized performance model"). In case neither partitioning nor network improvements can be found, one may decide not to implement the DS(Σ).

Conclusion

A DS consists of cooperating simulation programs, which are located on network-connected and geographically distributed hosts. Such programs may suffer of performance degradations originating from inadequate network capabilities and delays arising in the exchange of synchronization messages and data messages between the cooperating programs. Such performance degradations may however be compensated by the run-time gain (speedup) one may obtain with respect to the equivalent LS.

Software PE is a discipline to develop efficiently performing programs, in other words, programs that satisfy specific performance indices, such as program speedup in our specific case.

The M&S approach, that supports the development of a k-federate DS of a given system Σ, requires the development of a PV program (called PM[DS(Σ)]) that itself consists of k-performance submodels, which may in principle be run on a distributed platform.

In this chapter, PE is applied to predict, at design-time, the effect of program partitioning and network capabilities on the efficiency of DSs.

A PE decision procedure has been proposed to decide at design-time whether to remain on the LS version of the simulator or carry out the implementation of its DS version. The procedure is guided by a PM of the DS. The PM assumes that the DS is based on the HLA protocol standard and middleware.

References

D'Ambrogio, A. and Iazeolla, G. Steps towards the automatic production of performance models of web applications, *Computer Networks* 41(1): 29–39, 2003.

Fujimoto, R. M. *Parallel and Distributed Simulation Systems*, Wiley, New York, 1999.

Gianni, D., D'Ambrogio, A., and Iazeolla, G. A software architecture to ease the development of distributed simulation systems, *Simulation* 87(9): 819–836, 2011.

Gianni, D., Iazeolla, G., and D'Ambrogio, A. A methodology to predict the performance of distributed simulation, *PADS10 the 24th ACM/IEEE/SCS Workshop on Principles of Advanced and Distributed Simulation*, Atlanta, GA, 2010.

Iazeolla, G., D'Ambrogio, A., and Mirandola, R. Software performance validation strategies, In *System Performance Evaluation*, Gelenbe (Ed.), pp. 365–381, CRC Press, Boca Raton, FL, 2000.

Iazeolla, G., Gentili, A., and Ceracchi, F. Performance prediction of distributed simulations, Technical Report RI.02.10, University of Roma TorVergata, Roma, Italy, 2010.

Iazeolla, G. and Piccari, M. The Speedup in distributed simulation, Technical Report RI.03.10. University of Roma TorVergata, Roma, Italy, 2010.

Iazeolla, G. and Pieroni, A. Load Balancing and Resource Allocation in Distributed Simulation Systems. Eurosis Simulation Conference, MESM'2012, Muscat, Oman, 2012.

IEEE Std. 1516. IEEE Standard for Modeling and Simulation (M&S) High Level Architecture (HLA)—Framework and Rules, 2010.

Kuhl, F., Weatherly, R., and Dahmann, J. *Creating Computer Simulation Systems: An Introduction to the High Level Architecture* Prentice Hall PTR, New York, 2000.

Leveson, N. *Safeware*, Addison Wesley, New York, 1996.

Lyu, M. R. (Editor), *Handbook of Software Reliability Engineering*, McGraw-Hill, New York, 1996.

OMNeT++. Discrete event simulation v.4.0. User Manual. http://www.omnetpp .org, 2011.

Park, A. Master/Worker Parallel Discrete Event Simulation, College of Computing. PhD Thesis, Georgia Institute of Technology, Atlanta, GA, 2008.

Pieroni, A. and Iazeolla, G. On the speedup/delay trade-off in distributed simulations, *International Journal of Computer Networks* (IJCN) 4(5), 2012.

Pieroni, A. and Iazeolla, G. The Effect of the Speedup and Network Delays on Distributed Simulations. Winter Simulation Conference, Laroque, C., Himmelspach, J., Pasupathy, R., Rose, O., and Uhrmacher, A. M. (eds.), Roma, Italy, 2012.

Smith, C. U. *Performance Engineering of Software Systems*, Addison Wesley, New York, 1992.

Tolk, A. *Engineering Principles of Combat Modeling and Distributed Simulation*, Wiley, New York, 2012.

chapter nine

Reshuffling PDES platforms for multi/many-core machines

A perspective with focus on load sharing

Francesco Quaglia, Alessandro Pellegrini, and Roberto Vitali

Contents

Introduction

Simulation is an attractive and well-consolidated methodology to study real-world phenomena. In particular, parallel discrete event simulation (PDES) is a paradigm that has been extensively used (because of its modular and simple way of specifying simulation models) and has been proven as very effective in a wide set of fields, including physics, biology, and business-oriented processes (such as financial prediction or optimized system-configuration selection). A core aspect of PDES platforms is that they allow the exploitation of multiple computing units to give rise to a parallel execution of the simulation model, which can provide significant reduction of the time requested for delivering simulation outputs to end users or applications. Specifically, PDES has been shown to provide (large) speedups when compared to classical simulation paradigms where the execution of the simulation model takes place in a sequential fashion. This enables the wide usage of simulation in contexts where timeliness in the production of the simulation results plays a fundamental role, such as when employing simulation technology in time-critical decision processes.

The PDES paradigm dates back to the 80s and was originally thought as of a means for exploiting computing systems formed by clusters of machines relying on single-core central processing unit (CPU) technology. On the other hand, more recent technological trends lead to the proliferation and large diffusion of multicore hardware, where multiple computing units are hosted on the same machine and share several hardware resources, such as main memory. This unavoidably leads to the need for rethinking the organization of PDES platforms to make them perfectly suited for exploiting the computing power offered by modern multi/many-core machines.

In this chapter, we discuss some key aspects related to the reorganization process of these platforms and present in detail a recent literature approach exactly tackling this issue. The presentation is also targeted at showing how the approach, which is based on the symmetric multithreading software programming paradigm, can be suited for a change in the perspective on how to exploit computing resources for PDES applications in a balanced and effective manner. This is achieved via an innovative load-sharing paradigm suited for PDES systems run on top of multicore machines.

The chapter also considers a case study in the context of optimistically synchronized PDES platforms, which are based on speculative processing schemes and achieve causal consistency of the parallel run by means of rollback/recovery techniques. The case study reports the main outcomes of an experimental assessment of the suitability of the revised organization of PDES systems and of the load-sharing technique.

The remainder of this chapter is organized as follows. We initially provide background information in relation to the PDES paradigm, including some examples of PDES-compliant models, which may help the reader to fully understand the paradigm potential. Then we enter details related to the multicore technological trend, also discussing the motivations for its adoption. Afterwards, we outline the main challenges and opportunities related to the organization of modern PDES platforms to be run on top of multicore machines. Successively, we provide an overview of the aforementioned recent proposal particularly aimed at achieving balanced exploitation of the available computing resources when running PDES models on multicore systems. Finally, we present the case study on the design and implementation of an instance of PDES platform tailored for multicore machines, based on the optimistic synchronization paradigm, and report some results related to its runtime behavior.

Parallel discrete event simulation

PDES (Fujimoto 1990) is considered a de facto standard for developing simulation systems featuring high-performance executions of complex and generally applicable simulation models. In particular, PDES has been successfully exploited in differentiated contexts, including—but not limiting to—symbiotic systems or simulation-based (time-critical) decision making.

It is evident that the possibility to rely on a highly optimized PDES framework, able to exploit (large-scale) parallel/distributed computing platforms, provides various benefits:

- *Problem feasibility* is broadened. Very complex/large models, demanding large/huge CPU and/or memory resources, become tractable.
- *Simulation accuracy* is maximized. The simulation-model writer is provided with the ability to study complex phenomena with an always increasing level of detail, enabling researchers from different fields to enhance the quality of simulation results.
- *Timeliness of results* can be ensured. In this way, simulation results can be effectively used in scenarios where, for example, the outcome of simulations can drive a real/physical system (with proper timing constraints), to take the best decision, given the run-time variations of the surrounding environment.

The basic idea underlying PDES is to partition the simulation model into several distinct *simulation objects*, which are the core of the simulation process from the model writer's point of view. In fact, each object represents a portion of the real world being simulated, the evolution of which is described by object's state transitions, driven by a set of logical/mathematical properties. To represent real-world interactions, simulation objects can communicate with each other, by exchanging pieces of information in the form of *events*.

From a technical point of view, simulation objects are handled by *logical processes* (LP), which undertake the concurrent execution of simulation events. Traditionally, a PDES run entails *N concurrent* LPs, uniquely identified by a numerical code in the range [0,N – 1] and the overall simulation model keeps track of the evolution of the simulated world by relying on a global simulation state, which is partitioned into various LPs' private and disjoint simulation states. In particular, if we denote with S the global simulation state and with S_i the LPs' private simulation states, two properties hold:

$$S = \bigcup_{i=0}^{N-1} S_i \tag{9.1}$$

$$S_i \cap S_j = \phi, \forall i, j \in [0, N-1] \tag{9.2}$$

This means that, for two different LPs to interact, events must be exchanged in the form of explicit *message passing*, as it is not possible for a LP to update/modify any other LP's private simulation state.

In PDES, simulation events are *time stamped* and their execution is *impulsive*, meaning that there is no notion of time evolution during an event processing. The current simulation time at each individual LP is known as *local virtual time* (LVT), and can be expressed in any measure unit (i.e., one LVT unit can represent seconds, hours, or even years, depending on the actual simulation model). This notion of time is opposed to the *wall-clock time* (WCT), which is the actual notion of time that we—as human beings—are used to. Therefore, in one WCT unit, the LVT advancement can be of one or several units, depending on the actual complexity of the simulation model and on the efficiency of the simulation run.

During the execution of an event, other events can be generated—destined to any simulation object in the system—associated with a time-stamp value greater than or equal to the one of the inexecution event. This means that during the execution of the event E_x associated with the time stamp T_x a new event E_y associated with time stamp T_y can be generated and sent to any simulation object, ensuring that $T_y \geq T_x$. Therefore, the simulation execution evolves according to a causality pattern where the present cannot affect the past.

The execution of a PDES run occurs in two modes, namely *application-level mode* and *kernel-level mode*. The former execution mode is associated with actual simulation events processing, that is, the control flow is passed to some LP that schedules the involved simulation object. On the other hand, the latter entails all the procedures carried on by the simulation run-time environment to support the actual event execution (e.g., managing the LPs' queues or sending messages to remote kernel instances). Tasks executed while in kernel mode are usually referred to as *housekeeping* operations, and as the reader can imagine, enhancing the overall performance of a PDES execution involves minimizing the time spent in housekeeping (i.e., maximizing the time spent in event processing). Overall, LPs deliver simulation events to the hosted simulation objects via the invocation of proper event handlers. On the other hand, simulation-kernel instances take care of dispatching event-processing activities across the various LPs and of managing inter-LP communication. In particular, they handle the LPs' *event queues*, by reflecting the updates associated with incoming messages, and determine the best LP to be dispatched on a given computing unit to optimize specific execution metrics. A schematization of the traditional organization of a PDES platform is shown in Figure 9.1.

As depicted, LPs are hosted by different instances of the simulation kernel, whose interactions are carried out via message passing. As hinted, this type of organization has been originally devised in the 80s to exploit computing platforms based on clusters of commodity machines equipped with a single processing unit. This organization led to consider a tight 1:1 mapping between simulation-kernel instances and processing units as a de facto standard for the design/development of PDES platforms.

PDES-compliant simulation models

As mentioned before, a simulation model compliant to the PDES paradigm is structured in simulation objects that have disjoint simulation states. We want to provide here the reader with the description of two sample simulation models conforming to the PDES paradigm. In no way we mean this dissertation to be complete and thorough, yet we provide them for the sake of clearness, hoping they will eliminate residual doubts on the PDES paradigm, if any. In addition, we will use these models in the final part of the chapter to empirically show the benefits that derive by the forthcoming dissertations.

Personal communication service

Personal communication service (PCS) is a mobile phone simulation model (Vitali et al. 2012b) targeted at simulating the evolution of a wireless

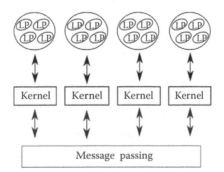

Figure 9.1 Traditional organization of parallel discrete event simulation platforms.

communication system adhering to Global System for Mobile (GSM) technology, where communication channels are modeled in a high fidelity fashion via explicit simulation of power regulation/usage and interference/fading phenomena on the basis of the current state of the corresponding wireless cell. The power regulation model is implemented according to the results given in Kandukuri and Boyd (2002).

In PCS, one simulation object models the evolution of a single wireless cell, and any LP is in charge of driving the execution of simulation events occurring in a single cell, namely at a single simulation object. The state of each simulation object is essentially a set of lists of records, keeping track of the current state of the wireless channels managed by the cell. Particularly, a channel can be either busy or free. In the former case, it means that some wireless communication is taking place via that channel, for which the state of the simulation object keeps track of the current value of the transmission power and of the current signal-to-interference ratio. It is clear that if no active wireless communication is handed off between adjacent cells, then each simulation object evolves along simulation time in a fully autonomous manner, given that any event occurring at that object (such as the setup of a call involving a mobile device within that cell) will give rise to future events only destined to the same object (such as the completion event for the previously mentioned call). This generates a so-called embarrassingly parallel PDES run, where no coupling at all ever takes place between the different objects included in the model. The overall scenario looks similar to one where each individual simulation object can be seen as a separate simulation model and the PDES platform would be in practice running multiple separate models concurrently. On the other hand, when simulating realistic settings mobile devices can switch cells, which may give rise to hand off of active calls. Hence, some level of coupling between pairs of simulation objects can arise, given that the simulation will evolve by generating some *leave* event of the ongoing call on the source cell and some *income* event for the ongoing call on the destination cell. This is the typical case to be faced by PDES platforms, which needs to be optimized in terms of run-time dynamics. In this scenario, the PDES platform is in fact solving a larger model (e.g., it is simulating the evolution of a large coverage area of the GSM system) by concurrently simulating the evolution of its individual, interacting components, according to a spatial decomposition of the simulated system into individual wireless cells. We note that the optimization of the run-time dynamics does not only involve optimizing the actual message exchange across the different simulation objects when cross scheduling of events takes place. Rather, several other aspects are involved, among which the one associated with the actual computing demand for the simulation of individual events at the different cells. More in detail, when the setup of a call is simulated, the CPU time required for determining the initial power assignment for the call depends on the current number of active channels within the cell and on the current power usage on these channels. The more channels are busy, the higher is the CPU cost for determining the minimum power to be assigned to the newly started call to meet some signal-to-noise ratio (according to the results given in Kandukuri and Boyd [2002]). Hence, the different simulation objects may exhibit different CPU requirements for simulating their events depending on the actual workload they sustain (e.g., in terms of frequency of call arrival for mobiles hosted by the corresponding cell), which may be different for objects simulating cells in different spatial regions. How to exploit the overall computing power to achieve balanced advancement of the simulation time across the interacting objects with different CPU requirements for the simulation of their events is another central issue to be dealt with by the PDES run-time environment.

Traffic

The Traffic benchmark (Vitali et al. 2012a) simulates a complex highway system (at a single car granularity), where the topology is a generic graph, in which nodes represent cities or junctions and edges represent the actual highways. Every node is described in terms of car interarrival time and car leaving probability, whereas edges are described in terms of their length. The highway's topology, in terms of its simulation objects, is distributed on a given number of LPs. Therefore, every LP handles the simulation of a node or a portion of a segment, the length of which depends on the total highway's length and the number of available LPs. Every LP simulates the cars that cross the junction or traverse the segment (depending on the actual nature of the LP). In addition, if the LP models a junction, the simulation model takes into account the entering paths, that is, in a junction cars can enter the system according to some specific distribution. Cars can join the highway starting from cities/junctions only, and are later directed toward highway segments with some probability, depending on the traffic pattern that needs to be simulated.

Whenever a car is received by a LP, which will correspond to some *car-arrival* event scheduled for that LP, it is queued in the LP's list of traversing cars, and its speed (for the particular LP it is entering in) is determined according to a Gaussian probability distribution, the mean and the variance of which are specified at start-up time. This allows capturing different speed limits that can be associated with different highway segments and the behavior of drivers in particular regions. Then, the model computes the time the car will need to traverse the node, adding traffic slowdowns that are again computed according to a Gaussian distribution. In particular, the probability of finding a traffic jam is a function of the number of cars that are currently passing through the node, and on the likelihood of accidents, which can be again specified according to some specific distribution. After having determined the exiting time from the node, a car-arrival event is scheduled toward the subsequent node to be passed through by the car, if any.

Traffic shows two peculiarities: On the one hand, there is a high coupling between the LPs, as the model intrinsically captures dynamics related to interactions across the LPs, which are naturally generated by the movement of vehicles within the highway; on the other hand, the events exchange patterns across the LPs are somehow constrained, as compared to PCS. In fact, from one LP an event (e.g., a car arrival) can be sent only to a specific other LP, whereas PCS gives more statistical freedom. Yet, with Traffic we still have great opportunities for parallelism, as in a large highway, traffic evolution within a given zone is unlikely affected by the one in a different (far) zone. On the other hand, events such as accidents lead to concentrate the simulated evolution of the highway in a few LPs, which may lead to an increase of the computing power needed for processing the simulation events for these LPs. As a consequence, the PDES environment should also in this case be able to react to such model execution dynamics to achieve effective exploitation of parallel execution schemes.

Current architectural trends: Multicores

The recent history in computer science has shown that, for software components to offer performance enhancements, the need for software optimizations were not necessarily a key factor. In fact, as *Moore's law* (1965) states, the total number of transistors on a microchip was doubling every 18–24 months. This electronic evolution was yielding a proportional increase in a single processor's clock speed, leading to an enhancement in computation efficiency that researchers, developers, and users were receiving for free.

Figure 9.2 shows (with reference to Intel processing units) how this trend has been almost the same until year 2003. Later on, although the number of transistors is still showing the same trend as before, clock speed increase has stalled. This is connected with the power curve. In fact, 130 W of power consumption in a processing unit is considered an upper bound (Sutter 2005), which generates a *clock-frequency wall*. This is related to physical constraints, as the power consumption (P) is proportionally related to the frequency (f) (Intel Corporation 2004):

$$P = CV^2 f \tag{9.3}$$

In fact, an increase in the clock frequency would produce unsustainable power consumption. Nevertheless, to face the continuously increasing demand in computing power, the industry's recent trend has brought the attention to multi/many-core architectures, where multiple processing units are packaged together into several interconnected processors. Of course, to efficiently rely on this emerging architecture, new programming

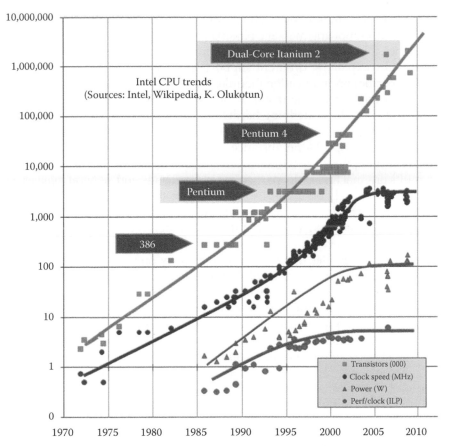

Figure 9.2 Intel CPU characteristics (updated August 2009). (From Sutter, H., *Dr. Dobb's Journal*, 30(3), 202–210, 2005. With permission.)

models have been developed. The high-performance simulation field is just recently starting to look into these new possibilities, which are the actual topic of this chapter.

PDES in the multicore era

In literature, optimizations targeted at PDES systems mostly address the historical scenario, characterized by the aforementioned 1:1 mapping scheme, where a simulation-kernel instance is given a single processing unit for the execution of the simulation.

To exploit most of these optimizations, the traditional PDES paradigm could be easily mapped onto modern multi/many-core architectures while still relying on a 1:1 mapping between kernel instances and CPU cores. However, this scenario would not allow the exploitation of all the potentials offered by the underlying architecture. For instance, although libraries targeted at supporting message passing are highly optimized on multicore machines, it would appear more promising to actuate cross-kernel communication by exploiting different supports such as shared memory and/or PDES-specific communication modules/schemes (Swenson and Riley 2012).

However, beyond the reshuffle of specific mechanisms to make them more suited for the underlying hardware's peculiarities, it looks convenient to relax the tight 1:1 mapping between kernel instances and processing units. In this way, a single kernel instance can run on top of multiple CPU cores (just like it happens for the kernel of modern operating systems targeted at multicore machines). In particular, a promising and general *symmetric* paradigm is to rely on a set of threads within the same simulation-kernel instance (referred to as *worker threads*), which are not bound to a specific activity (e.g., message passing) but that can undertake any operation in the system (i.e., both event processing or housekeeping operations) depending on the actual activities that are required to carry on the overall simulation. This gives rise to maximal flexibility in terms of exploitation of available hardware resources. Therefore, to support this paradigm, simulation kernels' internal architectures should be reshuffled by relying on multithreading, which poses a set of issues to be carefully addressed. In what follows, we initially discuss architectural issues involved in this type of reorganization, as presented by some literature results (Vitali et al. 2012b). Then, we focus on a specific and relevant aspect, which relates to how this architectural approach allows for an innovative way of (dynamically) getting a suited distribution of the simulation workload across the available CPU cores.

Paradigm shift toward multithreaded PDES

A paradigm shift toward the design/implementation of multithreaded PDES kernels, each one entailing multiple, symmetric worker threads that can concurrently run any of the LPs hosted on top of the same kernel instance, is nontrivial. In particular, the following two key aspects must be carefully addressed:

1. Avoiding synchronization phases while running housekeeping tasks in kernel mode to become a performance bottleneck
2. Avoiding loss of locality,* which might (unacceptably) degrade the efficiency of the caching hierarchy

---

* We explicitly refer here both temporal and spatial locality. In particular, if the same slice of data is repeatedly accessed along a given execution phase, then the overall performance will benefit from its residence in higher cache level. Accessing data along different threads may reduce the amount of relevant data kept in cache, and can therefore significantly affect the performance because of increased memory-fetch time.

As for point 1, there is one important difference depending on the execution mode of the simulation, namely *application mode* and *kernel mode*. In fact, when running in application mode (i.e., when the actual simulation model is being run within a particular LP), worker threads inherently execute in *data separation*[*] considering that, as mentioned before in this chapter, LPs handle their own application-level data structures in the form of private/disjoint states. On the other hand, when running in *kernel mode* (i.e., when the execution is under control of simulation-kernel modules for carrying on housekeeping operations) great care must be taken to avoid performance degradation.

Specifically, in typical PDES platforms, the services and data structures used at kernel level to support the simulation's execution are very reduced in number if compared, for example, to the ones used by common operating system's kernels. It is therefore more likely that, in a concurrent kernel-mode execution of multiple worker threads contention and synchronization would easily become a bottleneck, if ad hoc design mechanisms were not employed. Such a problem might be exacerbated in medium/fine-grain simulation applications, where control resides in application-level modules (for actual event processing) for limited WCT intervals, thus giving rise to frequent switches to kernel-level housekeeping modules.

Most notably, the data structures requiring frequent updates (to be performed coherently via proper kernel-level synchronization mechanisms) are the event queues of the LPs. Essentially, these data structures represent the core of cross-LP dependencies (we recall that the only way for different LPs to exchange information is message passing), thus involving update operations caused not only by activities executed by the worker thread currently taking care of running the *queue-owner LP*, but also by activities carried out by other worker threads. Figure 9.3 shows an example where two threads operating within the same kernel instance simultaneously attempt to deliver to the same event queue two new messages (two newly scheduled events) for a given destination LP. These messages might have been produced along the execution path of the concurrent worker threads by two LPs still residing on the same kernel instance, which might have scheduled the corresponding events for the same destination as a result of local event processing activities.

---

[*] Data separation is a property of parallel/concurrent programs where, for example, different threads can concurrently operate in isolation on a subportion of the overall data. This allows concurrent operations by a specific thread to be carried on regardless of the operations performed by other (concurrent) threads, without the presence of any critical section. This allows the parallel algorithm to produce correct results without relying on any locking primitive to protect critical sections, therefore giving rise to an improved execution speedup.

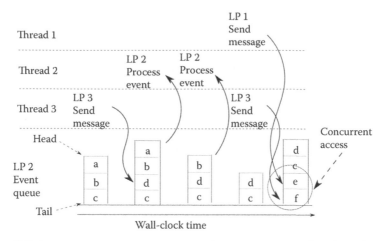

Figure 9.3 Accesses to a target event queue by concurrent worker threads.

Synchronizing accesses to the event queues via a conventional locking mechanism would give rise to scalability problems. Also, it would give rise to critical sections whose duration would depend on the actual time complexity of the queue update operation (which might even unpredictably depend on the specific event-time-stamp pattern).

As for point 2, a symmetric multithreaded simulation kernel allows virtual addresses related to both application- and kernel-level data structures to be, in principle, accessible by any worker thread (as all the worker threads associated with the same simulation-kernel instance operate within the same address space). However, such an unlimited access policy would cause frequent invalidation/refill of, for example, the top-level private caches of individual CPU cores, even when entailing processor affinity schemes involving the worker threads.

Addressing kernel-level synchronization

To reduce the synchronization costs while performing housekeeping operations, the architecture of the symmetric multithreaded simulation kernel can be organized as recently proposed in Vitali et al. (2012b).

According to the presented scheme, any housekeeping task potentially crossing the boundaries of individual LPs' data structures is dispatched according to the same rules employed to structure modern operating system drivers, by organizing it according to *top/bottom-half* activities. More in detail, any of these tasks is considered as a logical interrupt to be eventually finalized within a bottom-half module. Hence, on the interrupt occurrence, the task is not immediately finalized, and the target data structure is not immediately locked (or, the worker thread does not immediately enter a lock-waiting phase). Instead, a light top-half software module is executed, which registers the bottom-half function associated with the interrupt finalization within a per-LP bottom-half queue, resembling the Linux *task queue*.

The critical section accessing the bottom-half queue takes constant time as each new bottom half associated with the LP is recorded at the tail of the queue. Also, when the bottom-half tasks currently registered for a given LP are finalized, the corresponding chain of records is initially unlinked from the corresponding bottom-half queue, which is again done in constant time by unlinking the head element within the chain from its base pointer.

A scheme related to the above architectural organization is provided in Figure 9.4. Operatively, the architecture can rely on a *spin-lock array*, having one entry for each LP hosted by the multithreaded simulation kernel. The i-th entry is used to implement the critical section for accessing the bottom-half queue associated with the i-th LP hosted by the kernel, either for inserting a new bottom-half task, or for taking care of unlinking the current chain to process and flush the pending bottom-halves. To reduce the performance impact generated by spin-lock accesses, techniques such as the one described in Vitali et al. (2012b) can be used.

Overall, in this architectural organization, as soon as any worker thread becomes aware of a new message destined to the i-th locally hosted LP, it accesses the i-th bottom-half queue within a fast critical section that performs the insertion of the corresponding message delivery task. A similar situation occurs when the worker thread performs some received operation via the messaging layer, which delivers a message incoming from some remote kernel instance and destined to a locally hosted LP. As shown in Figure 9.4, the above circumstances are logically marked as interrupts, which will be finalized via the bottom-half mechanism.

Addressing locality

To cope with locality, a promising approach investigated in Vitali et al. (2012a) and Vitali et al. (2012b) is to devise the adoption of affinity mechanisms where any worker thread belonging to a given simulation-kernel instance is not allowed to run every LP hosted by that kernel at any time. Instead, it takes care of running a subset of these LPs, which are currently selected as being *affine* to the worker thread. In other words, for enhancing performance by locality related means, temporary binding mechanisms should be adopted, which associate a subset of the locally hosted LPs to a specific worker thread. In this case, this worker thread is the only thread taking care of running these LPs during a specific WCT window.

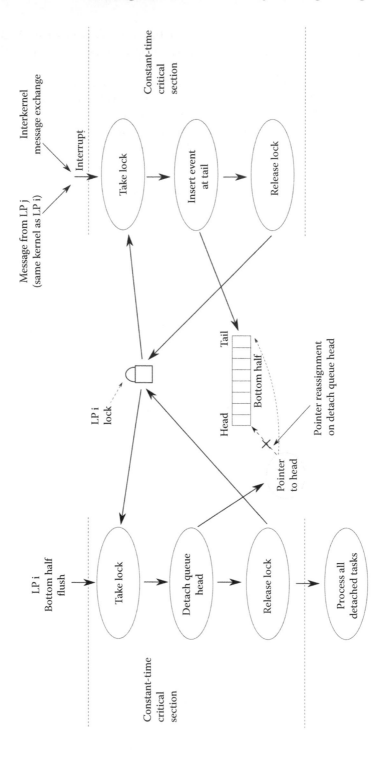

Figure 9.4 Interrupt handling subsystem.

Therefore, this worker thread should undertake the following activities to carry on the simulation:

- Flushing the bottom-half queues associated with its affine LPs
- Dispatching these LPs for event execution in time-interleaved mode

The binding of a specific LP to a worker thread is not meant to be fixed, but can change over time, also in relation to variations of the amount of worker threads belonging to a given kernel instance, as it will be further discussed in the Section "Load sharing model."

Overall system engineering considerations

By the above description it is clear that a few additional engineering aspects need to be taken into account when designing/developing a symmetric multithreaded PDES platform, when compared to a traditional single-thread organization. Here, we recall the already pointed out aspects and provide additional hints on the actual engineering process. As said, the base for the support of the top/bottom-half paradigm consists in having a nonblocking and asynchronous message notification system across threads within the same simulation-kernel process. The channels within the notification systems need to support multiplexing with selective operations based on identifying messages with the ID of the destination LP. As illustrated, such a nonblocking service can be instantiated via fast (and constant time) critical sections protected by spinlock, which also avoid paying costs associated with system calls access. On the other hand, other approaches could be employed such as software transactional memory layers, which allow nonblocking atomic operations in a seamless manner.

As for exploitation of the caching system, (temporal) affinity between LPs and worker threads is only one of the relevant aspects. Another aspect relates to the avoidance of false cache sharing problems. Particularly, platform-level data structures used to support housekeeping tasks should be explicitly separated per-thread and allocated in memory according to a cache aligned scheme. This can be achieved by either developing an ad hoc memory allocator with the desired properties, or by relying on standard cache aligned allocators. Details on how to cope with these aspects can again be found in Vitali et al. (2012a) and Vitali et al. (2012b).

Workload: Balance or share?

The traditional approach to the design and implementation of PDES platforms consists in having—as mentioned before—multiple LPs being run within a single-threaded simulation-kernel process. Therefore, the typical literature approach aimed at achieving effective simulation runs, in terms of efficient exploitation of the available computing resources, is *load balancing* (Boukerche and Das 1997; D'Angelo and Bracuto 2009; Glazer and Tropper 1993; Carothers and Fujimoto 2000; Reiher and Jefferson 1990; Meraji et al. 2010), a schematization of which is depicted in Figure 9.5.

This technique is based on migration of the application load* among different simulation-kernel instances (i.e., user-space processes) while the run is in progress. It is clear that, given the tight 1:1 mapping between kernel instances and CPU cores, no other means to dynamically rebalance the usage of resources can be employed, as each simulation-kernel process has a fixed amount of computing power allocated to it, namely one CPU core.

On the other hand, Figure 9.6 shows a different operational approach to maximize the fruitful resource utilization factor, which can be actuated thanks to the flexibility provided by the symmetric multithreaded organization of the simulation kernel—namely *load sharing*. In this situation, the constraint mapping one simulation-kernel instance to one CPU core is relaxed, allowing a single kernel instance to exploit a higher number of processing units to carry on the actual simulation run. This scenario enables a particular kernel instance to acquire/release computing resources, in the case of a

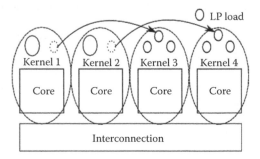

Figure 9.5 Load balancing.

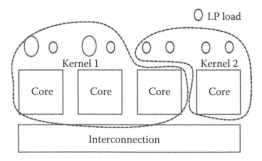

Figure 9.6 Load sharing.

---

* In the context of PDES, workload can be defined as the number and the granularity of pending events to be scheduled in the (near) future at each LP. Therefore, an *expected workload factor* can be associated with any LP according to this notion.

workload factor increase/decrease by the hosted LPs. This can be achieved by scaling up/down the number of worker threads operating within each kernel instance. Recall that these worker threads operate symmetrically (not according to sector-specific functionalities); hence, each of them is able to sustain the actual simulation workload (i.e., event processing).

This is clearly an orthogonal approach with respect to load balancing, considering that LPs are no longer migrated across various simulation-kernel instances. In addition, the two approaches can be merged to achieve an even better simulation performance, when the execution of a simulation run involves several different machines in a cluster or a geographically distributed system. In fact, while load sharing can be used to maximize the performance on a single node of the distributed environment, load balancing can be used at the same time to balance inter-node changes in the simulation workload.

This can be regarded as a promising approach, as it would allow making the best of the available computing power even in the case that the evolution of the system being simulated noticeably varies during the execution (giving rise to variations of the workload associated with individual LPs). This situation (unless for specific simulation model structures) can hardly be foreseen before the actual simulation is put in place.

Advantages of load sharing versus load balancing

As mentioned, load balancing relies on the migration of the workload from a given (overloaded) kernel instance to a (more unencumbered) one. Migrating workload is not a trivial task, as this entails moving LPs' states across different simulation kernels, rerouting events in the whole platform, and presents some technical details that can become a nonnegligible bottleneck in the execution of the overall simulation. Transparency issues also need to be considered as, unless for very limited cases where the simulation-kernel layer offers global memory–management schemes and automatic facilities for reinstalling the exact memory layout of the migrating LP's state onto the destination kernel (Peluso et al. 2011), the intervention of the application-level programmer is requested. Particularly, he might be requested to explicitly provide the software modules for correct management of the memory layout and/or content of the LPs' states during migrations.

To provide more details on some of these aspects, the task of migrating the information required to support the execution of the migrating LP onto the destination kernel instance exhibits latency Δ_m, the lower bound of which can be expressed as

$$\Omega(\Delta_m) = \delta_t \left[S_{state} + \sum_{i=1}^{N_p} S_{evt}^i \right] \qquad (9.4)$$

where δ_t is the average per-byte transfer time between source and destination simulation-kernel instances, S_{state} is the migrating LP's state size, N_p is the number of pending events for the migrating LP, and S_{evt}^i is the size of the i-th pending event for the migrating LP.

With the above lower bound, we do not intend to capture aspects associated with, for example, event-queue implementation and related scan/update costs. We do not even include the latency for transferring data needed to support correct recovery in case of rollback, which is proper when considering an optimistically synchronized PDES run. This data might entail, for example, already processed but not-yet-committed events—which might be required to be reprocessed in case of LP rollback after the migration phase—and state-log information to correctly reconstruct past snapshots of the LP's state onto the destination kernel.

Anyway, by Equation 9.4, there is a clear dependency between the actual cost for supporting rebalance and the complexity of the simulation model, for example, in terms of size of the state of individual LPs to be migrated. Also, in case of migration of multiple LPs, the above cost gets amplified (as it only expresses the per-LP migration overhead). Likely, this amplification could arise (or become relevant) in case of unbalanced execution scenarios of scaled-up models, possibly entailing a (very) large number of LPs.

On the other hand, for all the cases where load sharing is feasible, namely when aiming at the optimization of the (portion) simulation run hosted by a multicore machine, the above migration costs can be eliminated at all. In fact, the only additional paid costs relate to worker-thread suspension/reactivation, which are anyhow not directly dependent on the aforementioned simulation model's complexity (e.g., in terms of state size of individual LPs).

Load sharing model

To select the best-suited amount of CPU cores to be destined for worker threads' execution in a given kernel instance during a specific WCT window, several load-sharing schemes can be envisaged. A highly general and simple one, which has been introduced in Vitali et al. (2012a), is presented in what follows. Let us suppose that we are in a setting where K simulation-kernel instances are present, running on C CPU cores (hosted by the same multicore machine). Let us denote with $k_i, i \in [1, K]$, an individual kernel instance, with $numLP^{k_i}$ the cardinality of the set of LPs hosted by k_i and with LVT_l^e the LVT associated with event e stored in the event queue of $LP_l, l \in [1, numLP^{k_i}]$.

- Step 1: Each kernel instance k_i associates with each $LP_l, l \in [1, numLP^{k_i}]$, a *workload factor* L_l defined as the WCT for advancing the LVT of LP_l by one unit. The workload factor L_l is computed by k_i on the basis of the total number of simulation events currently registered as to be processed within the corresponding event queue, and having a time stamp that falls within a given distance in the future, normalized to the LVT advancement they would produce, weighted by the average CPU time for event processing by LP_l, that is

$$L_l = \frac{q_l \times \delta_l}{\text{LVT}_l^{q_l} - \text{LVT}_l^1} \tag{9.5}$$

In Equation 9.5, q_l denotes the amount of pending events within the event queue of LP_l with time stamps that fall in the interval of interest, LVT_l^i is the time stamp associated with the i-th pending event along the event queue, and δ_l the expected CPU requirement for event processing by LP_l along that chain of pending events. Among the above parameters, q_l and LVT_l^i are known in advance, as they are a function of LP_l input queue's current state. Instead, δ_l is not known in advance, as it expresses the expected cost for events that have not yet been processed. Anyway, it can be approximated by using an exponential mean over already processed events.

- Step 2: k_i computes its total workload as

$$L_k = \sum_{i=1}^{numLP^{k_i}} L_l \tag{9.6}$$

- Step 3: k_i determines the maximum degree of parallelism it can reach. This is done in relation to that each LP must execute events serially, that is, no two worker threads can simultaneously take on the execution of the same LP. Hence, the maximum degree of parallelism is determined by the number of knapsacks of LPs, such that the LPs within a same knapsack would globally induce the same workload as the LP associated with the highest workload factor. Obviously, one knapsack will consist of a unique LP, namely the one associated with the highest workload factor. This task is performed in several steps:
  - Workload factors for the LPs hosted by k_i are nonincreasingly ordered (let us call them in this order as $L_{l_1}, L_{l_2}, ..., L_{l_H}$).
  - As hinted, the first (i.e., the highest) factor L_{l_1} is taken as the reference value, and the knapsack formed by LP_{l_1} is defined.
  - The other knapsacks are built by aggregating the remaining LPs according to a *0-1 one-dimensional multiple knapsack* problem solving algorithm. This is an NP-complete problem, whose integral solution is nontrivial. However, a procedure can be followed where an ideally infinite set of J sacks, $J \in [1, \infty]$, is allowed to be used and for each of them, the greedy approximation approach

proposed by George Dantzig (1957) can be exploited, in which the constraint on the sack is relaxed, by allowing a maximum *overflow* of 30%. At each step of the algorithm, $\forall i \in [2, H]$ the j-th knapsack's size K_{n_j} is updated as $K_{n_j} = K_{n_j} + L_{l_i}$, and thus considered full if the size constraint is violated. In that case, a *new* sack is created (i.e., j is increased) and begins to be filled, until the total workload is distributed across the sacks.

- The well-suited number of worker threads required by simulation kernel k_i is $W^{k_i} = j$.
- Step 4: k_i notifies the tuple $\langle W^{k_i}, L^{k_i} \rangle$ to a *master kernel*. Generally speaking, a master kernel can be either identified via a distributed consensus protocol, or can be known a priori (e.g., by specifying it at compile time) depending on the actual simulation platform.
- Step 5: The master kernel computes the total system's workload:

$$L^{tot} = \sum_{i=1}^{K} L^{k_i} \tag{9.7}$$

- Step 6: A preliminary estimation of the number of cores to be allocated to k_i is computed as

$$T_{k_i} = \left\lfloor \frac{L_{k_i}}{L^{tot}} C \right\rfloor \tag{9.8}$$

enforcing at least $T_{k_i} = 1$.

- Step 7: A refined estimation is then calculated:

$$T'_{k_i} = \begin{cases} T_{ki} & \text{if } T_{ki} \le W^{k_i} \\ W^{k_i} & \text{if } T_{ki} > W^{k_i} \end{cases} \tag{9.9}$$

- Step 8: If $\sum_{i=1}^{K} T'_{k_i} < C$, then there are some CPU cores still available. In this case, all the kernels are nonincreasingly ordered by resource allocation reminder $R^{k_i} = (W^{k_i} - T'_{k_i})$, and until there are CPU cores still available, they are allocated in a round-robin fashion.
- Step 9: The master kernel notifies to each kernel k_i the tuple $\langle j, T'_{k_j} \rangle \forall j$.

At the same time, the work in Vitali et al. (2012a) also provides a model for determining the temporal binding of LPs to worker threads. Specifically, once the new amount of worker threads C_i to be employed by kernel k_i gets defined, a binding that allows balancing the whole workload related to local LPs onto the whole set of worker threads has been devised as follows. For the j-th LP hosted by kernel k_i, which we refer to as LP_i^j, the total amount of CPU time cpu_i^j required for processing its events during the last observation period can be evaluated. Again, the maximum cpu_i^j value across all the locally hosted LPs represents a reference knapsack, and the corresponding LP_i^j is assigned to a given worker thread. Then, the approximated knapsack algorithm can be run to determine which LPs must be assigned to the remaining worker threads.

Case study: Optimistic PDES

The optimistic PDES paradigm, as presented in Jefferson (1985), maintains the mapping of the simulation objects onto N uniquely identified LPs, mapped in their turn onto K simulation-kernel instances.

According to the optimistic synchronization protocol (Jefferson 1985), events are executed regardless of their safety. This means that, whenever an LP is ready for processing (i.e., it has at least one pending event within its event queue), it can be dispatched, although some other LP in the system might eventually generate a new event that had to be executed before the currently dispatched one. This approach allows for great exploitation of parallelism while executing the simulation model, but might lead to *causal inconsistencies*, because events might be no longer executed in time-stamp order.

The simulation kernel must therefore adopt some consistency-check procedure to ensure that the final outcome of the simulation is correct. In particular, optimistic simulation platforms rely on a *rollback recovery scheme*, which can be based on periodically saving the state of every LP (incrementally or not). The taken state logs can be reloaded in case of causality inconsistencies, so as to allow the resume of the LPs' evolution from a past, correct point along the simulation time axis.

Although this approach might seem cumbersome, it has been shown in literature to be incredibly effective, considering (among the other benefits) that it has been shown to provide run-time dynamics that are relatively independent of the simulation model's lookahead and of the delay according to which new messages (i.c., new events) are reflected into the event queues of the destination LPs (namely, the message delivery delay).

Beyond discussing basic principles underlying the optimistic paradigm, the seminal paper in Jefferson (1985) also provides a reference architectural organization for optimistic simulation systems, which we schematize in Figure 9.7. Specifically, we detail the suited set of data structures and functionalities/subsystems, which should be provided to implement a platform relying on the optimistic paradigm.[*]

Input and output message queues are used to keep track of simulation events exchanged across LPs, or scheduled by an LP for itself. Typically,

---

[*] By *subsystem*, we just mean a logical differentiation, not an execution by a separate thread/ process.

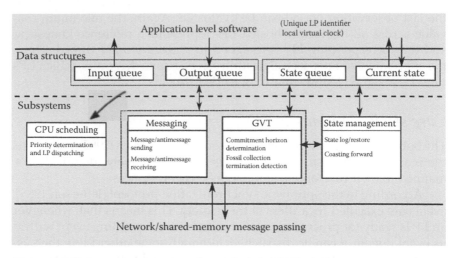

Figure 9.7 Reference architecture for optimistic PDES platforms.

they are separated for different LPs, so as to afford management costs. For the input queues, these costs are related to both event insertions and, for example, an event moved from the past (already processed) part to the future (not yet processed) in case of rollback of a specific LP. The input queue is sorted by message (event) time stamps, whereas the output queue is sorted by virtual send time, which corresponds to the LVT of the LP on the corresponding event-schedule operation. As discussed in several works (Vitali et al. 2010), the actual implementation of input queues can be differentiated (e.g., heaps vs. calendar queues), and possibly tailored to and/or optimized for specific application contexts characterized by proper event-time-stamp patterns (affecting the insertion cost depending on the algorithm used to manage the queue). On the other hand, output queues are typically implemented as doubly linked lists as insertions occur only at the tail (i.e., according to nondecreasing values of the LVT). Also, deletions from the output queues only occur either at the tail or at the head, the former occurring on a rollback operation that undoes the latest computed portion of the simulation at each LP. In particular, all the output messages (i.e., the generated events) at the tail of the output queue with send time greater than the logical time associated with the causality violation are marked, sent out toward the original destination in the form of *antimessages*—used to annihilate previously sent messages and inform the original destinations of the occurred rollback[*]—and then removed from the output queue. The latter are related to memory recovery procedures, which we shall detail later on in this section.

---

[*] Chained rollback can arise if the received events have already been processed by the destination LPs.

A *messaging subsystem* receives incoming messages from other simulation-kernel instances, the content of which will be then reflected within the input queue of the destination LP. Also, it notifies output messages (i.e., newly scheduled events) to LPs hosted by other kernel instances, or the aforementioned antimessages.

The *state queue* is the fundamental means for allowing a correct restore of the LP's state to a previous snapshot whenever a causality inconsistency is detected (i.e., the LP receives a message with time stamp lower than its current LVT, or an antimessage that annihilates an already processed event). The state queue is handled by the *state management subsystem*, the role of which is to save/restore state images, typically according to infrequent and/or incremental schemes (Vitali et al. 2010; Preiss et al. 1994; Quaglia 2001; West and Panesar 1996). Additional tasks by this subsystem are related to the following:

- Performing rollback operations (i.e., determining what is the most recent suited state that has to be restored from the log)
- Performing costing forward operations (i.e., fictitious reprocessing of intermediate events in between the restored log and the point of the causality violation)
- Performing *fossil collection operations* (i.e., memory recovery) by getting rid of all the events and states logs that belong to an already committed portion of the simulation.

The *Global Virtual Time* (GVT) subsystem accesses the message queues and the messaging subsystem to periodically perform a global reduction aimed at computing the new value for the commit horizon of the simulation, namely the time barrier currently separating the set of committed events from the ones that can still be subject to a rollback. This barrier corresponds to the minimum time stamp of not yet processed or in-transit messages/antimessages. In addition, this subsystem cares about termination detection, by either checking whether the new GVT oversteps a given predetermined value or by verifying some (global) predicate (evaluated over committed state snapshots [Quaglia 2009]) that tells whether the conditions for terminating the model execution are met. Finally, this subsystem is also in charge of starting the aforementioned fossil collection procedure.

In addition, a *CPU-scheduling approach* is used to determine which among the LPs hosted on a given kernel instance must take control for actual event processing activities. Among several proposals (Vitali et al. 2012a; Bauer et al. 2005), the common choice is represented by the lowest-time-stamp-first (LTF) algorithm (Hamada and Nitadori 2010), which selects the LP whose next pending event has the minimum time stamp, compared to next pending events of the other LPs hosted by the same kernel. Variants for LTF exist, among which a basic (stateless) approach relies on traversing the next pending events across the input queues of all the LPs, and a recent stateful approach (Ronngren et al. 1996) is based on reflecting variations of the priority of the LPs into the CPU-scheduler state, so that the LP with the highest priority can be determined via a query on the current CPU-scheduler state in constant time.

Load sharing in the context of optimistic simulation

In optimistic PDES runs, whenever a simulation kernel is hosting LPs with a load higher than the ones being hosted by other instances, the LVT associated with LPs in the other instances can advance more. This is related to the fact that, when every kernel instance hosts the same number of LPs, a more overloaded instance must execute more events per WCT time unit. As the advancement in LPs' LVT is just related to the number of events executed in a WCT time unit (we recall that events are impulsive), a less overloaded instance will process more events for all the hosted LPs, thus making their LVT advance more. This skew in the LVT values might induce, as a consequence, a higher amount of rollbacks (hence, an increase in the wasted computation). In fact, the probability that LPs hosted by the overloaded simulation kernel would create causal inconsistencies in the less loaded LPs gets higher.

A load-sharing system would be able to capture this imbalance, and assign a higher number of processing units to the more loaded simulation kernels. As a consequence, the LVT skew would be reduced, diminishing the amount of rollback operations and, in turn, increasing the overall efficiency of the simulation run.

Experimental evidence in relation to the above reasoning exists. Specifically, a set of measurements gauged from a real implementation of the innovative multithreaded architecture and of the overviewed load-sharing facilities within the open source ROme OpTimistic Simulator (ROOT-Sim) are available (High Performance and Dependable Computing Systems Research Group 2012; Pellegrini et al. 2011; Pellegrini and Quaglia 2013).

The experimental data refer to tests carried out on a 64-bits NUMA machine, namely an HP ProLiant server, equipped with four 2 GHz AMD Opteron 6128 processors and 64 GB of RAM. Each processor has 8 CPU cores (for a total of 32 CPU cores) that share a 10 MB L3 cache (5118 KB per each 4-cores set), and each core has a 512 KB private L2 cache. The operating system that has been used in the experimental study is 64-bits Debian 6, with Linux kernel version 2.6.32.5.

For assessing the validity of the proposal, the previously presented two real-world application benchmarks have been used, namely PCS, a mobile phone simulator and *Traffic*, a detailed highway simulator. Detailed descriptions for both these benchmarks are provided in Vitali et al. (2012a) and Vitali et al. (2012b). In the experimental study, the PCS benchmark has been configured to provide a constant workload across all the LPs during the whole simulation run. This has been done to measure the actual overhead of the load-sharing architecture, although not taking advantages from its ability to reallocate CPU cores, given the constancy of the workload. In this assessment, 1024 wireless cells (where one cell is modeled by an individual LP) have been simulated, each one handling 1000 wireless channels.

As for the Traffic benchmark, in the experiments carried out, the whole Italian highway system has been simulated (with the exclusion of the islands) by relying on 1024 LPs.

To effectively assess the proposal, the experimental study has been based on the observation of two metrics, whose related data are shown in Tables 9.1 and 9.2.

In the tables, we report results for a serial (traditional) DES execution based on a calendar-queue scheduler (which constitutes the baseline for the evaluation of the speedup provided by the parallel runs), a single-threaded classical optimistic execution, and several multithreaded executions with different amounts of kernel instances, giving rise to different multithreading degrees within each kernel instance. As the Traffic benchmark provides a more varying workload, in the corresponding table there is also a comparison with the orthogonal load-balancing approach (still offered by ROOT-Sim according to the description in Peluso et al. [2011]), whereas in the case of PCS this comparison is not shown, as that benchmark has been used solely for assessing the actual overhead of the presented architecture (with respect to a classical single-threaded

Table 9.1 Personal Communication Service Experimental Results

Configuration	Execution time (seconds)	Speedup
Serial	17,000	—
Single-threaded	44	386
Multithreaded (4 k)	58	293
Multithreaded (8 k)	54	315
Multithreaded (16 k)	52	327
Multithreaded (32 k)	52	327

Table 9.2 Traffic Experimental Results

Configuration	Execution time (seconds)	Speedup
Serial	17,500	—
Single-threaded	54	324
Load balancing	33	530
Multithreaded (4 k)	27	648
Multithreaded (8 k)	21	833
Multithreaded (16 k)	22	795
Multithreaded (32 k)	62	282

execution), given that it exhibits a balanced workload. The PCS bench-mark's execution involved 830 K events (overall), whereas the Traffic's one involved 400 K events (overall).

By the results of the experimental assessment, it can be seen that for the PCS benchmark, the configuration with 4 kernels shows the highest overhead (expressed by a WCT increase for completing the run), which is in the order of 20%. Increasing the number of simulation kernels (each one entailing a reduced amount of worker threads) provides a 13%–15% overhead reduction. These results have been achieved for relatively fine event granularity (about 30 μs for event processing), thus further support-ing the viability of the proposal, as applications exhibiting coarser grain events would absorb better the actual overhead induced by the symmetric multithreaded architecture, for example, relation to the activities required for handling the top/bottom half mechanism, which is not present in the traditional single-threaded organization. Also, the parallel approaches provide super-scalar speedup with respect to the serial DES execution, which indicates that they are actually competitive.

As for the Traffic benchmark, the experimental study shows that the parallel approaches provide super-scalar speedup. All the multithreaded versions of the simulation kernel provide speedup over the single-threaded one, which ranges between 40% (for the 4-kernel configuration) and 55% (for the 8- and 16-kernel configuration). The execution with 4-kernel instances shows a reduced speedup because of the following reasons:

- The rebalancing is more likely to map a worker thread on a core that is not actually sharing any level of cache.
- A worker thread can access remote memory with a higher probabil-ity (which, on NUMA machines, as the one used for the study, is very costly).
- Worker threads are more subject to false cache sharing effects.

As for the execution with 32 multithreaded kernels, the speed down is in the order of 15%. This is related to the fact that in this configuration no actual rebalancing is possible (in fact, each simulation kernel must have at least one worker thread to proceed in the simulation run). Therefore, this configuration again measures the architecture's overhead, with respect to the single-threaded organization, which is indeed comparable to the one shown when running the PCS (balanced) benchmark.

The last comparison is the one with respect to the load-balancing con-figuration. Although we note that this configuration provides speedup in the order of 40% with respect to the single-threaded approach, the deliv-ered performance is comparable with the 4-kernel multithreaded config-uration, while the 8- and 16-kernel configurations of the multithreaded architecture are still 30% faster than the load-balancing configuration.

Conclusion

In this chapter, we have discussed the issue of how to organize the architecture of PDES platforms to make them fully suitable for running on top of multicore machines. This is a relevant aspect related to the possibility to improve the efficiency of these platforms (thanks to the fruitful exploitation of scaled-up computing power), thus allowing them to increase the role they already have in engineering contexts where the timeliness of the delivery of simulation outputs plays a central role. We have presented general concepts related to the architectural organization, using the symmetric multithreading paradigm as the base for instantiating an archetypal organization of the PDES platform. Then we have shown how the symmetric approach is naturally prone to the achievement of balanced exploitation of the computing resources offered by a multicore machine. Finally, a real architecture of a PDES platform based on the above symmetric paradigm, and on the optimistic synchronization approach, has been presented as an example of a real system leading to full, balanced exploitation of such computing power. This chapter is intended to provide contributions on the side of showing how the multicore technological trend can be exploited for improving the timeliness according to which a simulation systems (in particular, a system based on the DES paradigm) can support the execution of general, complex, and dynamic models.

References

Boukerche, A. and Das, S. K. Dynamic load balancing strategies for conservative parallel simulations, In *Proceedings of the 11th Workshop on Parallel and Distributed Simulation*, Lockenhaus, Austria, 1997.

Bauer, D. W., Yaun, G., Carothers, C. D., Yuksel, M., and Kalyanaraman, S., Seven-O'Clock: A new distributed GVT algorithm using network atomic operations, In *Proceedings of the 19th Workshop on Parallel and Distributed Simulation*, Monterey, CA, 2005.

Carothers, C. D. and Fujimoto, R. M. Efficient execution of time warp programs on heterogeneous, NOW platforms, *IEEE Transactions on Parallel and Distributed Systems* 11(3): 299–317, 2000.

D'Angelo, G. and Bracuto, M. Distributed simulation of large-scale and detailed models, *International Journal of Simulation and Process Modelling* (IJSPM) 5(2): 120–131, 2009.

Dantzig, G. B. Discrete-variable extremum problems, *Operational Research* 5(2): 266–277, 1957.

Fujimoto, R. M. Parallel discrete event simulation, *Communications of the ACM* 33(10): 30–53, 1990.

Glazer D. W. and Tropper, C. On process migration and load balancing in time warp, *IEEE Transactions Parallel and Distributed Systems* 4(3): 318–327, 1993.

Hamada, T. and Nitadori, K. 190 TFlops astrophysical N-body simulation on a cluster of GPUs, In *Proceedings of the 2010 ACM/IEEE International Conference for High Performance Computing, Networking, Storage and Analysis*, New Orleans, LA, 2010.

High Performance and Dependable Computing Systems Research Group, ROOT-Sim: The ROme OpTimistic Simulator - v 1.0, 2012. [Online]. Available at http://www.dis.uniroma1.it/~hpdcs/ROOT-Sim/.

Intel Corporation, Enhanced Intel SpeedStep Technology for the Intel Pentium M Processor, 2004.

Jefferson, D. R. Virtual time, *ACM Transactions on Programming Languages and System* 7(3): 404–425, 1985.

Kandukuri, S. and Boyd, S. Optimal power control in interference-limited fading wireless channels with outage-probability specifications, *IEEE Transactions on Wireless Communications* 1(1): 46–55, 2002.

Meraji, S., Zhang, W., and Tropper, C. A multi-state Q-learning approach for the dynamic load balancing of time warp, In *Principles of Advanced and Distributed Simulation (PADS)*, Atlanta, GA, 2010.

Moore, G. E. Cramming more components onto integrated circuits, *Electronics* 38(8): 114–117, 1965.

Pellegrini, A. and Quaglia, F. The ROme OpTimistic Simulator: A tutorial, In *Proceedings of the 1st Workshop on Parallel and Distributed Agent-Based Simulations*, Aachen, Germany, 2013.

Pellegrini, A., Vitali, R., and Quaglia, F. The ROme OpTimistic Simulator: Core internals and programming model, In *Proceedings of the 4th International ICST Conference on Simulation Tools and Techniques*, Barcelona, Spain, 2011.

Peluso, S., Didona, D., and Quaglia, F. Application transparent migration of simulation objects with generic memory layout, In *Proceedings of the 25th Workshop on Principles of Advanced and Distributed Simulation*, Nice, France, 2011.

Preiss, B. R., Loucks, W. M., and MacIntyre, D. Effects of the checkpoint interval on time and space in time warp, *ACM Transactions on Modeling and Computer Simulation* 4(3): 223–253, 1994.

Quaglia, F. A cost model for selecting checkpoint positions in time warp parallel simulation, *IEEE Transactions on Parallel and Distributed Systems* 12(4): 346–362, 2001.

Quaglia, F. On the construction of committed consistent global states in optimistic simulation, *International Journal of Simulation and Process Modelling* 8(1): 172–181, 2009.

Reiher, P. L. and Jefferson, D. R. Virtual time based dynamic load management in the time warp operating system, *Transactions of the Society for Computer Simulation* 7: 103–111, 1990.

Ronngren, R., Liljenstam, M., Ayani, R., and Montagnat, J. Transparent incremental state saving in time warp parallel discrete event simulation, In *Proceedings of the 10th Workshop on Parallel and Distributed Simulation*, Philadelphia, PA, 1996.

Sutter, H. The free lunch is over: A fundamental turn toward concurrency in software, *Dr. Dobb's Journal* 30(3): 202–210, 2005.

Swenson, B. P. and Riley, G. F. A new approach to zero-copy message passing with reversible memory allocation in multi-core architectures, In *Proceedings of the 2012 ACM/IEEE/SCS 26th Workshop on Principles of Advanced and Distributed Simulation*, Zhangjiajie, China, 2012.

Vitali, R., Pellegrini, A., and Quaglia, F. Autonomic log/restore for advanced optimistic simulation systems, In *Proceedings of the Symposium on Modeling, Analysis, and Simulation of Computer and Telecommunication Systems*, Miami Beach, FL, 2010.

Vitali, R., Pellegrini, A., and Quaglia, F. A load-sharing architecture for high per-formance optimistic simulations on multi-core machines, In *Proceedings of the 8th IEEE International Conference on High Performance Computing*, Pune, India, 2012a.

Vitali, R., Pellegrini, A., and Quaglia, F. Towards symmetric multi-threaded opti-mistic simulation kernels, In *Proceedings of the 26th International Workshop on Principles of Advanced and Distributed Simulation*, Zhangjiajie, China, 2012b. D. West and K. Panesar, Automatic incremental state saving, In *Proceedings of the 10th Workshop on Parallel and Distributed Simulation*, Philadelphia, PA, 1996.

chapter ten

Layered architectural approach for distributed simulation systems
The SimArch case

Daniele Gianni

Contents

Introduction

Simulation systems incorporate diverse sets of general-purpose and application-specific functions that must be designed and integrated in simulation solutions satisfying the requirements of the systems engineering (SE) activity (Sage and Rouse 2009). When defining a simulation solution, developers may be unnecessarily exposed to the complexity originating from the design and implementation of the simulation software, particularly for solutions based on distributed environments. Consequently, SE activity cost and schedule may be affected as specialized know-how and development efforts are often needed to deliver the simulation solution (Boehm 1981; Grable et al. 1999). Furthermore, an initial simulation solution may need to be adapted during the SE activity to cope with the evolution of existing requirements as well as new emerging requirements. In software literature, the layered architectural pattern has been used to encapsulate groups of functions beyond stable interfaces, which hide the complexity of internal implementation details and decouple the implementation of each group from the adjacent ones (Buschmann et al. 1996).

The most prominent example of layered architecture is the Open System Interconnection (OSI), which underlies the principles of the Transmission Control Protocol/Internet Protocol networking software for the current Internet. This architecture has considerably contributed to the success of Internet by defining the integration standard for numerous software solutions at several layers, from the data link communication layer to the application layer. In the simulation domain, a layered architectural approach can lead to similar advantages to modeling and simulation–based SE activities. Particularly, a layered architectural approach can be applied to identify independent groups of simulation tasks that present low coupling and high cohesion (Sommerville 2006). As a result, this architectural approach can practically contribute to achieve the advantages of (1) hiding the complexity of the entire simulation system to the end user (e.g., simulation engineer or systems engineer), (2) supporting layers interchangeability for assembling layer implementations satisfying different sets of functional and quality simulation requirements, and (3) supporting simulation model portability across multiple platforms.

The above benefits may be of particular interest for parallel and distributed simulation (PADS) systems, in which the inherent complexity of the simulation infrastructure is further exacerbated by the intrinsic parallelism and communications of concurrent processes (Fujimoto 2000).

The chapter specifically addresses distributed simulation (DS) systems and presents SimArch (https://sites.google.com/site/simulationarchitecture/), an example layered architecture that is specifically tailored to discrete event simulation (DES) (Wainer 2009) and distributed discrete event simulation (DDES) (Fujimoto 2000). SimArch consists of four layers, each addressing only a macro functional requirement, and offers to the development of DES and DDES the previously mentioned advantages. Moreover, a SimArch implementation is available on the IEEE High Level Architecture (HLA) standard (IEEE 2010b), which platforms consequently can gain the above advantages. The standard has been shown to suffer from shortcomings such as high degree of complexity, lack of fine-grained component reusability, lack of inter-vendor interoperability, specialized know-how, and extra development effort with respect to the equivalent local simulation (LS) system (Gianni et al. 2011). Within the Simulation Interoperability Standards Organization (http://www.sisostds.org), the layered simulation architecture study group has started working on the definition of a layered solution to overcome the previously mentioned shortcomings and to enable the DS real-time execution using standard protocols such as the OMG Data Distribution Service (http://portals.omg .org/dds/).

The chapter is structured as follows. The Section "Background" provides an overview of the layered architectural pattern and of concepts related to DES, DDES, and the HLA standard. The Section "Layered simulation architecture for distributed simulation systems" presents the design process of a layered architecture for distributed simulation (DS) systems. The Section "Example application" provides evidence of the SimArch advantages of the effort and know-how reduction for the development of a DS system. Finally, the Section "Appendix" provides some implementation details of the SimArch architecture.

Background

The SimArch architecture has been defined using key concepts related to layered architectural pattern, DES, and HLA, which are summarized below for a rapid consultation.

Layered architectural pattern

This architectural pattern consists in the definition of loosely coupled layers that encapsulate the implementation of functional capabilities and that communicate only through specified interfaces. As such, *"the layered architectural pattern helps to structure applications that can be decomposed into groups of subtasks in which each group of subtasks is at a particular level of abstraction"* (Buschmann et al. 1996). In general, this pattern is applied to the design of large software systems that require the functional decomposition to support: incremental development, adaptability to new functional requirements, and software evolution for deploying new releases of algorithms, data structures, and new language technologies. This pattern has been proved to be of natural application in systems that inherently present functional groups at increasing abstraction level as the pattern brings several valuable advantages and only introduces often-irrelevant disadvantages.

Advantages and disadvantages

The layered pattern offers software developers mainly four advantages: (1) reuse of layers—layer implementations can be easily replaced to more conveniently achieve a wider compliance with the software requirements, (2) support for standardization—layer interfaces may be the subject of standardization aiming to support the previous advantage, (3) dependencies are kept local—software development and maintenance is improved as consequence of high cohesion and low coupling, and (4) interchangeability—legacy and noncompliant implementations can be easily encapsulated in new components implementing the layer interfaces. Differently, the pattern offers two advantages to the architecture end user: (1) platform independence, that is, isolating the most abstract aspects from the most concrete ones, and (2) raising the abstraction level of the software development by encapsulating the complexity of the lowest levels behind an interface specification. However, the pattern may also present three main disadvantages: (1) cascade of changing behaviors—severe problems may arise when a layer changes its behavior unexpectedly; (2) lower efficiency with respect to monolithic architectures—layer interdependencies and polymorphisms, both for the service and data specifications, can lead to considerable performance overhead when top layers heavily depend on lower layers; and finally, (3) difficulty of identifying layer and determining the right granularity—there is no general recipe for the layer identification, and domain-specific knowledge is often needed. Nevertheless, this last disadvantage has been mitigated with the introduction of domain-independent guidelines for the design of layered architecture.

Application guidelines for the layered pattern

Nine steps are generally developed when applying the layered pattern for the definition of a new software architecture (Buschmann et al. 1996):

1. Define abstraction criterion for grouping tasks into layers. The criterion often reflects the physical or logical nature of the software system being considered. Moreover, multiple criteria may be identified for each of the layers.
2. Determine the number of abstraction levels. The above criteria identify the abstraction layers, each generally mapping to a layer in the architecture. However, the mapping may not always be evident, and trade-off analysis may be needed to decide whether groups of tasks should be split or integrated in one architectural layer. The aim of the analysis is also to identify the right compromise between the

previously mentioned advantages and disadvantages, and more technically, the compromise on the software quality attributes, balancing cohesion, coupling, and performances.

3. Name the layers and assign tasks to each of them. The naming and assignment process generally starts from the top layer, which is the one exposed to the end user who perceives this layer as the one incorporating the capabilities of the entire system. Proceeding downward, the layers should be assigned the necessary tasks needed for the implementation of the above layer tasks. Finally, the name should provide a nominalization of the layer capabilities and services.

4. Specify the services. This is a central step in the entire design process. The services are the only access point for the tasks allocated within each layer. Ideally, only adjacent layers should be able to communicate. However, communication from nonadjacent layers could be permitted in favor of a more understandable and efficient architecture. A typical heuristics is that higher layers should define richer sets of services with respect to lower layers, for reasons that are related to the phenomenon known as "inverted pyramids of reuse."

5. For each layer, specify the interfaces offered to the adjacent layers. The interfaces should be defined to separate the internal content of the layer, which may be seen as a white box or black box from adjacent layers. However, the use of a black-box design approach is strongly recommended as this approach raises the layers coupling from content level to informational level (Fenton and Lawrence 1996).

6. Specify data interfaces. The data interfaces define the access methods for the objects and ensure the transparent communication of these objects across the entire architecture, independently from the layer implementations. As such, data interfaces contribute to further decouple layers. However, data interfaces are commonly defined for read-only access to the objects, and therefore data conversions mechanisms must be provided internally by the layer implementations requiring the modification of field values of these objects.

7. Structure individual layers. Although the interlayer relationships are central to the architecture definition, and the respective quality attributes, individual layers should also be accurately designed. In particular, individual layers should be considered for further decomposition into components that may possibly conform to established design patterns to enhance the internal quality of the layer.

8. Specify communication between adjacent layers. The interlayer interactions should be fully documented in terms of expected state evolution and possible interaction scenarios. Concerning the individual interactions, the interaction model generally used is the pull and push, in which a routine does not return the control to the caller layer until the routine has been entirely processed. However,

this interaction model has been shown to introduce dependencies between layers, and therefore the approaches based on the asynchronous callback can be preferred to return the service output and to decouple lower layers from higher ones.

9. Decouple adjacent layers. This step aims to ensure that the layer coupling is minimum or at least in line with the expected software quality of the entire architecture. Layer dependencies may not necessarily be symmetric and may still be suitable for the expected workload of the architecture, in both top-down or bottom-up directions. Nevertheless, further design considerations may need to be applied to minimize these interdependencies also in the allocation statements for the layer implementations.

The above steps constitute only a general guideline, which needs to be pondered against the design effort, the suitability of the architectural solution, and sound technical judgment in the specific context. In particular, an iterative development approach should be considered instead of a fully sequential one, if the iterative can more effectively lead to a suitable architectural solution. For example, compliance levels may be considered in the interface specifications to ensure the benefits of the layered approach for the most common functional requirements, while granting the possibility of introducing ad hoc solutions on custom functional requirements.

Discrete event simulation paradigm

The DES paradigm is used to reproduce the operations of a system that changes its state only in correspondence of events (i.e., changes of the state of the environment) at discrete times. However, this paradigm commonly serves as abstraction to model and simulate inherently continuous systems, permitting that the continuous properties can be either ignored or approximated to the occurrence of relevant events. The DES paradigm can be implemented in four different world views (Banks et al. 2005): activity scanning, double phase, event scheduling, and process interactions. In particular, we consider the process interaction view, which is based on well-established component-based software engineering (CBSE) concepts (Lau and Wang 2007): entity—that is, the process encapsulating the simulation logic consisting in the sequence of state changes and delivery/reception of external events; event—which represents the change of the environment state, related to either the simulation model (model event) or the simulation system (service event); port—which is an entity's socket on for sending (output port) or receiving (input port) events; and link—which conveys events from the output ports to associated input ports (Figure 10.1). These concepts are commonly used for the implementation of process interaction views in DES and DDES

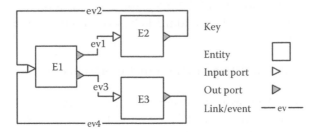

Figure 10.1 CBSE concepts for the process interaction view.

frameworks, for example, Scalable Simulation Framework (Nicol et al. 2011), D-SOL (Verbraeck 2004), and OSA (Dalle 2007).

Distributed simulation and the IEEE high-level architecture

DS consists in the execution of a simulation model on an interconnected set of processors, each hosting a segment of the model. DS can be classified in parallel and geographically DS. The former regards the model execution on a platform consisting of tightly coupled processors, which are provided with a shared memory for all the processors. Differently, the latter regards the model execution on a platform consisting of loosely coupled processors, each provided with a private memory area. Geographically DS can be further subclassified by the scale of the communication network, that is, LAN, MAN, or WAN. Regardless of the platforms, PADS systems present common solutions to similar problems, which are related to synchronization and communication issues deriving from the intrinsic concurrency. However, DS systems are further affected by problems related to software interoperability and adaptability, which inherently constitute a barrier to the widespread use of distributed computing platforms for simulation applications. These platforms are becoming more and more affordable and their availability would economically offer academics and industry several potential advantages, such as increased simulation performance, increased hardware and software reuse, interconnection of geographically distributed human actors, and software-as-a-service deployment of simulation models (Tsai et al. 2011; Bocciarelli et al. 2013).

The most predominant DS standard is HLA, which provides a general framework for designing, developing, and executing DS applications. The standard aims to promote interoperability and reusability of simulation components in different domains with the introduction of a process for simulation design and development (IEEE 2003); rules for the design and implementation of DS system (IEEE 2010b); interfaces for generic simulation oriented services (IEEE 2010a); a standardized data model that can be serialized in XML format (IEEE 2010c); and a verification and validation framework (IEEE 2007). The standard is based on the following concepts (Khul et al. 1999):

- Federate: A simulation program that represents the unit of reuse in HLA.
- Federation: A DS execution composed of a set of federates.
- Run-time infrastructure (RTI): A simulation-oriented middleware consisting of a RTI local component, which resides on the federate sites, and a RTI executive component, which is deployed on a central server.

With respect to previous protocol-oriented standards (e.g., IEEE Distributed Interactive Simulation [DIS] [IEEE 2012], Aggregated Level Simulation Protocol [Weatherly et al. 1991]), the major improvement HLA introduces is an API-oriented development approach of DS systems. This approach relieves developers from all concerns related to the communication and the synchronization within the DS system, obtaining considerable effort savings in the development process. Despite this improvement, HLA still suffers from three main shortcomings (Gianni et al. 2011): (1) the complexity of the API, (2) the strictly distributed orientation of the API, and (3) the absence of a standard communication protocol between RTI local and RTI executive. The API complexity derives from the wide set of generic simulation services (e.g., simulation life cycle management or data publishing and subscribing). As a result, considerable effort and specialized know-how is needed to develop HLA-based DS systems. Moreover, the API also suffers from an exclusively distributed orientation of the two HLA interfaces: RTIAmbassador, for the federate invoking HLA services; and FederateAmbassador, for the RTI local callbacks to the federate requests. As such, HLA does not provide any support for implementing the internal federate logic. In addition, the absence of a standard communication protocol affects the interoperability and reusability of HLA federates, which cannot practically be run in the same federation unless their versions of the RTI local and RTI executive components are released by the same vendor and are identified by compatible version numbers.

Layered simulation architecture for distributed simulation systems

The design of a layered architecture for DS systems—particularly HLA-based—begins with the identification of the motivations and exploitable advantages that the layered pattern can directly bring to the development of the DS systems, and transitively to the SE activity. A positive evaluation would lead to the architecture definition using the above-outlined guideline.

Motivation

The layered pattern can bring three types of advantages to HLA-based systems: general advantages, DDES and HLA-specific advantages, and side advantages related to the integration with ongoing trends in the software community.

The general advantages, which derive directly from those of the layered architectural patterns, are complexity hiding, layer interchangeability for different functional and quality requirements, and software reuse. Complexity hiding can bring the isolation of the simulation models from all implementation details of the underlying execution platform. This benefit is mainly related to the design effort reduction for the simulation model specification as developers are only concerned with model-specific issues and are raised from implementation-specific ones. Differently, layer interchangeability regards the possibility to immediately configure the entire DS system by selecting different layers implementations to meet specific functional requirements (e.g., distributed or local execution) or quality requirements (e.g., real-time execution). Both complexity hiding and layer interchangeability contribute to achieve the advantage of increased software reuse, which regards the reuse of existing layer implementations—such as local or distributed on the various standards (e.g., HLA, IEEE DIS)—and the portability of available simulation components over evolving or future DS standards.

The general advantages also underlay the DDES and HLA-specific advantages of effort saving and fine-grained HLA simulation component reusability. Particularly, the effort savings can be gained in three ways (Gianni et al. 2011):

- Reducing extra effort to acquire HLA knowledge and skills: 30% for a developer with average experience and 60% for a beginner
- Removing extra coding effort to create HLA federates: about 3.5 K extra lines of code (LOC) per federate
- Reducing extra design effort related to design evaluations, such as which federates to develop, which to reuse, which time advancement modality and simulation paradigm to adopt, which data to exchange, and which communication modalities to use

Component reusability is also enhanced as layer interchangeability contributes to refine the level of reuse unit from the federate to the individual simulation components. As such, HLA federations can be designed by selecting existing federates and by aggregating available simulation entities, and the associated support components, into newly ad hoc developed federates. Consequently, simulation interoperability also becomes a less impeding factor in multi-enterprise simulation experiments as these components can be effortlessly ported on compatible DS platforms, when using a layered approach (D'Ambrogio et al. 2007).

Finally, other side advantages can be obtained by leveraging on the increased level of abstraction offered to the developers, particularly in the area of DES paradigms, domain-specific languages (DSLs)—that is, languages that are specialized for a particular application domain (Mernik et al. 2005), and model-driven simulation engineering approaches (http://www.sel.uniroma2.it/Mod4Sim10). Building on the lower architecture layers, diverse upper layers may be integrated in multiparadigm simulation environments, such as agent-based and process interaction. Differently, the increased abstraction level also offers more easily reachable capabilities for the implementation of new DSLs for simulation, using the services from the top layer. For the same reason, the layered approach would contribute to ease the definition of automatic transformations between model specifications and newly defined high-level simulation languages.

Applying the layered pattern

The SimArch design proceeds as indicated in the above guidelines. First, we have identified the layering criterion "one functionality group per layer" as the one offering the highest level of cohesion within each individual layers and the relative lowest level of coupling between adjacent layers. Second, the layers can be intuitively identified after observing typical DS systems, which incorporate functions related to five groups: simulation model specification, simulation language implementation, local synchronized communication among the simulation components, coordination between a local and distributed environment, and finally, communication and synchronization across the distributed environment. Consequently, the criterion aimed to maintain these tasks is decoupled to provide reusable and interchangeable layers for different simulation languages implementations, simulation world views, simulation communication protocols, and simulation platform performances. As result, the SimArch architecture has been partitioned in the five layers shown in Figure 10.2.

Third, the layers roles and responsibilities can be identified as follows:

- Layer 0 is the distributed computing infrastructure, which can be either of simulation-oriented type (such as HLA or DIS), or of general purpose type (such as CORBA, Web Services, Grid, or Cloud), or else of mixed type (CORBA-HLA [D'Ambrogio and Gianni 2004]) or HLAGrid (Xie et al. 2005). Technically, this layer is outside the SimArch definition as it is the foundation on which the architecture is developed.

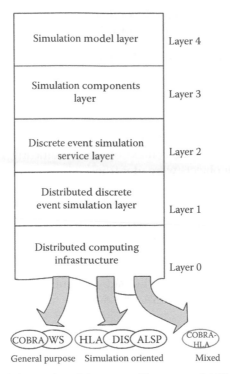

Figure 10.2 SimArch layered architecture. (Courtesy of ACM. Gianni, D. et al., *Proceedings of the First International Conference on Simulation Tools* (SIMUTOOLS'08), pp. 1–9, Marseille, France, 2008.)

- Layer 1 provides a DES abstraction on the distributed computing infrastructure, namely DDES. This layer provides the services for the synchronization with the distributed environment and for the delivery of model events to remote simulators.
- Layer 2 provides a transparent DES abstraction on either local or distributed environment. Similarly to Layer 1, this layer provides the services for the synchronization and the delivery of model events. However, at this layer, the services concern individual simulation entities. Moreover, these services also relieve the developers of Layer 3 simulation languages from the concerns of the execution environment, either local or distributed.
- Layer 3 contains the implementation of any DSL for simulation. For proof-of-concept purposes, we specifically provided SimArch with a Java internal DSL named jEQN (Gianni and D'Ambrogio 2008), for the extended queuing network (EQN) domain (Bolch et al. 1998). Nevertheless, other domains can be addressed and the respective simulation entities be implemented at this layer.

- Finally, Layer 4 contains the definition of the simulation model to be executed by the underlying simulation layers.

Next, the data interfaces are defined to ensure the fully decoupling of layers implementations.

Data interfaces

Data interfaces can be identified by listing the data objects that populate the domains of DES in the context of DS. Specifically, for each layer, the following relevant objects were identified (Gianni and D'Ambrogio 2007):

- Layer 3: entity (model) event, port, link, time, and name for the identification of the entity
- Layer 2: (model) event, time
- Layer 1: (model) event

The data interfaces can be subsequently defined as shown in Figures 10.3 and 10.4. SimArch defines the following interfaces:

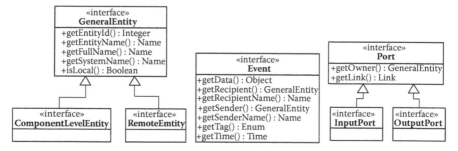

Figure 10.3 Specification of the data interface: Entity, Event, and Port.

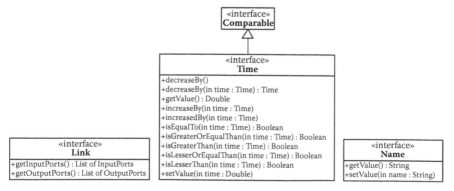

Figure 10.4 Specification of the data interface: Link, Time, and Name.

ComponentLevelEntity—which identifies local entities (including hierarchical composition), Event—which represents events scheduled between layers, GeneralEntity—which provides a reference to local and remote entities, InputPort; Link; Name—which represents entity names in SimArch, OutputPort; Port—which provides a common abstraction for InputPort and OutputPort, RemoteEntity—which provides a local reference remotely running simulation entities; and Time—which represents the simulation time.

The architecture definition proceeds with the development of the central steps in the entire process of the definition of the services and of the service interfaces.

Services specification

The services are inherently related to the above-mentioned layers descriptions. In particular, Layer 1 provides Layer 2 with the DDES service, for time-stamped distributed delivery of model events at discrete times. Similarly, Layer 2 provides Layer 3 with the DES service, which are transparent (local or distributed) time-stamped delivery of model events at discrete times. Layer 2 also provides Layer 1 with local accountability DES service, which enables Layer 2 to receive remotely generated DES events. Next, Layer 3 provides Layer 2 with the services for the activation and the suspension of a simulation entity, as well as the delivery of a DES event to an individual entity. Finally, Layer 3 provides Layer 4 with the services for the definition and execution of the simulation model. For each couple of adjacent layers, the service interfaces must be formally defined to ensure the interchangeability of different layers implementations.

Service interfaces

Similarly to the OSI architecture, the service interfaces define the only communication access points between adjacent layers, in both communication directions. Below, we adopt the notation LayerXToLayerY to indicate the interface from Layer X to Layer Y. Moreover, we purposely omit from this presentation the definition of the methods related to the cross-reference of layers instances, for the sake of conciseness.

The SimArch architecture is defined on Layer 0, which represent a distributed environment. Consequently, the Layer0ToLayer1 and Layer1ToLayer0 interfaces are specific to the distributed environment—for example, using HLA as Layer 0, Layer0ToLayer1 coincides with the FederateAmbassador and Layer1ToLayer0 coincides with the RTIAmbassador—and therefore these interfaces are outside the definition of the SimArch service interfaces.

Figure 10.5 shows the `Layer2ToLayer1` interface. The `initDis-tributedSimulationInfrastructure` and `postProcessing-DistributedSimulationInfrastructure` methods are general placeholders for system setup and recovery of initial state after a simulation execution, respectively. The remaining three operations provide the DES abstraction on the distributed environment: `sendEvent` delivers the event parameter to the recipient entity at the specified simulation time, `waitNextDistributedEvent` suspends the invoking thread until a remote model event is received, and `waitNextDistributedEvent-BeforeTime` suspends the invoking thread until either a remote model event is received or the specified time is reached by the DS system.

Figure 10.6 shows the `Layer1ToLayer2` interface, which enables Layer 1 to transparently schedule remote model events into the local system. To further decouple Layer 1 and Layer 2, the schedule service is accessible through the `scheduleEvent` operation, for any model event, and `scheduleSimulationEndEvent` operation, for the simulation end event only.

Figure 10.7 shows the `Layer2ToLayer3` interface, which offers access operation to the internal layer structures for the synchronization of the simulation entities. For example, the body service operation encapsulates the entity logic, which consists in the invocation of `Layer3ToLayer2` service operations (e.g., `hold` or `waitNextEvent`) and internal processing, on the entity state variables (e.g., queue length or busy/free state of a service center). This operation is invoked when the simulation is activated through Layer 4 service interface.

Figure 10.8 shows the `Layer3toLayer2` interface, which consists of two subinterfaces: `Layer3toLayer2DevelopersInterface`, detailed

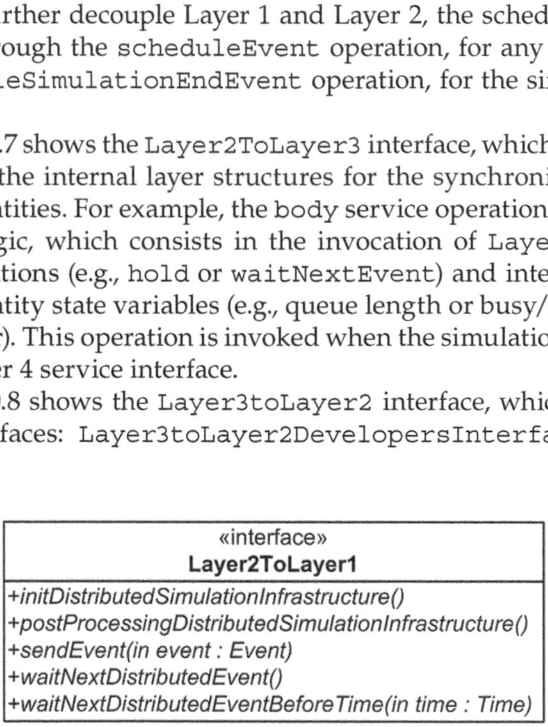

«interface»
Layer2ToLayer1
+*initDistributedSimulationInfrastructure()*
+*postProcessingDistributedSimulationInfrastructure()*
+*sendEvent(in event : Event)*
+*waitNextDistributedEvent()*
+*waitNextDistributedEventBeforeTime(in time : Time)*

Figure 10.5 Layer2ToLayer1 interface.

«interface»
Layer1ToLayer2
+*scheduleEvent(in event : Event)*
+*scheduleSimulationEndEvent(in time : Time)*

Figure 10.6 Layer1ToLayer2 interface.

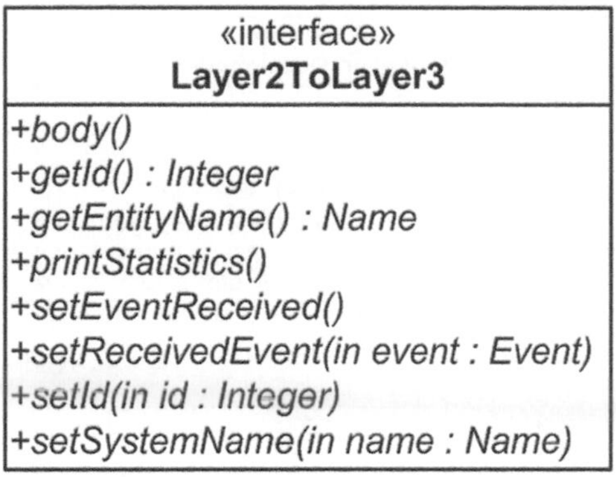

«interface»
Layer2ToLayer3
+*body()*
+*getId() : Integer*
+*getEntityName() : Name*
+*printStatistics()*
+*setEventReceived()*
+*setReceivedEvent(in event : Event)*
+*setId(in id : Integer)*
+*setSystemName(in name : Name)*

Figure 10.7 Layer2ToLayer3 interface.

«interface»
Layer3ToLayer2
+*getUserInterface() : Layer3toLayer2UserInterface*
+*getDeveloperInterface() : Layer3ToLayer2DeveloperInterface*

Figure 10.8 Layer3ToLayer2 interface.

in Figure 10.9, and `Layer3toLayer2UsersInterface`, detailed in Figure 10.10.

`Layer3toLayer2DevelopersInterface` is for accessing the DES service operations for the implementation of DSL components at Layer 3. Specifically, this interface offers the `send` operation to communicate with other simulation entities. Moreover, the interface also provides the operations to suspend the process execution for a specified simulation time (`hold`), until a model event is received (`wait`), or for a simulation time pending no model events are received earlier (`holdUnlessIncomingEvent`).

Differently, `Layer3toLayer2UsersInterface` is mainly for encapsulating Layer 3 service operations related to the configuration of the simulation system. Consequently, this interface is also the base for the definition of `Layer4ToLayer3Interface`, which enables developers to register the simulation entities in the simulation system (`registerEntity`) and to activate the simulation system within the simulation platform (`start`). For LS systems, the actual software execution will be achieved using the conventional operating system primitives for the execution of a software program. Conversely, DS systems require that the entire distributed platform is on execution before the local segment can be started (Gianni et al. 2011).

Communication between adjacent layers

The layer interactions can be specified in terms of static and dynamic aspects. The static aspects concern the specification of the data and service interfaces as these are critical to decouple layers. Differently, the dynamic

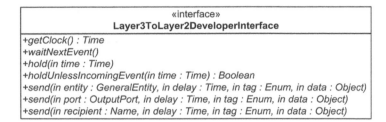

«interface»
Layer3ToLayer2DeveloperInterface
+getClock() : Time
+waitNextEvent()
+hold(in time : Time)
+holdUnlessIncomingEvent(in time : Time) : Boolean
+send(in entity : GeneralEntity, in delay : Time, in tag : Enum, in data : Object)
+send(in port : OutputPort, in delay : Time, in tag : Enum, in data : Object)
+send(in recipient : Name, in delay : Time, in tag : Enum, in data : Object)

Figure 10.9 Layer3toLayer2DeveloperInterface.

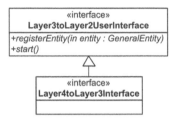

Figure 10.10 Layer4toLayer3Interface.

aspects can be specified in terms of (1) layers interaction patterns— represented with Unified Modeling Language (UML) sequence diagrams, and of (2) layers state evolution—represented with UML state diagrams.

Figure 10.11 shows the diagram of interaction patterns developing across the SimArch layers, for the DS type—the local ones can be easily inferred by the DS ones. In the diagram, each layer is represented as an actor that interacts with the adjacent layers through the service operations defined in the respective SimArch service interfaces. The diagram begins with the registration of all the locally instantiated simulation entities in the SimArch environment (registerEntity). Next, Layer 4 invokes the startEngine service to activate the simulation component execution. The startEngine method call is further forwarded to Layer 2, which is the execution container coordinating the simulation entities. Layer 2 thus enters in an initial loop in which all entities are activated simultaneously and made concurrently run until each reaches the invocation of a SimArch service operation hold, wait, or holdUnlessIncomingEvent. In this loop, each simulation entity may also invoke Layer 2 service operations for the delivery of model events (send), which are further processed by Layer 2 to determine whether the delivery can be resolved locally or should be forwarded to Layer 1 through the homonymous service operation. After all the entities are suspended on the above service operations, Layer 2 proceeds with a conservative advancement of the DS time. Specifically, if the local event list is empty, Layer 2 will request Layer 1 a time advancement to the next remote model event. Differently, if the local event list is full, Layer 2 requests Layer 1 to advance to the time of

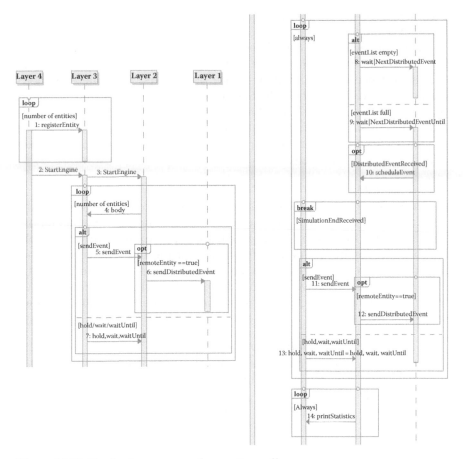

Figure 10.11 SimArch sequence of operation calls.

the earliest event in the list. If a model event is received, Layer 1 invokes Layer 2's `scheduleEvent` operation. In turn, Layer 2 will check whether the event is a simulation end event and, in case, invokes the Layer 3 service `printStatistics` before terminating all the simulation entities. Otherwise, when the time advancement can be granted for the next internal event, Layer 1 concludes the execution of the `waitNextDistributedEventUntil` and returns the control to Layer 2. In this case, Layer 2 continues the execution with the processing of the next event—n.b. remote model events are transparently scheduled by Layer 1 into Layer 2 as local model events— and informs the respective simulation process through Layer 3 services `setEventReceived` (for processes suspended on hold operation) and `setReceivedEvent` (for processes suspended on wait operations). Once the simulation logical process has been activated, the process will similarly continue executing the entity logic and occasionally invoking Layer 2 services until the process becomes blocked again and the loop reiterates, similarly.

The state evolutions complement the above architecture specification by displaying, for each layer, the possible states and their transitions in correspondence of the SimArch service operations. The state evolutions can be easily read from the state diagrams shown in Figures 10.12 through 10.14 for Layer 1, Layer 2, and Layer 3, respectively. Layer 4 diagram is omitted as this layer is substantially stateless as it would only consist in a sequence of the transient states: initial state, entity registration state, execution state, and final state.

Decoupling adjacent layers with the factory pattern

In SimArch, the service interfaces do not fully decouple the layers as the interfaces still require that each layer is created with the implementation-specific syntax defined by the layer constructors. In literature, this problem has already been solved with the use of the factory interfaces that hide the specific details related to the instantiation of a layer implementation. In SimArch, we have therefore defined the factory interfaces shown in Figure 10.15 and enriched the above service interfaces with accessory setter methods.

Designing error handling strategy

Currently, the entire architecture is defined so that layers do not catch local exception and do not attempt to recover from local errors. Rather, exceptions and errors are always propagated to the upper layer from the one they were raised, until their propagation reaches Layer 4. With the

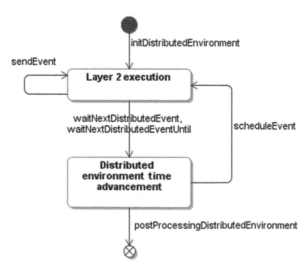

Figure 10.12 Layer 1 state diagram.

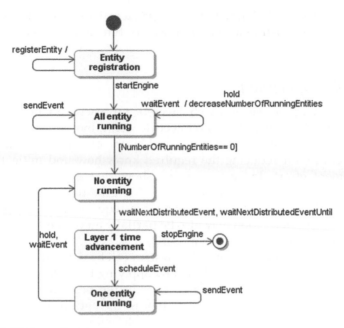

Figure 10.13 Layer 2 state diagram.

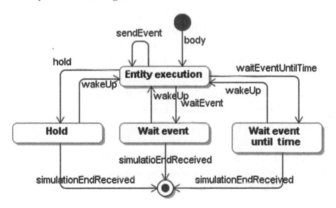

Figure 10.14 Layer 3 state diagram.

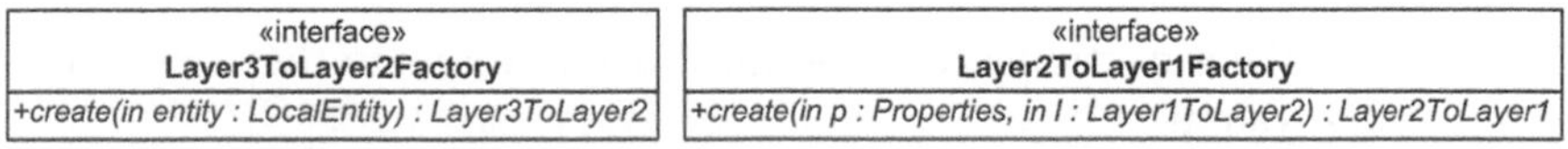

«interface» Layer3ToLayer2Factory	«interface» Layer2ToLayer1Factory
+create(in entity : LocalEntity) : Layer3ToLayer2	+create(in p : Properties, in l : Layer1ToLayer2) : Layer2ToLayer1

Figure 10.15 Layer factory interfaces.

objective of providing a scope and tracing the origins of exceptions and errors, each layer defines a base class Layer<X>Exception and a base class Layer<X>Error. The service operations are also associated to a specialization of one of these classes, depending on the identified types of exceptions or errors. For example, Layer 1 may trigger exceptions related to configuration errors of the distributed environment, or time issues related to the DS execution.

Example application

The advantages of the SimArch layered approach can be informally proved directly by the above architecture definition, for the complexity hiding and layer interchangeability, and by the definition of a mechanical procedure, for reducing the required know-how and mitigating the development effort. In particular, we show that a mechanical procedure can be used to derive a DS system from the respective LS system. Consequently, no specialized know-how is required to develop the DS solution, whose derivation may also be partially automated using model-driven approaches (Brambilla et al. 2012).

Starting from a LS system, a DS system can be derived by first specifying a model partitioning and then by applying the mechanical derivation procedure for each model partition. In the procedure description, two concepts are key: local entities—that is, the entities running within the considered model segment; and remote entities—that is, the entities running on other model segments. Moreover, remote entities can be distinguished in relevant and irrelevant remote entities. The relevant remote entities can be graphically identified as the remote entities whose incoming links originate from the model segment. Consequently, the irrelevant remote entities are all the remaining remote entities. Similarly, the auxiliary concepts of relevant and of irrelevant remote links can be defined between a local entity and a relevant remote entity and between a local entity and an irrelevant remote entity, respectively.

The procedure can be formulated in three steps:

1. Instantiate Layer 1 and Layer 2 implementations
2. Define and declare domain-specific entities:
   a. Local entities: code identical declarations as the ones in the LS system
   b. Relevant remote entities: allocate Layer 3 `RemoteEntity` interfaces
3. Connect entities:
   a. Local connections: code identical declarations to the ones in the LS system.
   b. Irrelevant remote connection: do not code and reduce the multiplicity of the respective local connection, if the irrelevant remote connection is a part of a multiple link.
   c. Relevant remote connections: code connections between local entities and the allocated `RemoteEntity` interfaces in substep 2b, using the local connection statements.

For example, let us consider the simple LS system on the top side of Figure 10. 16 and the respective DS system, which also includes the model partitioning in three segments, on the bottom side of the same figure. For the sake of clarity, we assume that submodel <i> is coded in federate <i>, where $i = 0$, 1, and 2.

Using the above procedure, the local entities can be immediately identified for federate 0 (Source, WaitingSystem0, ServiceCenter0, and Router), for federate 1 (WaitingSystem1, ServiceCenter1), and for federate 2 (WaitingSystem2, ServiceCenter2). The respective declaration statements can be immediately copied from the LS system into the three federates,

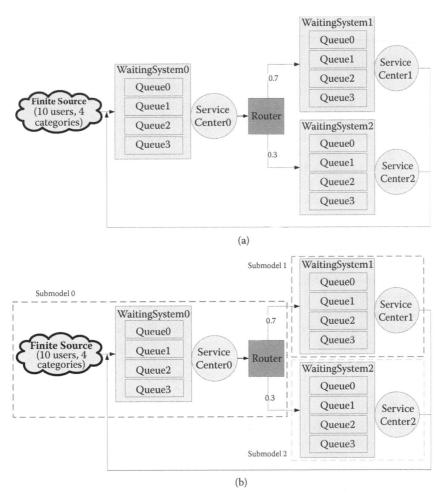

Figure 10.16 (a) Local simulation and (b) respective distributed simulation. (From Gianni, D. et al., *Simulation* 87(9), 819–836, 2011.)

respectively. Similarly, the relevant remote entities can be graphically identified, by following the respective definition, for federate 0 (WaitingSystem1, WaitingSystem2), for federate 1 (WaitingSystem0), and for federate 2 (WaitingSystem0). The respective statements can be coded allocating the `RemoteEntity` interfaces—specifically the `BasicRemoteEntity` class available in the SimArch implementation. Next, the local links can be immediately established by directly copying the respective statements, for federate 0 (Source-WaitingSystem0, WaitingSystem0-ServiceCenter0, ServiceCenter0-Router), federate 1 (WaitingSystem1-ServiceCenter1), and federate 2 (WaitingSystem2-ServiceCenter2). Differently, the relevant remote links need to be introduced ex-novo for federate 0 (Router-WaitingSystem1, Router-WaitingSystem2), federate 1 (ServiceCenter0-WaitingSystem0), and federate 2 (ServiceCenter2-WaitingSystem0).

Conclusions

The layered architectural approach has shown a promising potential to bring several advantages to the development of simulation solutions in support of SE activities. For example, a layered approach can enable systems engineers to assemble diverse layers implementations to develop a simulation solution that can most effectively satisfy the simulation requirements of SE activities. Moreover, the layered approach raises the developers from the complexity of the simulation solution, particularly in the case of DS solutions, which offer a number of benefits ranging from increased resource reuse to potentially reduced execution times. In the chapter, we have illustrated the application and the results of a design process for the definition of a layered simulation architecture, named SimArch, for DS systems—particularly the ones based on the IEEE High-Level Architecture (HLA) standard. The design process has consisted in a series of steps, which should be considered as general guidelines and which should always be calibrated to the specific application domain and to the objectives of simulation architecture. The SimArch advantages are exemplified through the design of a simple DS solution. A mechanical derivation procedure is also introduced to informally prove that the use of SimArch contributes to reduce the required know-how and to mitigate the development effort of a DS system with respect to the equivalent local one. However, we envisage further exploitations of the SimArch architecture in the areas of DSLs for DES and of model-driven simulation engineering (MDSE). Concerning DSLs, SimArch provides the facilities to define new internal DSLs using jEQN—a simulation language available in the current SimArch implementation—or directly using the SimArch Layer 2 services for the timestamped communication and synchronization among simulation logical processes. As a consequence, language developers are completely raised from the complexity of the simulation platform, particularly for DS systems. Similarly, the availability of new DSLs and SimArch would ultimately contribute to further lower the effort for the implementation of MDSE approaches for the generation of DS systems from visual specifications (e.g., UML or SysML). As SimArch provides a higher-level platform of simulation services, with respect to a general distributed environment or the HLA standard, MDSE approaches only have to define model transformations between the visual language and the DSL. In this direction, recent developments have been undertaken and initial MDSE approaches have been developed for the DS of process models in the Business Process Model Notation (http://www.bpmn.org), using jEQN as base DSL.

Appendix

SimArch implementation details

A SimArch implementation has been developed on the IEEE HLA-compliant Pitch pRTI 1516 software (http://www.pitch.se) and has been tested in LAN and WAN environments (Gianni and D'Ambrogio 2007). Below, we briefly present the internal details of the existing layers implementations. All the source code is available in Java from the project code repository (https://code.google.com/p/simarch/).

Layer 1: Distributed discrete event simulation over HLA

DDESoverHLA is the acronym that denotes SimArch Layer 1 implementation, whose internal view is shown in Figure 10.17. DDESoverHLA consists of four main components: DDESoverHLAEngine; DDESoverHLAEngineAmbassador; FederationManager; and HLAEvent.

The DDESoverHLAEngine implements the interface Layer2ToLayer1 using the standard services provided by the HLA interface RTIAmbassador. For example, the waitNextDistributedEvent service operation is implemented as follows:

```
waitNextDistributedEvent():
do
      tfuture = tcurrent + tadvancingStep;
      timeAdvanceRequest(tfuture);
      wait(timeGrant);
while not (eventReceived);
sendReceivedEventToUpperLayer();
```

This implementation consists of a do-while cycle that performs time advance requests to a future time until a model event is received from

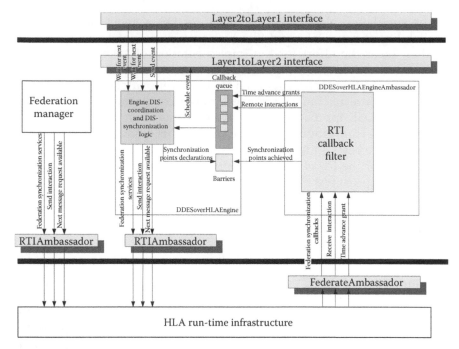

Figure 10.17 DDESoverHLA internal view.

the RTI executive. The future time is computed by adding the current time to the system parameter advancement step. The `timeAdvanceRe-quest` statement is used to request a time advancement at the specified time and the `wait(timeGrant)` statement suspends the current process until a `timeGrant` RTI callback is received. While waiting, the engine automatically processes incoming RTI callbacks from the `FederateAmbassador`. The expected types of callbacks are of the following types:

- `timeAdvanceGrant`
- `receivedEvent`

The `timeAdvanceGrant` callback is dispatched by updating the internal clock time to the one associated to the callback. The `receivedEvent` callback is dispatched by updating the flag `eventReceived` to true and by storing the received model event in a temporary buffer. According to the HLA standard, with the current time settings, an `interaction-Received` notification is always followed by a `timeGrant` notification, which confirms the safe advancement to the interaction time. The sendReceivedEventToUpperLayer statement follows the do-while cycle and invokes the Layer 2 service method schedule, to transparently schedule remote model events into the local system. Furthermore, the DDESoverHLAEngine also maintains the consistency between the local and distributed environments by preventing Layer 2 from processing unsafe local events (i.e., events that present a simulation time greater than the current DS time).

Differently from the DDESoverHLAEngine, the DDESoverHLA EngineAmbassador implements the interface FederateAmbassador and filters relevant RTI callbacks, which are processed by invoking DDESoverHLAEngine methods, as better detailed in (Gianni et al. 2011).

Next, the FederationManager is an independent federate that regulates the simulation execution and ensures simulation reproducibility and causality in the initial and final phases of the federation life cycle (Khul et al. 1999). In particular, the FederationManager ensures that the simulation starts and ends only when all the federates are ready to start and finish the execution, respectively.

Finally, the HLAEvent is an auxiliary component that encapsulates the conversion between the internal model event format and the corresponding HLA data structure. This component provides methods to convert internal model events into HLA events, and vice versa, and to send HLA events using the RTIAmbassador methods.

Layer 2: Local and distributed simulation engines

SimJ is the acronym that denotes SimArch Layer 2 implementation, which is based on a hierarchy of simulation engines, that is, the component responsible for the coordination of the simulation entities. The engine hierarchy is structured as to offer reuse of the engine code, for both the implementation of different world views (e.g., process interaction or event scheduling) and of the local/distributed versions. The local engine was inspired by common Java-based local simulation frameworks, such as sim-java (Howell et al. 1998) or JavaSim (http://j-sim.cs.uiuc.edu/). Differently, the distributed engine was designed ex-novo. Figure 10.18 shows the engine architecture, including the main internal data structures (local clock, event list, references to the local simulation entities) and the flows of the SimArch service operations from and to the adjacent layers.

Layer 3: Domain-specific languages for simulation

Layer 3 implementation is split into two sublayers: sublayer 3.1, named Basic Sub-Layer 3, and sublayer 3.2, named jEQN. Basic Sub-Layer 3 provides the common and reusable domain-independent code that further

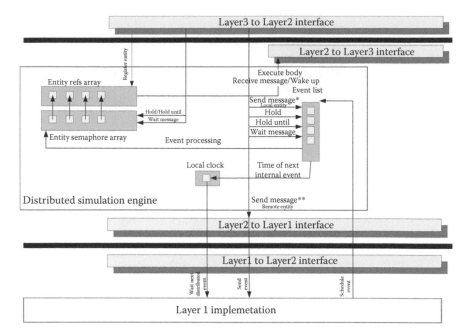

Figure 10.18 SimJ internal details: distributed version.

eases the implementation of DSLs with the introduction of local state variables and semaphores for tracing the entity states and for synchronizing the entity execution, respectively. This Sub-Layer mainly consists of two classes[*]: `BasicLocalEntity` and `BasicRemoteEntity`, whose relationship with the SimArch interfaces is shown in Figure 10.19.

The `BasicLocalEntity` implements the Layer 3 state diagram in Figure 10.14 and provides the framework for the definition and implementation of application-specific state machines within the execution state of the state diagram in Figure 10.14. This class also implements the `Layer3ToLayer2Interface`'s service operations `sendEvent`, `hold`, `wait`, and `waitUntilNextEvent` as own methods, incorporating the lock statement on a class-private semaphore. Conversely, the unlock statement is defined on the same semaphore within the implementation of the `Layer2toLayer3Interface`'s service operation `setReceivedEvent`. Similarly, the `BasicRemoteEntity` implements the common and domain-independent statements for the access of internal attributes, such as `entityName` and `systemName`, for local stubs of remotely running entities.

---

[*] `BasicLink` and `BasicPort` classes are also defined to implement the respective SimArch interfaces. However, these classes do not incorporate any software logic and their implementation is limited to provide the concrete data structure for linking ports.

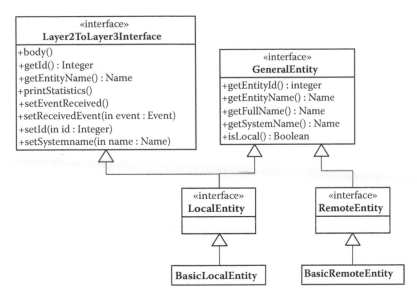

Figure 10.19 Domain-independent sublayer 3.1.

The design of the jEQN language has been preceded by the definition of a conceptual model for the EQN domain (Gianni and D'Ambrogio 2008). The conceptual model aimed to distinguish EQN concepts in EQN entities, which incorporate simulation logic, and in EQN parameter, which support the definition of the EQN entities. Using this model, the following types of simulation entities can be identified:

- User sources, for the generation of users
- Waiting systems, for the storage of users while waiting for a service center to become available
- Service centers, the processing unit for users
- Routers, for directing the users toward other entities
- Synchronization nodes, for representing parallelism (join and fork nodes) and limited capacity resources (e.g., pool of tokens, allocate token, and release token)

Similarly, standard software engineering principles, such as separation of concerns (Mernik et al. 2005), have been applied for the definition of the support components ranging from the decision policies for controlling the simulation logic (e.g., termination policy for the user generation in a source entity) to the sequence of numbers for pseudorandom generators (Gianni et al. 2009). As result, the language has been straightforwardly implemented from the domain model as shown by the following excerpt of the jEQN implementation, for the source entity:

```
#1  while (terminationPolicy.getDecision().booleanValue()) {
#2      try {
#3              nextUserBornTime = timeFactory.
                    makeFrom(interarrivalTime.getNext());
#4              User u = usersGenerator.getNextUser();
#5              u.setBornTime(getClock());
#6              send(outPort, nextUserBornTime, Events.NEW_
                    INCOMING_USER, u);
#7              hold(nextUserBornTime);
#8      } catch (TimeAlreadyPassedException ex) {
#9              ex.printStackTrace();
#10             throw new JEQN TimeException(ex);
#11     } catch (UnlinkedPortException ex) {
#12             ex.printStackTrace();
#13             throw new JEQN ConfigurationError(ex);
#14     }
#15 }
```

In statement #1, terminationPolicy determines the condition for the generation of the users. Statements #3 through #5 are domain-specific and concern the instantiation of the user to be created, including the definition of user name, user born time, and so on. Statements #6 and #7 are central in the domain logic definition: the former dispatch the user to the recipient EQN entity that is statically linked in the model definition at Layer 4; and the latter suspends the entity execution for a simulation time that corresponds to the interarrival time between the generation of two users. The remaining statements can be ignored for the purpose of this section.

References

Banks, J., Carson, J., Nelson, B. L., and Nicol, D., *Discrete-Event System Simulation*, 4th Ed. Prentice Hall, Upper Saddle River, NJ, 2005.

Bocciarelli, P., D'Ambrogio, A., and Gianni, D. 4SEE: A model-driven simulation engineering framework for business process analysis in a SaaS paradigm, In *Proceedings of the Symposium on Theory of Modeling & Simulation-DEVS Integrative M&S Symposium*, p. 31. Society for Computer Simulation International, San Diego, CA, 2013.

Boehm, B. W., *Software Engineering Economics*. Prentice Hall, Upper Saddle River, NJ, 1981.

Bolch, G., Greiner, S., de Meer, H., and Trivedi, K., *Queueing Networks and Markov Chains*. Wiley, Hoboken, NJ, 1998.

Brambilla, M., Cabot, J., and Wimmer, M., *Model-Driven Software Engineering in Practice*. Morgan & Claypool San Rafael, CA, 2012.

Buschmann F., Rohnert H., and Sommerlad P. *Pattern-Oriented Software Architecture: A System of Patterns*, Vol. 1. Wiley, Hoboken, NJ, 1996

Dalle, O., The OSA project: An example of component based software engineering techniques applied to simulation, *The 2007 Summer Computer Simulation Conference* (SCSC'07), pp. 1155–1162, San Diego, CA, 2007.

D'Ambrogio, A. and Gianni, D., Using CORBA to enhance HLA interoperability in distributed and web-based simulation, In *Proceedings of International Symposium on Computer and Information Science 2004* (ISCIS04), pp. 696–705. Springer-Verlag, Antalya, Turkey, 2004.

D'Ambrogio, A., Gianni, D., and Iazeolla, G., Software technologies for the interoperability, reusability and adaptability of distributed simulators, In *Proceedings of the 2007 European Simulation Interoperability Workshop* (EuroSIW-07), Santa Margherita Ligure, Italy, 2007.

Fenton, N. E. and Lawrence, S., *Software Metrics*. Pfleeger, London, UK, 1996.

Fujimoto, R., *Parallel and Distributed Simulation Systems*. Wiley, New York, NY, 2000.

Gianni, D. and D'Ambrogio, A., A domain-specific language for the description of extended queueing networks model, In *Proceedings of the IASTED International Conference on Software Engineering* (SE08), Innsbruck, Austria, 2008.

Gianni, D. and D'Ambrogio, A., A language to enable the distributed simulation of extended queueing networks, *Journal of Computer* 2(4): 76–86, 2007.

Gianni, D., D'Ambrogio, A., and Iazeolla, G., A layered architecture for the model-driven development of distributed simulators, In *Proceedings of the First International Conference on Simulation Tools* (SIMUTOOLS'08), pp. 1–9, Marseille, France, 2008.

Gianni, D., D'Ambrogio, A., and Iazeolla, G., Ontology-based specification of simulation sequences, *International Journal of Simulation, Systems, Science and Technologies, special issue on Internet and Web Technologies* 10(2): 67–78, 2009.

Gianni, D., D'Ambrogio, A., and Iazeolla, G., A software architecture to ease the development of distributed simulation systems, *Simulation* 87(9): 819–836, 2011.

Gianni, D., Iazeolla, G., and D'Ambrogio, A., A methodology to predict the performance of distributed simulations. In *2010 IEEE Workshop on Principles of Advanced and Distributed Simulation* (PADS), pp. 1–9, Atlanta, GA, 2010.

Grable, R., Jernigan, J., Pogue, C., and Divis, D., Metrics for small projects: Experiences at SED, *IEEE Software* 16(2): 21–29, 1999.

Howell, F. and McNab, R., simjava: A discrete event simulation package for Java with applications in computer systems modelling, In *Proceedings of the First International Conference on Web-based Modeling and Simulation*. SCS International, San Diego, CA, 1998.

IEEE: Standard for Modeling and Simulation (M&S) High Level Architecture (HLA)—Federate Interface Specification. Technical Report 1516.1-2010, IEEE, 2010a.

IEEE: Standard for Modeling and Simulation (M&S) High Level Architecture (HLA)—Frameworks and Rules. Technical Report 1516-2010, IEEE, 2010b.

IEEE: Standard for Modeling and Simulation (M&S) High Level Architecture (HLA)—Object Model Template (OMT) Specification. Technical Report 1516.2-2010, IEEE, 2010c.

IEEE: Recommended Practice for High Level Architecture (HLA) Federation Development and Execution Process (FEDEP). Technical Report 1516.3-2003, IEEE, 2003.

IEEE: Recommended Practice for High Level Architecture (HLA) Verification, Validation and Accreditation (VV&A) of a Federation—An Overlay to the High Level Architecture Federation Development and Execution Process. Technical Report 1516.4-2007, IEEE, 2009.

IEEE: Standard for Distributed Interactive Simulation—Application Protocol, Technical Report 1278.1-2012, IEEE, 2012.

Khul, F., Weatherly, R., and Dahmann, J., *Creating Computer Simulation Systems: An Introduction to High Level Architecture.* Prentice Hall, Upper Saddle River, NJ, 1999.

Lau, K. K. and Wang, Z, Software component models, *IEEE Transactions on Software Engineering* 33(10), 709–724, 2007.

Mernik, M., Heering, J., and Sloane, A. M., When and how to develop domain-specific languages, *ACM Computing Surveys* 37(4): 316–344, 2005.

Nicol, D. M., Jin, D., and Zheng, Y., S3f: The scalable simulation framework revisited, In *Proceedings of the Winter Simulation Conference*, pp. 3283–3294. Winter Simulation Conference, Phoenix, AZ, 2011.

Sage, A. P. and Rouse, W. B., *Handbook of Systems Engineering and Management.* Wiley, Hoboken, NJ, 2009.

Sommerville, I., *Software Engineering.* Addison Wesley, Boston, MA, 2006.

Tsai, W. T., Li, W., Sarjoughian, H., and Shao, Q., SimSaaS: Simulation software-as-a-service, In *Proceedings of the 44th Annual Simulation Symposium*, pp. 77–86. Society for Computer Simulation International, San Diego, CA, 2011.

Verbraeck, A., Component-based distributed simulation, In *Proceedings of the 18th Workshop on Parallel and Distributed Simulation* (PADS'04), pp. 141–148, Kuftein, Austria, 2004.

Wainer, G., *Discrete Event Simulation: A Practitioner Approach.* CRC Press, Boca Raton, FL, 2009.

Weatherly, R., Seidel, D., and Weissman, J., Aggregate level simulation protocol, n *Proceedings of the 1991 Summer Computer Simulation Conference*, Baltimore, MD, 1991.

Xie, Y, Teo, Y. M., Cai, W., and Turner, S. J., Service provisioning for HLA-based distributed simulation on the grid, In *Proceedings of the 19th Workshop on Principles of Advanced and Distributed Simulation*, pp. 282–291, Monterey, CA, 2005.

chapter eleven

Reuse-centric simulation software architectures

Olivier Dalle

Contents

Introduction

Computer simulation is used for many purposes and is one of the first applications of computer programming. Given this long history and the many intended use of the simulation software written so far, writing about simulation software architectures without an additional strong point of focus would certainly turn into a challenging and endless exercise. In this chapter, our additional point of focus is placed on reuse. In general purpose software engineering (as opposed to simulation software engineering), the motivations for reuse have long been advocated and demonstrated: lower risks of defects, collective support of potentially larger user community, lower development costs, and so on. In simulation software architectures, we can also cite business-specific motivations, such as providing a better reproducibility of simulation experiments, or avoiding a complex validation process. In practice, although it is rarely discussed, reuse is a problem that may be considered in two opposite directions: reusing and being reused. Accordingly, this chapter is divided into two parts, each geared at one of these two directions. This dichotomy also reflects the dual nature of modeling & simulation (M&S), with modeling on one side, which requires domain-specific support and aims at being reused, and simulation on the other side, which requires a number of generic software elements and aims at reusing general purpose solutions. Hence, the first part of this chapter demonstrates how some of the main features required for building the simulation engine can be found in existing general purpose software solutions, while the second part of this chapter demonstrates how specifically tailored techniques and solutions have to be used to solve critical issues related to the modeling aspect. The two parts of this chapter also target different audiences: the first part is primarily intended for first-time experimenters and beginners that look for tips and ideas on how to start a new simulator development project without wasting too much time on reinventing the wheel with graphical user interfaces, data base, and so on; the second part is rather intended for experienced simulation software developers, that look for ideas and techniques to further improve the design of their more mature simulation software.

Reusing

Building new simulation software is a very popular exercise. Indeed, when considering all the applications of simulation, such as gaming, scientific studies, military applications, and many more, a significant number of computer scientists and software engineers have certainly been involved in the development of a simulator at some point in their programming experience. Unfortunately, what might appear as an easy project at the early stage of the development often ends up being endless source difficulties and eventually a time-consuming (and cost-consuming) exercise.

Underestimation is certainly not the only reason that explains why starting such a development is so popular. Even though reinventing the wheel is often considered a waste of time, it actually turns to have some benefits (Oshri et al. 2007). For example, reinventing the wheel is an excellent way of learning how things work, and why some solutions are better than others. Out of the many reasons that might motivate this decision, a common one is that, at some point, existing simulators are not good enough, either because the problem that motivates the simulation is new or different, or because new techniques have emerged from software engineering and M&S that made the old (-fashioned) simulators obsolete.

In this chapter, we discuss ideas for efficiently building yet another good *simulation* software based on reuse. More precisely, the scope of this discussion is limited to intertwined aspects of software engineering and simulation. Indeed, building a good simulation product is also a question of design, ergonomics, and is dependent on the functional coverage of the software: the less a software has functionalities, the less it takes the risk of disappointing its end users (although there is an obvious lower limit to this principle).

This first part is therefore highly focused on the simulation aspect of M&S, that is, the activity of running a program that executes a simulation model onto a computer, while the activities related to the other aspect, modeling, are mainly addressed in the second part of this chapter. More precisely, in this part we mainly consider models as black-box entities that need execution support and produce data, according to a given workflow. This approach might seem arguable due to the fact that modeling and simulation are tightly coupled. Nevertheless, dividing a problem of higher complexity into smaller ones is a simple application of Descartes' approach: even though some tight connections exist between modeling and simulation, our claim is that if any of the elements required for building a simulator, the software in charge of the simulation, can be reused off-the-shelf, then the option of reusing these elements can save tremendous development efforts, and therefore help focus on more M&S domain-specific issues.

The approach discussed hereafter consists in building a new product by reusing already existing products and implemented techniques. Reuse has long been popular in the simulation community (see, e.g., discrete event systems [DEVS], proposed by Zeigler in 1976), and many efforts have been made to build products that allow for reuse. Component-based modeling and simulation (CBMS), such as with DEVS, is generally accepted as a good way to achieve reuse, but other techniques exist, such as agent-based simulation, middleware run time infrastructures such as high-level architecture (HLA) or, more recently, the web-based mash-ups (Harzallah et al. 2008).

In the case of CBMS, the underlying idea is that many products implement CBMS so that the component developed within such a product can be reused by others in the same product, or even in other products. However, even with a nonambiguous formalism, such as the one used in DEVS, reuse of DEVS components is still an open issue in the community, despite the fact that the formalism has existed since the 1970s. Going one step further, Simulation Interoperability Standards Organization (SISO) did put a significant effort in standardizing simulation model components with the Basic Object Model (BOM) standard. BOM is a significant contribution that includes a number of interesting ideas, but it still does not solve fully the issues of reuse, in particular on semantics aspects (Mahmood 2013).

The reuse approach promoted in this part of the chapter is slightly different: it consists in making abstraction of models or simulation domain-specific code to allow the reuse of general purpose products or techniques that are already implemented, widely available, used, and well documented. Reusing a mature, general purpose product has plenty of benefits. First, because it is mature, it certainly went through several iterations of bug fixes and improvements and already has a community of experienced users. Second, because it is designed to be general purpose, it may come with its own self-contained philosophy, which might be different from what would first come in mind when implementing a simulation-specific product, but prove to proficiently serve the product.

Assuming this approach is worth to try, the next question is what good products and implemented techniques are available for reuse and how can we reuse these products to build a (good) simulation product? The answer to the latter question is certainly not unique and depends a lot on the current trends. In the following, a few examples of timely available techniques or products are given.

The wide variety of software products that fall under the appellation of simulation software makes it difficult to accurately define the minimal set of functionalities required in simulation software. However, a reasonable set of features that might be found in modern simulation products could include the following:

1. Front-end user interface: A graphical user interface to support user interactions
2. Project management: A system for building complex projects with dependencies and versions
3. Management of workflows: A machinery to support simulation methodology through workflows
4. Online database: A solution for implementing the persistence of models, experiments, and results and make them available to the community
5. Component-based modeling: A framework for the composition of software entities, and in particular the simulation models

This list could easily be augmented with other important features, such as visualization, or validation and testing, but the purpose of this part is to be demonstrative rather than exhaustive. For each of the aforementioned features, one ready-to-use solution is suggested. Here again, more solutions might exist, but the point of this discussion is to show that at least one such solution exists and provides a potentially good support for building simulation software.

Front-end user interface

Using and extending an already existing integrated development environment (IDE) is certainly not a novel idea. On the contrary, many simulators already reuse or extend existing general purpose IDEs. The Eclipse

IDE (available from http://www.eclipse.org) is particularly popular for this purpose for two reasons: first, it is highly customizable because of its all-as-a-plug-in architecture, and second, it leaves the choice to the end users of configuring their own environment according to their needs. Before pursuing further the discussion about Eclipse's interesting features, it is worth stressing that in the following discussion, we use Eclipse as an example among others, and in many cases (if not all), the features we mention about Eclipse can actually be found in competitor IDEs, such NetBeans, IntelliJ, Emacs, Xcode, Kdevelop, and many others.

The all-as-a-plug-in approach allows Eclipse users to select the set of plug-ins they want to use among a large (and continuously expanding) collection of available plug-ins. More importantly, it allows users to remove the plug-ins and features they do not like. Indeed, in terms of productivity, being able to skip or replace badly designed features can already result in significant improvements.

The ability of the Eclipse platform to deliver domain-specific functionalities can even be further extended to the point where the IDE itself could be entirely dedicated for a particular application. In this case, the platform becomes a so-called *Rich Client*, whose aim is no longer to let people add or remove plug-ins, but to provide a fully integrated solution, prepared and optimized for a particular application, usually by a vendor. The Omnet++ simulator is an example of such an extreme customization, as shown in Figure 11.1.

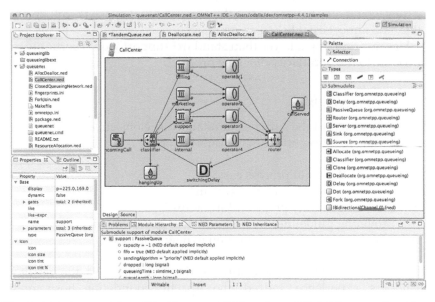

Figure 11.1 A screenshot of the Omnet++ Eclipse-based Rich Client user interface.

Hence, from a simulation product designer point of view, Eclipse offers an interesting range of possibilities: at one extreme, one can start modestly by contributing a single plug-in covering a particular aspect of the domain-specific application, while at the other extreme, one can create a fully dedicated, self-contained application. It is also worth mentioning that Eclipse plug-ins offer extension points, which means that existing general purpose plug-ins can be extended or specialized for more specific purposes. For example, a general purpose Extensible Markup Language (XML) editor could be extended to support an XML-based domain-specific language.

Eclipse also comes with a rich-featured Plug-in Development Environment (Eclipse PDE) and a number of frameworks, such as the Eclipse Modeling Framework, Graphical Editing Framework, and Graphical Modeling Framework, that allow the rapid development of eclipse extensions and plug-ins.

Coming back to our discussion, what are the benefits of reusing an IDE such as Eclipse compared to building a dedicated user interface from scratch? First of all, considering its large number of existing plug-ins, Eclipse provides an extensive support for almost all the existing programming languages. Furthermore, this support can be extended, adjusted, or replaced with a number of alternatives, and new specific support can be built from existing plug-ins, which speeds up the development. Eclipse also supports well graphical modeling and notations and provides a way of building specialized views for particular tasks. Without an IDE such as Eclipse, nonetheless all these features would have to be redeveloped, but they would also have to suffer the comparison: Eclipse is well known, stable, and almost established as standard.

Hence, providing a lower than standard support would certainly not contribute to the popularity of a new product.

Project management

Maven is a project of the Apache Foundation. The objectives of Maven are the following (quoting http://maven.apache.org):

- "Making the build process easy."
- "Providing a uniform build system."
- "Providing quality project information."
- "Providing guidelines for best practices development."
- "Allowing transparent migration to new features."

What makes Maven interesting is the fact that it is the result of developers' experience. Indeed, the Apache Foundation hosts many developments, some of which require a complex building machinery and project management. Hence, realizing the limitations of ant, the traditional building tool for Java-based projects, a new tool was built internally to serve some of the Foundation's projects. Realizing the value of their tools, the developers of Maven decided to release it officially as an Apache Foundation project. At the time Maven was first released, it had already gone through a number of improvement iterations resulting from its actual use in major development projects.

Delving a bit more into the details, the starting point in Maven is an XML-based project description file called a Project Object Model (POM). Such a POM file (named pom.xml) is associated to each subcomponent of the project (Listing 11.1).

```xml
<project xmlns="http://maven.apache.org/POM/4.0.0"
  xmlns:xsi="http://www.w3.org/2001/XMLSchema-instance"
  xsi:schemaLocation="http://maven.apache.org/POM/4.0.0
      http://maven.apache.org/xsd/maven-4.0.0.xsd">
  <modelVersion>4.0.0</modelVersion>
  <groupId>com.mycompany.app</groupId>
  <artifactId>my-app</artifactId>
  <packaging>jar</packaging>
  <version>1.0-SNAPSHOT</version>
  <name>Maven Quick Start Archetype</name>
  <url>http://maven.apache.org</url>
  <dependencies>
    <dependency>
      <groupId>junit</groupId>
      <artifactId>junit</artifactId>
      <version>3.8.1</version>
      <scope>test</scope>
    </dependency>
  </dependencies>
  <build>
    <resources>
      <resource>
        <directory>src/main/resources</directory>
        <filtering>true</filtering>
      </resource>
    </resources>
  </build>
  <properties>
    <my.filter.value>hello</my.filter.value>
  </properties>
</project>
```

Listing 11.1 A sample Maven Project Object Model file (copied from the tutorial at http://maven.apache.org).

This file contains all the necessary information to deal with the component life cycle: how to generate its documentation; what helpers are needed to build its code, run tests, produce reports (e.g., test coverage); describe its dependencies to other subcomponents; define the current version number; deploy the code; deploy the documentation on a dedicated website; save updates on a source code management system (e.g., SVN or CVS); interface with a bug-tracking system, and so on.

Although this latter enumeration is quite disorganized, Maven is all the contrary: it provides a strict file layout for each project (new types of projects being defined by means of *archetypes*) and everything in Maven is designed so as to avoid unexpected situations that could cause a failure to build the project. For example, Maven does not allow a (sub) project to be locally customized such that it becomes permanently linked to a particular execution context (e.g., to a specific host machine or user account).

This comes at a price in terms of flexibility, because users are not given the option to customize an existing project as much as they might want. However, it is worth noting that (1) in real life, the same price has to be paid when working collectively under the supervision of a (human) project manager (i.e., a project manager is expected to enforce common rules in the development team), and (2) Maven fully supports dependencies, which makes it easy to derive a new project from an existing one.

Maven also includes a bit of magic: it downloads and compiles all the required dependencies, including most of itself from online repositories. In practice, the only things required to build a Maven project from scratch are the Maven command mvn, or its integration plug-in for Eclipse, and the source files (including the POM file). Everything else, including compilers and tools, is downloaded and possibly recompiled by Maven on the fly. This makes the first compilation a bit lengthy, because it may trigger the downloading and recompilation of tens of packages, but it leads to another astonishing observation: Maven downloads and recompilation are pretty fast. Furthermore, a local cache significantly improves the performances of subsequent recompilations of a project.

Compared to a proprietary integrated solution developed from scratch, or even compared to a solution based on classic development tools such as ant or make, the benefit of using Maven is to cover the project management at large. Indeed, make, ant, and the likes are only providing support for building a project. Most of them do not even provide the automatic analysis and download of dependencies, although the configure scripts help at least to find out what is missing. Nonetheless, Maven finds what is missing but it finds and installs it properly and quickly. And it provides many more, as mentioned earlier. In the particular context of scientific simulations, the fact that Maven is a network-centric tool is also invaluable, because it provides an easy and well-tested means for putting the simulations online, and help to reproduce scientific results.

Management of workflows

For the management of workflows, the choice can be made between either a workflow management system for business process, such as Bonita, or a scientific workflow management system, such as Taverna (http://www.taverna.org.uk) or the myExperiment-associated web service (http://www.myexperiment.org). Both solutions have their pros and cons: Bonita's business process offers advanced controls over the process, whereas Taverna hides all the workflow logic in services, which are simply connected by means of input and output ports.

Bonita is a project of the ObjectWeb (OW) Consortium. The OW developments are open source projects geared at providing middleware solutions for grid, cloud, and general purpose distributed computing (see http://www.ow2.org). Bonita is a set of three tools available as open source projects, but also released for the market under the name Bonita Open Solution (BOS). The three tools included in BOS are the following: (i) a studio to design workflows using the Business Process Model and Notation (BPMN) formalism, (ii) an engine to run the BPMN workflows, and (iii) a web-based user interface for users to manage and interact with workflows (albeit, this latter interface might not be necessary in a simulation environment if the workflow is supposed to be hidden). With Business Process Execution Language, BPMN is one of the standardized formalisms for the specification of business processes; version 2.0 of the standard was approved by the Object Management Group in January 2011.

Let us have a closer look at BPMN. As shown on Figure 11.2, BPMN is a flow-chart notation that offers four kinds of elements: flow objects (activities, events, gateways), connectors (sequence flow, message flow and associations), artifacts (data objects, text annotations, groups), and swimlanes (a workflow is contained in a pool that can be divided into multiple lanes representing flows progressing in parallel, like swimmers in a pool). Bonita's connectors give the ability to keep the users in the loop and let them interact easily with the workflow when required, or let the workflow proceed automatically when some end-user feedback is not necessary. New connectors can be added to match the particular requirements of the application. For example, such connectors could be linked to electronic mail, social messaging, file storage, web services, enterprise content delivery systems, and so on.

Taverna/myExperiment is a project initiated by multiple research teams and sites in the United Kingdom and the Netherlands. The idea is to build workflows of services that can run either locally on a workstation, or distributed on high performance or cloud computing facilities. The project has collected a significant number of contributed workflows and services, in particular in the bioinformatics community, but not only.

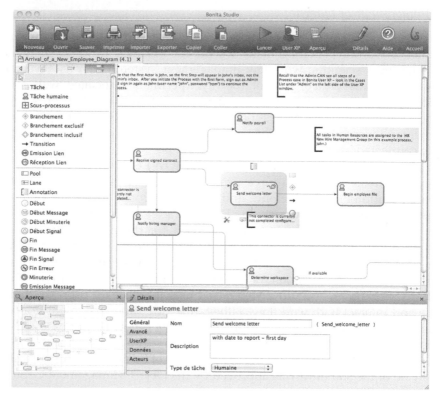

Figure 11.2 A screenshot of the Bonita Studio interface.

Compared to a solution developed from scratch, Bonita and Taverna already provide all the necessary support for expressing and managing workflows. Nonetheless, using such a product saves a significant amount of development time, but it also prevents the development from suffering from a number of failures that would undoubtedly result from the use of a new and not-as-much-tested one.

Indeed, coming back to simulations, it should be kept in mind that a single error in a workflow engine might invalidate thousands of hours of computation. Despite Bonita and Taverna might still suffer from a few bugs, their already long history gives them a clear advantage compared to a newly developed solution, in addition to the development time saved by reusing the product.

Online database

Ruby on Rails (RoR) is a popular framework for developing RESTful databases. As its name suggests, it is based on the Ruby language, which is not so common, but has actually some advantages. Indeed, Ruby (as well as Python and a few others), is a so-called *prototyping-language* because it was specially designed and optimized for the rapid prototyping of applications. With such a language, a prototype of a new application, including its graphical user interface, can be developed in a very short time. Ruby is very well designed, and supports advanced features such as object-oriented programming (OOP), mixins, functional programming, and many more. The Rails framework, that gave its name to RoR, adds to Ruby advanced database support.

Rails implements the well-known model–view–controller design (MVC), which connects a database to a user interface through a set of controllers: the user interface is made of views, which only contain the code for presenting the data to users; the controller contains the logic of the interactions of the users with the model, that provide an object-oriented front-end to query the relational database.

To MVC, which is not a new idea, Rails adds a collection of design ideas and principles that have recently emerged from the experience of developers in software industry. The first of these design principles is *DRY*, which means "Don't Repeat Yourself." Indeed, in Rails everything is done so that you do not have to repeat again and again the same coding idioms. This is partly achieved thanks to a set of generator scripts that automate the generation of recurring code when it cannot be factorized.

Once the code is generated, we enter into the scope of another interesting principle of RoR: promote conventions over configurations. RoR provides reasonable defaults for everything: nonetheless, the code of the application is automatically generated, but it is immediately ready for production without any configuration. Then, of course, the default behavior can be changed wherever it is deemed necessary.

The third principle is to rely on a RESTful web interface. REST is a design pattern for web applications that fully exploits the potential semantics of uniform resource identifiers (URIs) and hypertext transfer protocol (HTTP) verbs (Fielding 2000). In brief, REST relies on a generic abstraction called a resource and provides a simple means to implement actions (including the classical create, read, update, and delete database operations, but not only) on these resources using HTTP URIs and verbs (GET, PUT, POST, DELETE).

For example, the *GET* verb applied to the URI "server.org/customer/2" would trigger a view action on the second record of the table that stores the resource *customer* in the database, while the *POST* action on the same URI could be used to trigger an update action on that same record. Therefore, interacting with a database becomes almost as easy as typing a URI in a web browser (at least for GET actions).

An interesting feature of RoR is certainly its ability to support incremental development. RoR provides a *migration* mechanism that contains the necessary *up* code for migrating the database schema from one version to the next one. This mechanism also contains the *down* code to undo the last modification and rollback to the last version of the database. This incremental approach dramatically changes the development process. Indeed, instead of spending months to achieve a perfect database design before starting to code, RoR lets the developer play with database schema and make changes to the application on the fly, as new bugs or missing parts are discovered. Furthermore, this migration mechanism is also designed to properly update data that already populates the database (in case it is already in production): first, the migration takes the form of a Ruby program, automatically generated, but that can easily be changed to implement advanced migration strategies; and second, it provides a number of default strategies for common patterns, such as adding, renaming, or removing a column to a table. For example, when removing a column to a table, the option is offered to either keep the data in place (the column becomes invisible in the model) or to definitely remove the data.

RoR also includes many invaluable features. For example, it comes with three databases: one is for development, another is for the code already in production, and the last one is a testing database. Altogether, these three databases allow the developer to put the application into production early in the development process, while developments are still ongoing, because new untested features are safely kept separated from the production version, in the development database.

RoR can be used as is, but it is still a bit tedious, and a number of extensions (plug-ins) and additions are available. For example, Hobo is a framework built on top of Rails that adds even more DRY concepts: in Rails, the views of the MVC pattern are implemented using erb files, which are HTML page snippets with embedded Ruby code, that is, Ruby code that is dynamically interpreted. In Hobo, the erb files are replaced with *DRYML* files, an XML-based language that is specially optimized to enforce the DRY and convention over configuration principles. In practice, the results are astonishing, and a few lines of DRYML code are usually enough to make all the necessary changes required by an application.

Compared to a proprietary database solution developed from scratch, RoR and Hobo provide a means for the rapid creation of a RESTful database. Furthermore, this creation process can be made incremental, while a traditional database development usually requires to explicit the full database schema at the beginning of the development, to explicit the relations that need to backed-up in the supporting code.

Last but not least, a traditional development of a database that has a web interface may require the tedious writing of numerous HTML forms. With Hobo, this step is much easier, if not anecdotic, although there is no magic, because this comes at the cost of a long and steep initial learning curve. Given the abundant literature on Ruby and RoR, this learning phase is not unreasonable, and certainly a great source of inspiration for later developments.

Component-based modeling

The Fractal Component Model (FCM) (Brunneton et al. 2006) is another project of the OW consortium. FCM provides support for building component-based applications. It also provides means for applying the Separation of Concerns (SoC) software engineering principle.

First, it provides an Architecture Description Language (ADL) and advanced mechanisms for building component architectures, such as factories or template components. Thanks to the ADL, the concern of building the topological description of the hierarchy of components is separated from other concerns. Factories are special components that can dynamically instantiate new components. Therefore, the way components are instantiated may be implemented in a self-contained component, which is a means of separating the instantiation concern from others. The default FCM ADL parser is a hierarchical factory component. Template components are special factory components that may be used to build a generic model of hierarchical components. Such template models may then be used to instantiate homomorphic copies of the model.

Second, FCM offers good support for nonfunctional concerns. This framework consists of embedding each component into a software membrane: the content part of the component implements its functional concerns, and the membrane part implements its nonfunctional concerns. The membrane consists of several controllers, each of which is responsible for a nonfunctional concern (Figure 11.3). The framework allows for the construction of new membranes by assembling new or existing controllers. The selection of which membrane to associate with which content may be specified using the ADL.

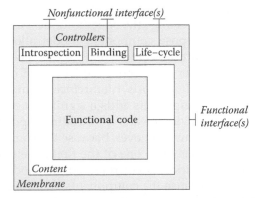

Figure 11.3 Anatomy of a Fractal component. In this example, the membrane contains three controllers that offer the introspection, binding, and life-cycle nonfunctional services.

Interestingly, FCM provides shared components. In a hierarchical component model, a shared component is a component that has more than one parent in the component hierarchy. To the author's knowledge, very few component models do effectively support the shared component feature: the FCM does explicitly support sharing whereas some others, such as JainSLEE (Lim and Ferry 2002) provide proxying techniques, which is a practical way of implementing sharing.

Unlike most other component models, FCM is not linked to a particular programming language. Indeed, FCM is a specification that may be implemented in multiple languages. Several implementations are already available or are under development, in different languages, such as Java, C++, C, or SmallTalk. No actual middleware implementation exists for the particular purpose of coupling Fractal components developed in these various languages; however, non-FCM-specific solution could be envisioned, such as through the use of web-services connectors between components.

Several solutions are available for the distributed execution FCM applications. For example, the FCM site provides a Java library called FractalRMI that transparently implements proxy/stub coupling between distributed components. Another library, called Fractal-BF (Binding Factory), allows the definition of advanced coupling between components, including distributed coupling. Because FCM is a specification, some integrated solutions such as the ProActive platform (distributed by ActiveEon) provides support for distributed high-performance computing on the grid; ProActive implements the FCM specifications while offering advanced distributed services such as FireWall bypassing strategies or dynamic load balancing of workloads between computing nodes.

Building a proprietary component specification for a particular purpose requires a careful design, because once the design starts to be used, it becomes very difficult to change the specifications. Furthermore, using a well-known general purpose component model allows to reuse the wide range of tools and libraries contributed for this model, as is the case for FCM. Compared to other component models, FCM is one of the rare existing specifications that supports hierarchical components. Indeed, supporting hierarchical components adds a significant complexity to the model, because it must provide means for building, updating, and navigating through the hierarchy. Moreover, because the component model is the base for modeling, it has to be one of the first developments: as long as the component specification is not released, no model can be developed. On the contrary, because the components are designed to separate the business logic of the models from the technical logic of simulation, the development of models can start almost as soon as the component specification is ready.

Being reused

This second part elaborates on two designs for building reusable simulation software:

1. A traditional component-based approach, with two popular and complementary variants, namely the plug-in-based approach and the component-object (CO) approach.
2. A layered component-based approach, as found in the Open Simulation Architecture (OSA), that allows for a maximum reuse by enforcing a strict SoC. Using this approach, the whole simulation is built as a composition of many layers that can be transparently replaced and reused depending on the needs.

Component-based architectures

Plug-in-based approach

Development of software is in general considered to be an expensive and error-prone task. Consequently, already early in the development of software, developers started to create libraries that provided reusable methods to reduce the amount of code to be written and thereby reducing the number of potential bugs. But libraries turned out not to be *the one solution*. A classical problem related to libraries is that of different library versions and applications running on the same machine requiring different versions thereof. Further on, classical libraries contain methods (e.g., methods to compute the square root of a number) but no classes to create objects with an internal state and so on. This limitation was one of the ideas behind component libraries, which provide no longer single methods but more complex objects fulfilling well-specified tasks (e.g., buttons for a graphical user interface, access to databases, report generation). Applications can be created faster based on such libraries, but they are still limited to the components (and the versions of these) available during compilation time.

Modern applications to be executed efficiently on current hardware (making use of many of the features provided by the actual hardware it is executed on) need to be more flexible than a precompiled monolithic software can be. This is a well-known fact and reflected in many important software packages, for example, operating systems. More recently, this idea is adapted by more and more software packages—containing an adaptable middleware. Such software packages can adapt themselves to the hardware they are used on. One of the mechanisms to provide the required degrees of freedom is that of plug-ins (also known as hot spots, etc.). Basically, a plug-in-based architecture resembles the idea of well-defined interfaces for which alternative implementations can be provided. These alternatives are described according to a predefined pattern and a central management element of the plug-in system can be queried for fitting instances. The advantages of plug-in-based architectures are manifold:

- Users can integrate the functionality they need.
- The software can select (automatically) from the available plug-ins to adapt itself to the problem and the hardware.
- Single plug-ins can be updated without the need to rerelease the overall software.
- Plug-ins can be provided by third parties.

Plug-in-based architectures require a well-defined environment that can, for example, be provided by the Open Services Gateway initiative (OSGi) plug-in architecture. This architecture is being used, for example, in Eclipse, and has proven to be sufficiently mature for use in real-life applications.

In our opinion, such an architecture is a *good* one for M&S software. M&S software needs to adhere to a number of principles, needs to provide many techniques (for which alternatives might exist), might need to deal with quite differing complexity challenges (model/experiments with the model), might need to run on developers machines as well as on more powerful systems of computers, and although most parts of any M&S software are intrinsically identical, modelers usually need a special way to model (a modeling formalism or language). These aspects are a good argument for a plug-in-based architecture as such an architecture provides the flexibility required to create a highly reusable M&S software that can cope with these.

JAMES II is the first M&S software that has been created based on a plug-in-based architecture (also known as plug'n simulate) (Himmelspach and Uhrmacher 2007). This architecture is composed of a pure plug-in-based architecture, a front-end–back-end architecture, and a service-based architecture. It has been created as a framework to provide the maximum reuse possible (see Himmelspach [2012]). In expression, JAMES II can be used to create specific M&S software, thereby reducing the overall coding efforts significantly and by instantly allowing a more complex analysis of the resulting software than it would most likely not be (that easily) possible without JAMES II in the background (see, e.g., Zinn et al. [2013]).

The core of JAMES II provides a plug-in registry, a number of predefined plug-in types, and the experimentation layer (Ewald et al. 2010; Himmelspach et al. 2008). The experimentation layer controls the execution of experiments with models (comprising parameter generation, replications, etc.). All the functionality used during the course of an experiment computation is thereby provided by plug-ins.

JAMES II is being under development for a decade and there are already more than 600 plug-ins available providing support for different modeling means (e.g., cellular automata, DEVS, PDEVS, species reactions, ml-rules, and stochastic pi), experimentation control means (e.g., optimization algorithms and trajectory analyses), data storing (e.g., csv files, databases, and streams), and many more.

Component objects and inversion of control

Plug-ins offer a coarse-grain composition model in which components can be used and composed for building many different simulations, but within one such simulation, they are used (instantiated) once.

On the contrary, the CO model allows a component to be instantiated multiple times, which leads to a finer grain of composition. These two forms of components are not exclusive: COs are typically used at the modeling level while plug-in components are used at the simulation level.

The idea of a CO model is to combine the best of both OOP and component-based software engineering (CBSE).

Compared to objects in OOP, COs have stronger isolation properties: Following strictly the black-box model, from outside, the content of the CO is hidden such that other COs cannot directly interact with its internal state, nor make any assumption on its internal structure. Following the CBSE principles, COs can only interact with each other through designated points, which may be called interfaces or ports, depending on the model. Furthermore, those interfaces or ports usually rely on a binding mechanism that connects COs to each other, either statically, or dynamically, as new instances are created.

Another popular idiom of CBSE is Inversion of Control (IoC), which prohibits one component to know a priori which other components it is connected to: on the contrary, with IoC, the overall topology of the application is not known from each individual component, but rather described by a global specification, typically expressed using an ADL. Therefore, the dependent components of a given component are discovered during execution. This way, because no a priori hidden dependency exists between components, components can be easily replaced, and therefore reused, provided they match.

As mentioned earlier, COs are well suited for implementing the model part of a simulation. Let us recall that the model is the part of the simulation that describes the structure and behavior of the system or object of the study. Often, this system is complex, mixes many aspects, and may involve many copies of a given element, for example, cells in a biological system simulation. In most extreme cases, a system may be so complex that it has to be developed concurrently by different teams, each having their own expertise area. In this case, the internals of the components developed concurrently in each team are not supposed to be known or even understood by the other teams, which makes the need for IoC even more stringent.

Examples of such COs models used in simulation include DEVS (Zeigler et al. 2000), BOMs (SISO 2006), and OSA (Dalle 2007).

DEVS is a formal specification for discrete event systems with many implementations, in various languages, including Java, C++, and Python. DEVS is a general purpose formalism in which each component implements a timed state machine. In addition, DEVS supports hierarchical constructions, for which closure under coupling has been proved (i.e., a DEVS component assembly can be seen—and replaced—by a single DEVS). BOMs is a SISO standard that defines a component as a unit that contains both an implementation and a description of its semantics, which is expected to make the reuse less prone to errors, such as reusing a component in a wrong context. OSA is built atop the FCM, which is introduced in the Section "Component-based modeling:" in OSA, the COs are FCM components equipped with added functions for simulation, such as event scheduling; in OSA, as will be further described in the Section "Multilayered component architecture," components are used not only for modeling, but also for many other activities related to the simulation methodology, such as observation, instrumentation, distribution, or scenario description.

Multilayered component architecture

Computer simulations are intrinsically multilayered, although the software tools do rarely provide explicit support for multiple layers. Furthermore, as explained hereafter, this layering may be found in multiple dimensions.

Dimensions

In a typical simulation, as shown on Figure 11.4, layers can be found in two orthogonal dimensions: in the simulation functional dimension (methodology) and in the modeling dimension.

In the functional dimension, a simulation can be seen as the superposition of multiple layers, such as the following (example given for a typical in silico experiment built for computer simulation):

- A model layer that describes the system under testing (SUT).
- An experimental frame (EF) layer that describes how an experiment is built around the SUT, what stimuli are sent to the model, and when and what to observe in response to those stimuli. The EF layer is made itself of multiple sublayers:
  - A scenario model that augments the SUT model with new elements used only for building the history of an experiment.
  - An observation layer that describes which elements of the model(s) should be observed, how to produce data samples (statistics computations are made in a separate layer).

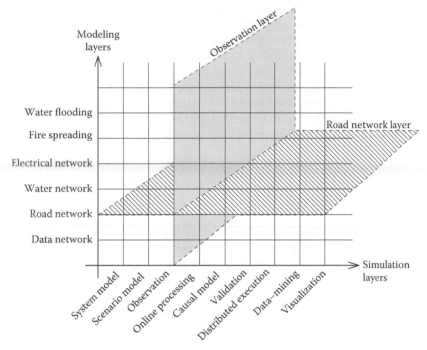

Figure 11.4 An example of multidimensional and multilayered simulation architecture.

- An online data-processing layer that handles the previous samples produced by the observation and decides of which processing to apply (if any): save to traces, compute statistics, and so on.
- A causal model, that links the scenario inputs to the simulation outputs. For example, such a model is needed if we want to compute metrics like the average time needed by the SUT to process an input stimulus coming from the scenario. In this case, the input stimulus must be time-stamped by the causal model before it is applied to the SUT and at the other end, the result produced by the SUT must deliver an observable response containing this initial time stamp.
- A simulation control layer (start, stop, pause, save). This layer might be in charge of deciding for consistent states in which the simulation can be stopped and restarted, possibly after being saved in a persistent storage facility.
- A distribution layer (in case of a distributed execution on multiple computing nodes). In a data-parallel approach, this layer is typically in charge of producing a spatial decomposition of the model that could efficiently be distributed on multiple hosts.

- An off-line data-processing (or data-mining) layer that processes data postmortem, after one or several simulations have finished.
- A visualization layer that includes production of graphics and/or might include the ability to build animations, replay sequences of interest, and so on.
- A reference layer that provides means for identifying the elements of other layers in a nonambiguous manner such that elements used in different contexts can be referred to (e.g., allow various teams working in different lab to refer to a particular configuration, model, or result).
- A validation layer that helps to validate a model, or assess the correctness of use (e.g., detect when model is subject to values out of its validated domain.

In the modeling dimension, a complex model can be seen as the superposition of multiple models. For example, in a complex city model, one might be interested in modeling the interactions between the following model layers for studying, for example, emergency situations:

- The road network, to identify congestion points
- The electricity network, because it controls operation of multiple other systems, including the roads (traffic lights), street lights, cell phone, and so on
- Telecommunication networks
- Water network(s), especially important in case of flooding situations
- Fire spreading model
- Data network(s)

These layers might in turn be decomposed in multiple sublayers. For example, a road network might be decomposed in a highway network, urban network, and/or secondary roads network.

Technical issues

The two main issues in layered simulations are (1) how to connect layers together, and (2) how to merge all the layers into a single executable entity.

In many simulators, many concerns are mixed altogether in the same source-code units. For example, it is common to find statistics calculation mixed with the code instructions that implement a model. This mixing of concerns makes the code hard to reuse and maintain: the initial statistics computation might not be relevant in other studies, and multiple versions often end up living in the same code base, which may have a negative impact on many aspects, for example, breaking a previous validation of the model.

The software engineering term that describes this mixing problem is SoC (Aksit 1996). Various solutions have been proposed in the software engineering community to solve this SoC problem. The solution we present is the one used in the OSA project (Dalle 2007). It is based on two elements: the FCM, already introduced earlier, and aspect-oriented programming (AOP) (Filman et al. 2004; Kiczales 1996).

In FCM, components are hierarchical (a component may include other components) and their interaction is very similar to object interaction in OOP. The Java implementation of FCM, in particular, makes extensive use of the Java Interfaces and object-oriented features (in short, an FCM Java component is a *heavy* Java object).

AOP provides a powerful means for establishing strictly controlled additional connections between objects/components.

In OSA, each layer is made of Fractal components. Layers are then glued together using AOP and the FractalADL. FractalADL is a well-integrated contributed part of the FCM implementation in Java language. In the remainder of this chapter, we illustrate using two examples, first presented in Ribault and Dalle (2008) and in Dalle and Mrabet (2007), how these two techniques, AOP and FCM/ADL, can help to separate concerns in multiple layers.

Example of layer separation using AOP

In this example, we will use AOP for separating the instrumentation and modeling concerns in distinct layers. A very common motivation for reproducing the behavior of a system in a computer simulation is to be able to virtually observe (collect data about) the behavior of that system. The Instrumentation Framework (IF) refers to the software pieces and programming instructions required in such computer simulations to implement observations.

Computer simulation programs usually mix a generic (reusable) technical part, often referred to as the simulation kernel, and more specific parts that depend on the system under study, usually simply referred to as the model implementations or modeling code.

In many simulators, the IF is deeply interleaved in both parts. When no or little instrumentation support is provided by the simulator environment (within the kernel, or thanks to dedicated libraries), the coding part of the instrumentation is left up to the model developer.

Let us first consider a sketching example without using AOP, shown in Listing 11.2: the proportion of instructions that appear in bold font in the listing represent the instrumentation part, whereas the lines in normal font represent the modeling part of the example. Clearly, in this example, the instrumentation represents a significant part of the code. Although no effort was done here to minimize the set of instructions used for instrumentation, it clearly demonstrates that without special programming techniques such as AOP, it is nearly impossible to avoid mixing of

```
class MobileObject {
  private float X;
  private float Y;
  private SimTime last_time;
  private Sampler PosSampler =
      new Sampler("POS");

  public void newpos(float dX, float dY) {
    int sample = 0;
    X = X + dX * (Now() - last_time);
    Y = Y + dY * (Now() - last_time);
    sample = sample +1;
    PosSampler.write("X"+sample+"="+X);
    PosSampler.write("Y"+sample+"="+Y);
  }
  ...
```

Listing 11.2 An instrumentation example, without using aspect-oriented programming.

modeling and instrumentation instructions, because the instrumentation directly refers to the modeling instructions, for example, the successive values of variables X and Y.

Indeed, instrumentation instructions typically start, at the lowest level of the instrumentation, by collecting raw data from the model. Because the model developer implements the model, he/she is in the best position for implementing this collection. For this purpose, and depending on the services and libraries provided by the simulator being used, he/she may use existing libraries or services for generating traces, or implement his own trace generation or data collection mechanism. For example, a common practice is to offer support in the simulation application programming interface for the declaration of observable (state) variables within the modeling code.

In any case, these common practices imply that whatever the goals of the study are, all the possible observations for a given model need to be decided (and hard-coded) at the time the model is implemented. As a consequence, two strategies may be envisaged: a lazy one in which the model developer only provides a minimalistic set of observable variables, or an exhaustive one in which the model developer tries to identify the exhaustive set of potentially useful observable objects.

In terms of reuse and performance, both approaches have opposite benefits and drawbacks: assuming that the execution overhead of the observation is related to the cardinality of the set of observable objects, then we may expect a better performance from the lazy approach than from the exhaustive one, but when trying to reuse the modeling code for another study, the risk to not find the needed objects in the set of observable objects is higher with the lazy approach, which means that there is a high probability that the code will need to be modified. Modifying the code of an existing model simply for instrumentation purposes is a questionable practice. For example, it raises the problem of supporting potentially conflicting instrumentations released by uncoordinated parties or simply the problem of properly switching from one instrumentation to another in a simulation study where the same model is reused many times in various contexts.

The OSA Instrumentation Framework (OIF) is a separate layer of the OSA architecture. The OIF is connected to the existing component hierarchy with a two-level structure: a lower sampling level and a higher sample-processing level, as shown on Figure 11.5. The sampling level is in charge of collecting data samples in the model components during the simulation runs. Data samples correspond to successive values taken by the observed objects during a simulation run. Each sample is time-stamped with the current simulated time or a sequence number.

At the lowest level, the data samples are automatically generated every time an observed object state changes, thanks to AOP.

The samples are directed to a dedicated controller, called the probe, which is automatically added by the FCM in component that owns at least one observed object. Thus, modeling components that are not instrumented does embed any instruction related to instrumentation, which avoids unnecessary overhead.

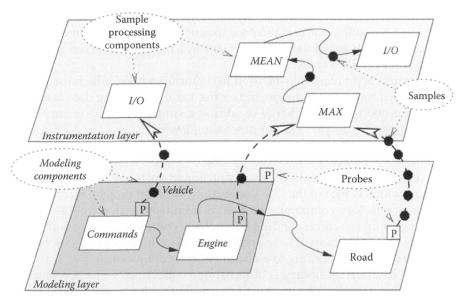

Figure 11.5 Layers of the open simulation architecture instrumentation framework.

The probes in turn forward their samples to the upper layer. The components of the upper layer(s) are in charge of the online statistical processing and implements input/output policies of samples. For this purpose, we integrated in OSA the COSMOS Fractal component framework (Conan et al. 2007).

The specification of which variables to observe is taken from an instrumentation descriptor file. This file is an XML file that obeys the OSA ADL specification, which is an extension of the Fractal ADL.

Without entering the details (see Miles [2004] for more details), the most important additional constructions offered by an AOP language such as AspectJ are the following:

- Aspect: This construction resembles a Java class definition, with attributes and methods declaration, but with additional constructions, such as the Inter-Type Declaration (ITD) and Advice briefly described hereafter.
- ITD: This construction is used to extend an existing type declaration (a Java class or Interface) with new elements (new attributes or methods).
- Advice: An advice is a piece of code that executes every time a particular instruction of the initial program is reached. This particular instruction is specified using an AspectJ pattern-matching construction called a pointcut. An advice may be executed prior, in place of, or after the instruction identified by the pointcut.

In OSA, AspectJ advices are used to trap the successive values of the observed variables. For this purpose, pointcuts need to be defined such that, for each observed variable, the cases in which the observed variable value is set can be trapped.

Then, this pointcut may be used to define an advice, which forwards the trapped value to the probe, indicating the origin of the data with its id. The probe is then in charge of adding a simulation time stamp, and forwarding the sample to upper processing layers.

Layer separation using Fractal ADL

The Fractal ADL (FractalADL) is a contributed software library, written in Java, which is part of the OW Consortium's Fractal project. FractalADL provides a factory component that reads architecture descriptions from files and build the corresponding hierarchical component-based software architecture in memory. These architecture descriptions are provided as XML definitions, according to a document type definition.

The FractalADL library is built using a collection of Fractal components. Interestingly, the component assembly that forms the FractalADL factory component is built recursively: it reads its own architecture description (i.e., the architecture of the hierarchical components used to implement the FractalADL factory) using a hard-coded bootstrap component architecture. Thanks to this flexible, reflexive architecture; the FractalADL components can be extended at will, which in turn allows to extend the ADL itself, and therefore the language definitions it is able to recognize. This flexibility might seem excessive, but it is consistent with the Fractal philosophy described earlier, in which the nonfunctional services provided by the component can be customized and extended at will. This noticeable feature is used extensively in OSA, in particular to allow the scheduling of exogenous events directly within a (model) architecture definition, or to specify the points in the modeling code where to collect data samples for the instrumentation framework.

Although almost all the content of the factory could be reengineered, and therefore almost all its functional specifications could be changed, a typical FractalADL factory supports the following constructs:

*   Definition of a component, which is a container for more definitions, specifying its name and source (either binary code or another ADL definition file)
*   Specification of component interfaces (services offered and used)
*   List of components bindings (how services offered by some components are connected to services used by others)
*   Component execution location (on which host to deploy the component instance for execution)
*   Component content (list of subcomponents in case of a hierarchical component)
*   Component special features (e.g., the capability of scheduling simulation events as in OSA)

Listing 11.3 illustrates the previous basic constructs through a simple client–server example, as shown on Figure 11.6: at the top level, the application is composed of two components, a client, named *Client* and a server named *Server*. Because the semantics of each XML tag word is self-explanatory, it is not to be further explained.

The client and server components each define an interface of type *Service*, which is named *cli* with role client on the client side and named *srv* on with role server on the server side. These two interfaces are then bound to each other using a binding instruction.

Let us now illustrate how the FractalADL features can be used to create a new layer for building a scenario on top of this model layer. Our scenario consists in adding a new component in between the client and the server, to simulate a man-in-the-middle (MITM) attack on the server, as shown on Figure 11.7. The key idea of this new layer is to fully implement

```xml
<?xml version="1.0" encoding="ISO-8859-1" ?>
<!DOCTYPE definition="skipped... " >
<definition name="ClientServeurApp">
  <component name="Client">
    <interface name="cli" role="client"
          signature="Service"/>
    <content class="ClientImpl"/>
  </component>
  <component name="Server">
    <interface name="srv" role="server"
          signature="Service"/>
    <content class="ServerImpl"/>
  </component>
  <binding client="Client.cli"
        server="Server.srv"/>
</definition>
```

Listing 11.3 FractalADL definition of a client-server model.

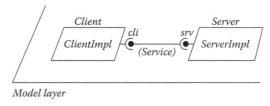

Model layer

Figure 11.6 A simple client–server model architecture.

the scenario incrementally, without making any change to the original model specification.

This modification done using FractalADL is shown on Listing 11.4: first, notice that the definition of the surrounding container, named *ClientServerAppMitm*, is declared as an extension of the previous definition given in Listing 11.3. This extension is twofold: first, a new *MITM* component is added, with two interfaces of (the same) type *Service*, named *mcli* and *msrv* with role client and server, respectively; Second, two bindings are added such that the MITM component is connected on one side to the client, and on the other side to the server. Because these two binding are using the same interface as in the original model, they brake and replace the original bindings, as shown on the Figure 11.7.

Such a layering approach comes with noticeable benefits in terms of reuse: because layers are well separated, they can be reused themselves,

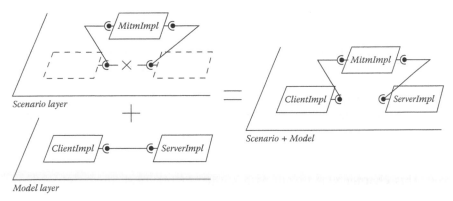

Figure 11.7 A scenario layer is added to the model, which results in a new architecture with an added man-in-the-middle component.

```xml
<?xml version="1.0" encoding="ISO-8859-1" ?>
<!DOCTYPE definition="skipped... " >
<definition name="ClientServeurAppMitm" extends="ClientServeurApp">
  <component name="MITM">
    <interface name="mcli" role="client"
            signature="Service"/>
    <interface name="msrv" role="server"
            signature="Service"/>
    <content class="MitmImpl"/>
  </component>
  <binding client="Client.cli" server="MITM.msrv"/>
  <binding client="MITM.mcli"  server="Server.srv"/>
</definition>
```

Listing 11.4 FractalADL definition to build the scenario architecture incrementally using the man-in-the-middle component and its new bindings.

which adds a coarser level of reuse to the finer level of the component. For example, following on the simple client—server use case, one could reuse the MITM scenario combined with another model, for example, two clients and one server, as shown on Figure 11.8. In this particular case, however, a slight change is required in the definition of the scenario layer, to reflect the fact that the extension mechanism is to be applied to another model definition.

Conclusion

This chapter discussed reuse-centric approaches for building M&S software. We advocated that reuse works in two directions: reusing software from others and contributing reusable software for others. Even though the previous statement seems to follow an obvious logic, it is actually almost never presented following this dual approach in the literature. Indeed, the direction to choose is dependent on the maturity of the project and the experience of the developer, and depending on the targeted audience, one direction or the other is more relevant. In this chapter, we intended to address both audiences: we covered the first reuse direction, reusing from others, in the first part of this chapter, targeting more specifically new comers to the field of M&S, whereas targeting developers having more experience in M&S was covered in the second part of the chapter.

This chapter was meant as a source of practical ideas rather than an exhaustive scientific review of the literature on the topic of reuse, abundant both in the general field of software engineering and in M&S. Nevertheless, following the excellent suggestions of early readers of this chapter, we are happy to provide a few references hereafter for further reading, especially in the more specialized field of M&S: Dahmann 1999; Harkrider and Lunceford 1999; Pratt et al. 1999; Petty and Weisel 2003; and Tolk et al. 2013.

We also deliberately left apart the reuse approach based on HLA (IEEE 1516 standard), which would require its own specific chapter or even more. Further reading related to this matter can still be found in this book in Chapter 10 by Daniele Gianni: "A layered architectural approach for distributed simulation systems: The SimArch case."

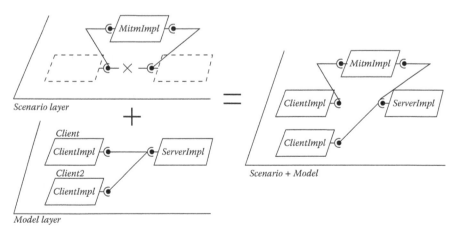

Figure 11.8 An example showing the reuse of a scenario layer with another model layer.

References

Aksit, M. (1996). "Separation and composition of concerns in the object-oriented model." *ACM Computing Surveys*, 28A (4), p. 148.

Bruneton, E., Coupaye, T., Leclercq, M., Quéma, V., and Stefani, J. B. (2006). "The fractal component model and its support in java." *Software – Practice and Experience*, 36(11,12), 1257–1284.

Conan, D., Rouvoy, R., and Seinturier, L. (2007). "Scalable processing of context information with COSMOS." In *Proceedings of 7th IFIP International Conference on Distributed Applications and Interoperable Systems*, June 2007, Paphos, Cyprus, pp. 210–224.

Dahmann, JS. (1999). "High level architecture interoperability challenges." *Presentation at the NATO Modeling & Simulation Conference*, October 25–29, 1999, Norfolk, VA. NATO RTA Publications.

Dalle, O. (2007). "The OSA Project: An example of component based software engineering techniques applied to simulation." In *Proceedings of the Summer Computer Simulation Conference (SCSC'07)*, July 2007, San Diego, CA, pp. 1155–1162. Invited paper.

Dalle, O. and Mrabet, C. (2007). "An Instrumentation Framework for component-based simulations based on the Separation of Concerns paradigm." In *Proceedings of 6th EUROSIM Congress (EUROSIM'2007)*, September 9–13, 2007, Ljubljana, Slovenia.

Ewald, R., Himmelspach, J., Jeschke, M., Leye, S., and Uhrmacher, A. M. (2010). Flexible experimentation in the modeling and simulation framework JAMES II—implications for computational systems biology. *Briefings in Bioinformatics*, 11(3), 290–300.

Fielding, R. T. (2000). Architectural Styles and the Design of Network-based Software Architectures. PhD thesis, University of California, Irvine.

Filman, R., Elrad, T., Clarke, S., and Aksit, M. (2004). *Aspect-Oriented Software Development*. Addison-Wesley, Boston, MA.

Harkrider, S. M. and Lunceford, W. H. (1999). "Modeling and simulation composability." In *Proceedings of the Interservice/Industry Training, Simulation and Education Conference*, November, 29–December 2, 1999, Orlando, FL.

Harzallah, Y., Michel, V., Liu, Q., and Wainer, G. A. (2008). "Distributed simulation and web map mash-up for forest fire spread." In *Proceedings of the 2008 IEEE Congress on Services - Part I*, pp. 176–183, July 2008.

Himmelspach, J., Ewald, R., and Uhrmacher, A. M. (2008). "A flexible and scalable experimentation layer." In *Proceedings of the 40th Conference on Winter Simulation*, Austin, TX, pp. 827–835. Winter Simulation Conference.

Himmelspach, J. and Uhrmacher, A. M. (2007). "Plug'n simulate." In *Simulation Symposium*, Norfolk, VA, 2007. ANSS'07. 40th Annual, pp. 137–143. IEEE.

Kiczales, G. (1996). "Aspect-oriented programming," *ACM Computing Surveys*, 28 (4es), 154.

Lim, S. B. and Ferry, D. (2002). *Jain SLEE 1.0 Specification*. Sun Microsystems Inc. & Open Cloud Ltd, Santa Clara, CA.

Mahmood, I. (2013). "A Verification Framework for Component Based Modeling and Simulation." PhD thesis, KTH Royal Institute of Technology, Sweden.

Miles, R. (2004). *AspectJ Cookbook*. O'Reilly Associates, Sebastopol, CA.

Oshri, I, Newell, S., and Pan, L. S. (2007). "Implementing component reuse strategy in complex products environments." *Communications of the ACM*, 50: 63–67.

Petty M. D. and Weisel E. W. (2003). "A composability lexicon." In *Proceedings of the Spring Simulation Interoperability Workshop*, March 30–April 4, 2003, Orlando, FL, pp. 181–187.

Pratt, D. R., Ragusa, L. C., and Von Der Lippe, S. (1999). "Composability as an architecture driver." In *Proceedings of the Interservice/Industry Training, Simulation and Education Conference*, November 29–December 2, 1999, Orlando, FL.

Ribault, J. and Dalle, O. (2008). "Enabling advanced simulation scenarios with new software engineering techniques." In *20th European Modeling and Simulation Symposium* (EMSS 2008).

SISO. (2006). *Base Object Model (BOM) Template Specification*. SISO-STD-003-2006.

Tolk, A., Diallo, S. D., Padilla J. J., and Herencia-Zapana H. (2013). "Reference modelling in support of M&S—foundations and applications." *Journal of Simulation* (2013) 7, 69–82.

Zeigler, B. P., Praehofer, H., and Kim, T. G. (2000). *Theory of Modeling and Simulation: Integrating Discrete Event and Continuous Complex Dynamic Systems*. Academic Press, San Diego, CA.

Zinn, S., Himmelspach, J., Uhrmacher, A. M., and Gampe, J. (2013). "Building Mic-Core, a specialized M&S software to simulate multi-state demographic micro models, based on JAMES II, a general M&S framework." *Journal of Artificial Societies and Social Simulation*, 16(3), 5.

chapter twelve

Conceptual models become alive with Vivid OPM

How can animated visualization render abstract ideas concrete?

Dov Dori, Sergey Bolshchikov, and Niva Wengrowicz

Contents

Introduction

Conceptual modeling is increasingly recognized as a vital stage in the process of developing any complex system. It allows for expressing the meaning of terms and concepts used by domain experts to discuss the problem and to find correct relationships between different concepts (Fowler 1997). Conceptual modeling explains not only the structure of the system under study but also, not less importantly, the dynamic aspect of the system—its behavior. Conceptual models are inherently abstract, requiring higher-order thinking capabilities to overcome the layers of abstraction represented in a conceptual model by the entities with their symbols and relations among them. Such abstract thinking is indispensable for understanding conceptual models, but it is removed from facts of the *here and now* and from concrete examples of the concepts being considered.

Marton (1981) made a fundamental distinction between a first-order perspective, which pertains to describing various aspects of the world, and a second-order perspective, aimed at describing various aspects of people's experiences of various aspects of the world. Analogously, at a different level—the conceptual model level—we distinguish between the conceptual model itself and the way people understand and interpret it. While abstract thinkers have the cognitive tools to process well-prepared models with relative ease, concrete thinkers require a certain amount of scaffolding to facilitate the understanding of conceptual models.

Abstract thinkers can reflect on ideas and relationships separate from the objects that share those relationships. For example, a concrete thinker can think about this particular object, whereas an abstract thinker can think about the class of those objects. This difference in abstraction capabilities poses a challenge for the majority of people, who are not used to conceptual modeling and whose abstract thinking is not geared for examining and comprehending such models. This is the main motivation behind the development of the conceptual model visualization and concretization mechanism, Vivid object-process methodology (OPM), which we developed and assessed, as described in this chapter.

Background

Object-process methodology

OPM (Dori 2002) is a compact holistic conceptual modeling approach to the representation and development of complex systems, which is simple and intuitive while also being formal. OPM, the emerging ISO 19450 standard, is a formal yet intuitive paradigm for systems architecting, engineering, development, life cycle support, communication, and evolution. The application of OPM ranges from simple assemblies of elemental components to complex, multidisciplinary, dynamic systems. OPM was originally aimed for use by information and systems engineers for knowledge management and representation of multidisciplinary man-made sociotechnical industrial and information systems. OPM notation supports conceptual modeling of systems. Its holistic approach includes refinement mechanisms using a single kind of diagram and equivalent text to describe the functional, structural, and behavioral aspects of a system.

OPM provides two semantically equivalent modalities of representation for the same model: graphical and textual. A set of interrelated object-process diagrams (OPDs) constitutes the graphical model, and a set of automatically generated sentences in a subset of the English language constitutes the textual model expressed in the object-process language (OPL). In a graphical-visual model, each OPD consists of OPM elements, depicted as graphic symbols. The OPD syntax specifies the consistent and correct ways to manage the arrangement of those graphical elements. Using OPL, OPM generates the corresponding textual model for each OPD in a manner that retains the constraints of the graphical model. As OPL's syntax and semantics are a subset of natural English, domain experts can easily understand the textual model.

OPM has formal syntax and semantics, and this formality serves as the basis for model-based systems engineering. An OPM model consists of a set of hierarchically structured OPDs, whose arrangement mitigates a systems' complicated expression. Each OPD is obtained by refining (in-zooming or unfolding) of a thing (object or process) in its ancestor OPD, thus providing a more elaborated context for that thing. Any diagram can include new things and display things from other diagrams, where some or all of the details, which are unimportant in the context of the specific diagram, are not presented. It is sufficient for some detail to appear once in some OPD for it to be true for the system in general even though it does not appear in any other OPD.

The domain-independent nature of OPM enables the entire scientific and industrial community to model, develop, investigate, and analyze systems in their domains, providing for people with different skills and competencies to employ a common intuitive yet formal framework. The OPM approach provides a framework for system lifecycle management. In particular, it facilitates a common view of the system under construction, test, integration, and daily maintenance, facilitating a multidisciplinary environment. Using OPM, companies can improve their overall, big-picture view of the system's functionality, flexibility in assignment of personnel to tasks, and managing exceptions and error recovery.

OPM is founded on two elementary building blocks. These are (1) stateful objects—things that exist, possibly at one of their states, which represent the system's structure and (2) processes—things that happen to objects and transform them, representing the system's behavior. Processes transform objects by (1) creating them, (2) consuming them, or (3) changing their states. The two semantically equivalent modalities, graphical and textual, specify each OPM model. The graphical modality is a set of one or more self-similar and interrelated OPDs. The textual modality of the model is a corresponding set of sentences in OPL—a subset of natural English. This is a verbal mirror image of the graphical model that can facilitate its comprehension by nonexpert viewers, as all it requires is understanding simple English. In an appropriate software environment, such as the Object-Process CASE Tool (OPCAT) (Dori et al. 2010), the graphical model is automatically translated into a textual model while it is being developed and edited by the modeler.

OPM things: Objects and processes

Elements, the basic building blocks of any system modeled in OPM, are of two kinds: things and links. The modeling elements object (possibly in a specific state) and process are OPM things. The modeling element link designates an association between two things.

An OPM object is a thing that exists or can exist once constructed, physically or informatically. Associations among objects are structural relations. They constitute the object structure of the system being modeled, that is, the static, structural aspect of the system. Objects may exist temporally. Once created, they exist in their entirety, until their existence terminates by consumption (disappearance, destruction, or elimination).

An OPM process is a pattern of object transformation—a thing that expresses the behavioral, dynamic system aspect: how processes transform objects in the system and how the system functions to provide value. Processes transform objects by creating them, consuming them, or changing their state. Thus, in an OPM system model, processes complement objects by providing the dynamic, procedural aspect of the system.

A nontrivial process comprises a hierarchical network of subprocesses. Every level of the process hierarchy induces a partial order on the processes, that is, some processes are sequential, such that one process must end before one or more other processes start, while other processes may occur in parallel or as alternatives. A function is a process that provides value, that is, benefit at cost. At any level in the process hierarchy, a process in a system should provide value as part of its ancestor process.

Emergence, created by combining objects and processes, gives rise to a function that exceeds, and is not a simple sum of, the functions of its parts. The structure-behavior combination of the system, which attains its function, is the system architecture. The underlying idea behind the embodiment of the system architecture is the system concept.

We know of the existence of an object when we can name it and refer to its unconditional, relatively stable existence. However, the object's transformation, that is, its creation, change over time, or consumption, cannot occur unless a process acted on that object.

An object state is a particular situation at which an object can be at some point during its lifetime. At any given point in time during the life of an object, the object is in one of its states or in transition between two of its states.

Unification of function, structure, and behavior

The OPM structure model of a system is an assembly of the physical and informatical (logical) objects connected by long-lasting associations among them, called structural relations. During the lifetime of a system, creation and destruction of aggregation-participation relations—a kind of structural relations—may occur.

The OPM behavior model of a system, referred to as its dynamics, reflects the mechanisms that act on the system over time to transform systemic (internal to the system) and/or environmental (external) objects. At any point in time, the state of a system is the aggregation of the states of its constituent objects and processes. The combination of system structure and behavior—the system's architecture—enables the system to perform a function, which is the value the system delivers to at least one stakeholder, who is the system's beneficiary.

OPM integrates the functional (utilitarian), structural (static), and behavioral (dynamic) aspects of a system into a single, unified model, which is nonetheless expressed bimodally in both graphics and text. In each OPD of the OPM model, structure and behavior can coexist without highlighting one at the expense of suppressing the other. Maintaining focus from the viewpoint of overall system function, this structure-behavior unification provides a coherent single frame of reference for understanding the system of interest, enhancing its intuitive comprehension while adhering to formal syntax and deep semantics.

Function-as-a-seed OPM principle

The top-level value-providing process of a modeled system expresses the function of the system as perceived by the system's main beneficiary or beneficiary group. Modeling with OPM begins by defining, naming, and depicting the function of the system as its top-level process. The structure and behavior of the system emerges from this function. Identifying and labeling this top-level process, the system's function, is a critical first step in the methodology part of OPM, which prescribes how an OPM model should be constructed. An appropriate function name clarifies and emphasizes the central goal of the modeled system and the value that the system is expected to provide for its main beneficiary. Such a deliberation, which often provokes a debate between the system architecture team members at this early stage, is extremely useful, as it exposes differences and often even misconceptions among the participants regarding the system that they set out to architect, model, and design.

By using a single holistic and hierarchical model for representing structure and behavior, clutter and incompatibilities can be significantly reduced even in highly complex systems, thereby enhancing their comprehensibility. OPM has proven to be better in visual specification and comprehension quality for representing complex reactive systems compared to object model template, a unified modeling language (UML) predecessor (Peleg and Dori 2000). OPM is supported by OPCAT (Dori et al. 2003; Dori et al., 2010), a software environment that is used in this work to model the transcription case study presented in the Section "Experimental design: The Vivid OPM evaluation system." OPM operational semantics are defined by a translation into a state transition system (Perelman et al. 2011), and a related OPCAT simulation environment was developed (Yaroker et al. 2013). OPM is in final stages of ratification as ISO Standard 19450. OPM main elements with their semantics and examples from the biology domain are presented in Table 12.1.

OPM operational semantics

The OPCAT simulation environment supports concurrent, synchronous, and discrete time execution. The execution we used in this work is qualitative in nature with one instance defined for each object (e.g., molecule) and each process. This enables analyzing the behavior and qualitative underlying mechanisms of the system under study. Although multiple instances can be defined in OPCAT simulation and quantitative aspects can be inspected, these are out of the scope of this work.

Processes are executed in a synchronous manner, one after the other, according to a defined timeline. The default timeline, within the context (in-zoomed frame) of each process, is from top to bottom. Alternative scenarios or loops, which override the default timeline, can be modeled using an invocation link. Concurrency is supported, and processes whose ellipse topmost points are located at the same height in the diagram are executed concurrently.

Table 12.1 shows the OPM entities with their symbols, definitions and operational semantics. An example from the biology domain is added, based on the work by Somekh et al. (2012).

Table 12.2 shows OPM procedural links: links connecting an object or its state with a process, including their symbols, definitions, and operational semantics. Table 12.3 shows OPM structural links: links connecting an object with an object or a process with a process, including their symbols, definitions, and operational semantics.

Each process has a (possibly complex) precondition and a postcondition. A process is triggered (attempted to be activated according to its place in the timeline), and its precondition is then checked. Only if the precondition is satisfied, the process is executed. On normal process termination, the postcondition must hold. The precondition of a process is

Table 12.1 OPM Entities with Their Symbols, Definitions, and Operational Semantics

Entity name	Entity symbol	Definition and operational semantics	Biological example
Object: Systemic	Object	An object that consists of a matter or a piece of information	mRNA — mRNA is a molecule
Object: Environmental	Object (dashed)	An object that is external to the system, randomly generated	Temperature (dashed) — Signals and environmental effects on system
Process	Process	A pattern of transformation that objects undergo	Transcription — Transcription is a biological process
State: Initial/Regular/Final	Object [state 1] [state 2] [state 3]	An **initial** (state 1)/**regular** (state 2)/**final** (state 3) situation at which an object can exist for a period	Complex — mRNA [nonexistent] [nascent] [capped] — mRNA has three states: nonexistent (initial state), nascent, and capped

Table 12.2 OPM Procedural Links: Links Connecting an Object or State with a Process

Link name	Link symbol	Definition and operational semantics	Biological example
Instrument/ Condition link represents precondition	(A) Process —○ Object (B) Process —○ Object state 1 (C) Process —○ Object state 1	A link denoting a condition required for a process execution. The condition can be the existence of an object (A) or the existence of an object in some state (B). The condition is checked when the process is triggered.(C) When a "c" inside the circle is added, then if the condition does not hold, the process is skipped and the next processes (if any) try to execute. (A), (B) if the condition does not hold, the systems halt (a mode used for checking model consistency)	RNA Polymerase II —○ Transcription The RNA polymerase II being existent is a precondition for transcription process to be executed
Consumption link represents precondition and postcondition	(A) Process →← Object (B) Process →← **Object** state 1	A link denoting that a process consumes an object (A) or an object at some state (B). The object (A) or object's state (B) existence is a precondition for process execution. Their nonexistence is the process postcondition	Nucleotide Set → Transcription The nucleotide set being existent is a precondition for transcription process to be executed. After execution nucleotide set is consumed (nonexistent)

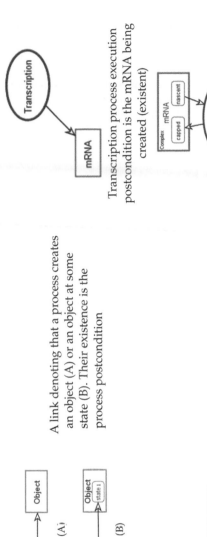

Creation link represents postcondition

(A)

A link denoting that a process creates an object (A) or an object at some state (B). Their existence is the process postcondition

(B)

Transcription process execution postcondition is the mRNA being created (existent)

Changing object state links (input and output links) represents precondition and postcondition

Links denoting that a process changes an object from state 1 to state 2. The input link consumes state 1 and the output link creates state 2

The mRNA nascent state is a precondition for the nascent RNA capping process execution. The change to the capped state is a postcondition for the process execution

Table 12.3 OPM Structural Links: Links Connecting an Object with an Object

Characterization		A fundamental structural relation representing that an element exhibits an attribute object
Participation (consists of)		A fundamental structural relation representing that an object (whole) consists of one or more objects (part[s])
General structural link		A unidirectional association between objects that holds for a period, possibly with a tag denoting the association semantics

expressed by its preprocess object set—the set of objects, which must exist, some possibly in specific states, for the process to start. The postcondition is defined similarly by the postprocess object set.

Logical expressions (AND, OR, XOR) between objects in the pre- and postprocess object set can be defined. By default, the logical relation between the objects in the pre- or postprocess object set is a logical AND, meaning that all the objects in the preprocess object set must exist in their defined states for the process precondition to be true. It is possible to change this default definition by using the XOR and OR relation between various objects. Process execution can also depend on random signals. To model this, we connect the process to an environmental object without or with a specific state. This environmental object is added to the preprocess object set, defining the process precondition. OPM semantics also include event links, for modeling reactive systems, and time exception links.

Unified modeling language

The UML (Fowler 2004; Object Management Group 2014) is a standardized visual specification language for object modeling. UML is software oriented and supports the object-oriented paradigm. The language comprises a graphical notation with strong aspect-separated views used to create an abstract model of a system. Even the behavioral aspect in UML is spread over several different types of diagrams: sequence, collaboration, activity, and statecharts, based on Harel's work (1987). This separation makes it difficult to comprehend the system's overall picture and to keep the different views consistent. Both the numerous types of diagrams (UML 2 has 13) and the software orientation make the notation hardly appropriate as a basis for the task at hand of translating a conceptual model to an animated clip.

xUML (Mellor and Balcer 2002) is an executable modeling language that profiles UML with a semantics extension called action specification language (ASL). The extension aims to capture the language determinism and ensure its executability. Several vendors support their own ASL, but no standard semantics has been accepted.

Petri nets

Petri nets (also known as a place/transition nets or P/T net) is one of several mathematical representations of discrete distributed systems. As a modeling language, it graphically depicts the structure of a distributed system as a directed bipartite graph with annotations. As such, a Petri net has place nodes, transition nodes, and directed arcs connecting places with transitions. Petri nets do not describe explicitly the data flow or objects in the model and this characteristic makes it inappropriate for the general task of conceptual modeling.

IBM Rational Rhapsody

IBM Rational Rhapsody lets systems engineers capture and analyze requirements quickly and then design and validate system behaviors. A Rational Rhapsody systems' engineering project includes the UML and SysML and allows the following (IBM Corporation 2009):

- Performing system analysis to define and validate system requirements
- Designing and specifying the system architecture
- Systems analysis and design
- Software analysis and design
- Software implementation
- Validation and simulation of the model to perform detailed system testing

Rational Rhapsody enables the visualization of the conceptual model via simulation. Simulation is the execution of behaviors and associated definitions in the model. To simulate a model, it requires statecharts, activity diagrams, and textual behavior specifications, which capture the behavior of the model. Structural definitions such as blocks, ports, parts, and links are used to create a simulation hierarchy of subsystems. The preparation for the simulation is relatively complicated process containing the following steps:

- Creating a component
- Creating a configuration for the component
- Generating the component's code
- Building the component application
- Simulating the component application

MagicDraw

MagicDraw is the business process, architecture, software, and system modeling tool with teamwork support, based on UML 2. It is designed for business analysts, software analysts, programmers, quality assurance engineers, and documentation writers, and provides analysis and design of object-oriented systems and databases (No Magic 2013b).

MagicDraw has Cameo Simulation Toolkit, a separate add-on providing simulation and animation capabilities. It extends MagicDraw to validate system behavior by executing, animating, and debugging UML 2 State machines and activity diagrams in the context of realistic mock-ups of the intended user interface (No Magic 2013a). These industrial solutions have the following disadvantages:

- Backed by UML, they provide for simulation and animation of software and hardware system models, but not of conceptual models of systems in general.
- They require several types of UML diagrams (activity and statecharts as a minimum) to operate, compared with OPM, which has only one type of diagram.
- These solutions require a complex sequence of steps to build the animation.

Vivid OPM

OPM explicitly addresses the system's dynamic-procedural aspect, which describes how the system changes over the time. The single OPM model provides for clear and expressive animated simulation of the OPM model using OPCAT, which greatly facilitates design-level debugging. However, the resulting model is basically static, and as such, it does not fully reflect the behavior of the system being architected or designed. The simulation module provides animation and enables visual verification of the OPM model, but the animation is at the abstract, conceptual level. It shows the model entities—process ellipses, object boxes, and state rountangles—changing colors as processes transform objects while red dots slide along the links. The underlying model is still conceptual; it represents concepts as animated geometric shapes. Humans, however, grasp moving pictures of the real things in the model much more intuitively than looking at their symbols. The more visual and dynamic a model, the more intuitive and deeper is humans' understanding of the system.

Motivation and benefits

To enhance complex system dynamics comprehension, we wish to decrease the abstraction level of models of those systems and bring the dynamic aspect closer to reality by actually playing what the conceptual model expresses formally but not intuitively enough.

Our conjecture when setting up to design and implement Vivid OPM was that creation of a visual dynamic model from a static one, which represents changes of objects in space along time by mimicking the actual system's behavior, is likely to enhance comprehension of the system's behavior without requiring knowledge of any specific modeling language. To realize this visual dynamic model rendering, we have developed Vivid OPM—a software module that takes an OPM conceptual model and plays a video clip of the dynamics of the system under development or research. The produced video clip does not require any knowledge of modeling language or technical background, which might prevent stakeholders from understanding the conceptual model. Figure 12.1a depicts a screenshot of OPCAT simulation in action. Figure 12.1b shows the corresponding Vivid OPM

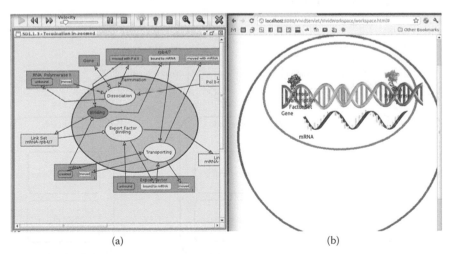

(a)   (b)

Figure 12.1 Vivid OPM animation with OPCAT simulation. Part (a) describes the OPCAT simulation and part (b) describes the Vivid animation.

animation, which is driven by the OPM model in Figure 12.1a. This figure demonstrates the drastic difference in what it takes a biologist to understand the conceptual model and its visualization.

Architecture

Vivid OPM (Bolshchikov 2011; Bolshchikov et al. 2011) is a software program that uses a conceptual OPM model of a system to generate an animation of the systems, bringing the conceptual model closer to reality. As noted, understanding a Vivid OPM animation does not require knowledge of OPM or any other modeling methodology.

Vivid OPM comprises three main parts (see Figure 12.2). On the one side is OPCAT—the OPM-based systems modeling environment with built-in simulation and the Vivid OPM plug-in that controls the simulation. On the other side is the visual web-based client—a GUI platform that visualizes the system's dynamic aspect driven by the OPCAT-resident OPM conceptual model of the system. In the middle is a Tomcat Apache server, which has two purposes: (1) it provides servlets, which are Java classes used to extend capabilities of an Oracle server (Oracle 2013), for the client side to upload files and run the animation; (2) it exchanges messages between the client side and OPCAT regarding the animation status.

Two communication channels connect the system's three parts. The client side is connected with the Apache Tomcat server via asynchronous http requests using a polling strategy for animation. The client side sends periodically an update request to the server. The server side and OPCAT talk over Java Messaging Service, which enables exchanging messages between Java applications.

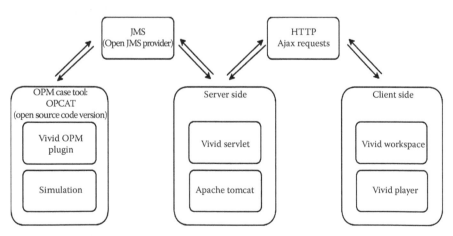

Figure 12.2 Architecture of Vivid OPM.

Vivid OPM modeling workflow

The workflow of Vivid OPM consists of two stages: configuration and animation. Preparing for animation requires configuration to provide missing information, which is not specified in the conceptual model, for example, object images and their initial spatial layout.

The configuration stage starts as the user uploads the OPM model file to the server for which the Vivid OPM component needs to be added. The Vivid OPM script then extracts all the objects in the model from the file and displays them on a canvas. The modeler can then upload a desired icon for each object. By default, all the states of an object have the same icon, but this can be changed so there would be a different icon for each state. Using a dragging mechanism, the user specifies the spatial arrangement for an object's states. As soon as the configuration is ready, it is saved into a separate file on the server. If a model has been previously configured, Vivid OPM will read the file instead of parsing the model again. Thus, the configuration step needed to be done only once per model, and later on the file can be reused.

Once the configuration is completed, Vivid OPM can be launched for animation. The OPM model and the Vivid OPM animation operate simultaneously. Triggering the Java-based simulation process in OPCAT on the server side, Vivid OPM periodically executes update requests arriving from the client side. On receiving an update from the server, a Java plug-in pauses the simulation of OPCAT for Vivid OPM to complete the current animation step. The animation executes by transitioning an object between its states. The JavaScript code executes over time the spatial transition effect: moving of an object, resizing it, changing its color, and so forth. Whenever an object's graphical transition is completed, the client side notifies the server to proceed with the Java-based simulation execution of OPCAT. This synchronization mechanism ensures that the simulation of OPCAT is aligned with the execution of the graphic effects of Vivid OPM.

Evaluation of Vivid OPM

The objective of developing and implementing Vivid OPM was to provide modelers with a tool that can enhance the understandability of the OPM conceptual model, which is often problematic for most people who are not experienced with conceptual modeling in general and OPM-based conceptual modeling in particular.

To assess the value of the animation that Vivid OPM plays from a given OPM model in terms of enhancing model comprehension, we have designed and conducted a controlled experiment, whose goal was to assess the effect of adding Vivid OPM to an OPM conceptual model in terms of its contribution to better understanding of the OPM model.

This section is divided into five subsections. The Section "Experiment population and background" describes the research population. In the Section "Experimental design: The Vivid OPM evaluation system," we describe the experimental design, which includes description of the application that we built particularly for the experiment, as well as description of the two OPM models used in the experiment. In the Section "Experiment hypothesis and method," we state the research hypothesis and methods. The Section "Data analysis and results" contains the results and their interpretation. Finally, the Section "Conclusions and discussion" states the conclusions of the experiment.

Experiment population and background

The experiment was carried out in the fall semester of 2011–2012 at Technion, Israel Institute of Technology. A total of $n = 154$ students from the faculty of Industrial Engineering and Management took part in the research, which was conducted in three different courses. We refer to these groups, respectively, as Course A ($n_1 = 13$), Course B ($n_2 = 132$), and Course C ($n_3 = 9$).

The first group consisted of $n_1 = 13$ students who took course A, titled "mini-project in industrial engineering." These were freshmen with almost no prior knowledge of OPM. Throughout the course they constructed two OPM models of a real-life project (Erkoyuncu et al. 2013). The second, major group, included $n_2 = 132$ students in course B, "analysis and specification of information systems." Their background included basic knowledge of OPM, UML, and SysML. The last group $n_3 = 9$ included third- and fourth-year undergraduate students, as well as several graduate (MSc) students in the Information Management Engineering program, who took course C, "methodologies in information systems development." These students had relatively extensive knowledge of conceptual modeling in OPM, which they had applied during the course in real-life projects, such as the EU FP7 TALOS—border control system project. Each of the three groups received a 45-minute lecture during the semester about Vivid OPM, describing its use, architecture, and capabilities.

The experiment was carried out using the Vivid OPM evaluation system. As described in the Section "Experimental design: The Vivid OPM evaluation system," this is a web-based system, which was especially designed and developed for the sake of this experiment. Upon login to the Vivid OPM evaluation system, each student was automatically and arbitrarily referred to the one out of two possible groups: the experimental group ($n = 76$), who received the OPM model along with the Vivid OPM animation and the control group ($n = 78$), who received the OPM model without the Vivid OPM animation. Table 12.4 presents

Table 12.4 Distribution of Students between the Test Group
and the Control Group in Each Course

Course	Course name	Experimental group (with vivid OPM)	Control group (without vivid OPM)
A	Mini-project in industrial engineering	6	7
B	Analysis and specification of information systems	63	69
C	Methodologies in information systems development	7	2

the distribution of students between the experimental group and the control group in each course.

Experimental design: The Vivid OPM evaluation system

This section describes the Vivid OPM evaluation system (Bolshchikov 2011) that was designed and built especially for the purpose of conducting the experiment. This application was built for several purposes:

- Managing the experiment that was performed concurrently by sub-groups of 21 students
- Dividing participating students equally into the experimental and control groups
- Storing the questionnaire responses
- Providing basic analytics regarding the amount of correct responses

Figure 12.3 depicts the Vivid OPM evaluation system architecture.

As Figure 12.3 shows, the Vivid OPM evaluation system consists of four modules: Registering, Models, Questionnaire, and Analysis. As soon as the application is launched, the Registration module opens. Here, the user fills in his/her ID and name. The Registering module then assigns each student to a group, based on prior information about how many people need to be in each group.

The experiment was conducted on two OPM models from two disparate domains: one from the biological domain—mRNA life cycle and the other from the banking domain—an automated telling machine (ATM). The rationale behind this choice was to test a sample of the diverse domains in which OPM can be applied and to cancel out possible variability in the levels of students' familiarity with each domain. Specifically, an OPM model and its Vivid OPM animation of an ATM might not provide

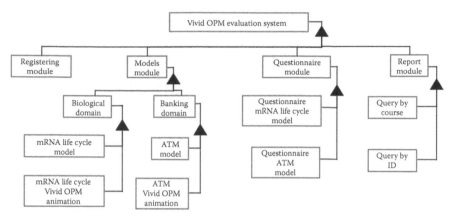

Figure 12.3 Architecture of the Vivid OPM evaluation system.

reliable results as most of the students are familiar with its operation from their daily lives, but they are much less familiar with a molecular biology system.

Figure 12.4 shows the system diagram (SD; top level) and the next level down (SD1) of the mRNA life cycle system as presented to the experimental students within the Vivid OPM evaluation system. Our Vivid OPM evaluation system automatically assigned the mRNA life cycle system as "OPM Model One" and "Vivid OPM One" to half of the experimental students and the ATM model system as "OPM Model Two" and "Vivid OPM Two" to the rest of the experimental students. A similar assignment was made for the control group students. This took care of cancelling out any potential learning that might occur during the experiment and cause bias in the outcomes.

Figure 12.5 shows the menu that a control group student sees. Comparing this to the menu in Figure 12.4, we see that a control group student cannot access Vivid OPM Model One and Vivid OPM Model Two.

As Figure 12.4 shows, each OPM model was presented using its two modalities, namely, each OPD was accompanied beneath it with its corresponding, automatically generated text of OPL paragraph, composed of OPL sentences. The presentation of the textual modality next to each graphical modality model piece facilitates the understanding of the OPM model even before adding the potential enhancement of the Vivid OPM part, increasing the challenge of proving that the addition of the Vivid OPM component is of value in terms of enhancing model understandability.

Each of the models was provided by a prerecorded Vivid OPM animation clip of the two systems used in the experiment. These are represented by the two objects at the bottom of Figure 12.3: mRNA life cycle Vivid OPM animation and ATM Vivid OPM animation. The two Vivid OPM

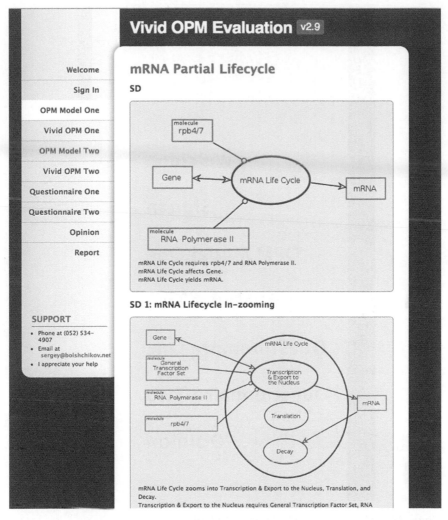

Figure 12.4 mRNA life cycle system as presented to the experimental and control students within the Vivid OPM evaluation system.

animation clips were uploaded to the website of the Vivid OPM evaluation system, such that each experimental group student could watch it as many times as she or he deemed necessary. Figure 12.6 shows a snapshot of the preprepared clip of the ATM system as presented to the experimental students within the Vivid OPM evaluation system.

The questionnaire module contained two blocks: the ATM model questionnaire and the mRNA life cycle model questionnaire, each composed of nine questions. Each question can have more than one correct answer.

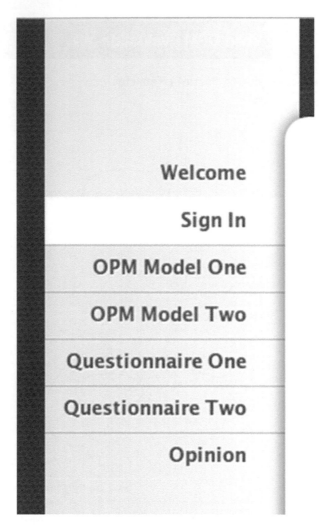

Figure 12.5 Menu that a control group student sees.

Figure 12.7 shows a screenshot of the top part of the ATM system questionnaire, which, in this case, happens to be "Questionnaire Two." As an example, the first question from the ATM system questionnaire is: "Who initially has Cash?" with the following response options:

- ATM
- Bank
- Account
- Customer

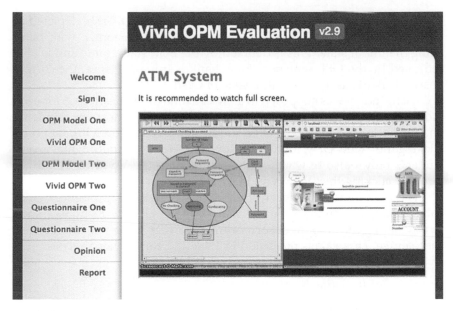

Figure 12.6 A snapshot of the preprepared movie of the ATM system as presented to the experimental students within the Vivid OPM evaluation system.

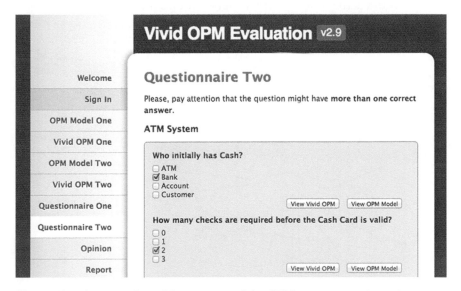

Figure 12.7 A screenshot of the top part of the ATM system questionnaire.

Examining Figure 12.8, one can see that the object Cash, depicted at the bottom of the OPD, has two states: bank and customer, with bank being the initial state. This is denoted by the bold state frame and is also indicated by the OPL sentence "State bank is initial." Hence, the correct answer is Bank, as it is the initial state of Cash.

Each student was supposed to answer a total of 18 questions, 9 for each model. As Figure 12.7 shows, while answering the questions, students had the option to browse the OPM model diagrams without limitation by clicking the "View OPM Model" button, and—for the experimental group students only—also to watch without limitation the Vivid OPM animations, by clicking the "View Vivid OPM" button.

The last module is the Analysis module. It is invisible to the experiment subjects, as it is intended for initial analysis of the subjects' responses, such as calculating the amount of correct and incorrect answers of students in each course. This data is presented in Table 12.5 for further analysis.

Experiment hypothesis and method

Our research hypothesis was that there would be significant differences in students' achievements, as reflected in their responses, between the experimental group, who used Vivid OPM, and the control groups, who did not use Vivid OPM. The differences would be such that the Vivid OPM groups would achieve significantly higher grades than the control groups.

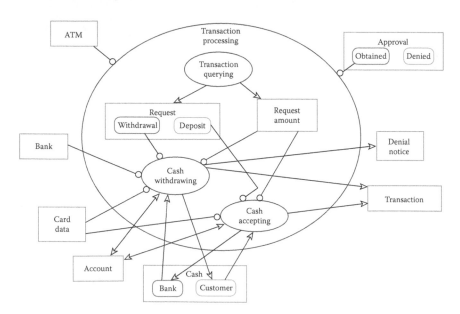

Figure 12.8 Transaction querying of the ATM system in-zoomed.

Table 12.5 Students' Means and Standard Deviations of
Correct Answers for the Four Question Categories

Question category	M	SD
Structure	3.13	1.2
Behavior	1.75	0.82
Change of state	3.98	1.09
Interaction	2.48	1.18
Total	11.34	2.92

For data collection, we used a qualitative method. We conducted an achievement test, in which we examined the students' outcomes in the course. The achievement test contained 18 closed questions that tested their ability to read and understand complex system models. This test was validated by three systems engineering experts.

Each one of the 18 questions belonged to one of the following four categories, listed in increasing order of difficulty level:

1. Structure: This category relates to structural aspects of the system. It included five questions regarding issues such as the state of an object and static relations between objects. An example of a question from this category is: "What is the relation between molecules General Transcription Factor Set and RNA Polymerase II?"
2. Behavior: This category relates to behavioral aspects of the system. It included four questions regarding processes and their impact on objects. An example of a question from this category is: "What process breaks RNA Polymerase II apart from rpb4/7 and Gene?"
3. State change: This category relates to aspects pertaining to state changes. It included five questions about the change of object states. An example of a question from this category is: "Which molecule returns to its initial condition?"
4. Interaction: This category relates to relative complex mutual actions and reactions between objects. It included four questions concerning interactions between objects through processes. An example of a question from this category is "Which molecules move with mRNA?"

Data analysis and results

Exploratory factor analysis with varimax rotation was performed for the 18 test questions. The analysis indicated four distinct content worlds that match the four test question categories: structure, behavior, change of state, and interaction. The total explained variance was 41.67%. We summarized the number of correct answers in the test as a whole and the correct answers in each aspect for each student. Means and standard deviations are presented in Table 12.5.

Our research hypothesis was that there would be significant differences in course grades between students who were given the system and questions relating to it with the OPM model and the additional Vivid OPM and those who were given the system and questions relating to it with the OPM model only and without the Vivid OPM component. We expected the grades of the experimental students, who used Vivid OPM, to be significantly higher.

To test this hypothesis, we first used one-way analysis of variance (ANOVA) to test if there were differences in the number of correct answers between the three courses. The analysis of variance showed that the effect of course was not significant, $F(2151) = 0.652$, $p > .05$, indicating that there were no differences between the three courses. This enabled us to coalesce the experimental and control groups in each course into one large experimental group (with Vivid OPM) and one large control group (without Vivid OPM).

Further, the four question categories, namely, structure, behavior, change of state, and interaction, were subject to a one-way multivariate analysis of variance (MANOVA) with the two research groups, experimental (with Vivid OPM) and control (without Vivid OPM). The students' mean of correct answers in each aspect and their standard deviations by research group are presented in Table 12.6.

The MANOVA result was significant for research group by aspect, $F(4149) = 3.57$, $p < .01$, $\eta_p^2 = 0.09$, indicating that there were significant differences between the experimental and control groups in number of correct answers.

To detect the source of the differences, we conducted a separate ANOVA for each of the four question categories. The results indicated significant differences only in the interaction question category: $F(1152) = 14.11$, $p < .001$, $\eta_p^2 = 0.09$. No significant differences were detected in the

Table 12.6 Students' Means, Standard Deviations, and *F* Results of Correct Answers for Differences between Experimental and Control Groups for the Four Question Categories

Question category	Experimental group (with Vivid OPM)		Control group (without Vivid OPM)		
	M	SD	M	SD	F(1152)
Structure	3.20	1.24	3.06	1.17	0.471
Behavior	1.75	0.85	1.74	0.80	0.002
State change	4.08	1.00	3.88	1.17	1.22
Interaction	2.83	1.09	2.14	1.24	14.11**

**$p \leq 0.001$.

structure question category, behavior question category, and change of state question category.

Conclusions and discussion

Our experimental results indicate that, in line with our research hypothesis, Vivid OPM enhances model understandability. Refining this conclusion by looking into each one of the four question categories reveals that Vivid OPM did not enhance model understandability in three of the four aspects: structure, behavior, and change of state. This is likely so due to the simplicity and user-friendly presentation of OPM even without the additional animation that Vivid OPM provides. Only when really complex questions are posed, such as those belonging to the interaction category, Vivid OPM, with its intuitive interface and animated actions, does provide a significant added value. It is this complex side of the question spectrum where the real benefit of Vivid OPM is exposed: When the question to be answered involves understanding a complex situation, in which several objects interact through processes, Vivid OPM contributes to providing a correct answer, indicating its benefit as a means of disambiguating complicated situations and aiding in making the model more comprehensible. Although OPM with its graphic and textual modalities is sufficient for correctly answering questions of the first three categories—structure, behavior, and state change—when it comes to the really difficult questions, those of the interaction category, Vivid OPM is of real and significant help in providing a correct answer.

Summary and future work

In this chapter, we have developed, presented, tested, and evaluated a visualization, in addition to OPM, called Vivid OPM. Vivid OPM enables animation of a conceptual OPM model, making it closer to reality than the original OPM conceptual model. The OPM model presents both the structure and the behavior of the system under study using a static set of diagrams. Although the diagrams can be animated, the animation is of the geometric shapes representing objects and processes in the diagrams. In this type of diagram animation, symbols of objects (denoted as boxes or rectangles), states (rounded-corner rectangles within the object boxes), and processes (ellipses) change colors as objects become existent and change states, and as processes are being executed. At the same time, red dots run along procedural links connecting objects or states with processes, symbolizing the passage of time.

In contrast, the type of animation that Vivid OPM provides is more concrete. It involves the dynamics of the object icons themselves, not of their symbols. This is a principal change: We animate icons that look like the objects themselves, not their conceptual symbols with the names recorded inside them. This is a big step toward making the abstract conceptual model concrete. Furthermore, the object icons perform movements, rotations, sizing, and color changes, as dictated by the OPM model and the Vivid OPM extension, making the dynamic aspect of the model closed to *the real thing* so it is more concrete and more comprehensible.

We tested OPM and Vivid OPM against OPM only on a research population of 154 information systems engineering students using a specially developed system for this task, which is also described in detail in the Section "Experimental design: The Vivid OPM evaluation system."

We found that there is a significant advantage in the addition of the Vivid OPM component to the OPM model when it comes to being able to respond to questions related to interactions, which are the most complex category of the four question categories we defined. The fact that Vivid OPM is most helpful in answering the most complex questions is a testimony of its value. When the questions are simple enough, there is no significant added value to Vivid OPM, because OPM itself is sufficiently intuitive and self-explanatory. However, as the question complexity increases, the real discriminatory added value of Vivid OPM becomes evident.

Future work in the direction of model-driven animation includes extending the two-dimensional Vivid OPM to three dimension using computer-based augmented virtual environment (CAVE) technology. A vision for this system is presented in the work by Dori (2012). In this vision, an intelligent multimedia device dubbed computer-based augmented virtual environment for realizing nature (CAVERN) is proposed as a quantum leap in molecular biology research. CAVERN is a system that leverages state-of-the-art technologies that include CAVE, supercomputing, electron microscopy, conceptual modeling, and biological text mining. A new notion of the human spatiotemporal comfort zone is presented along with a fourth multimedia learning assumption—the limited spatiotemporal comfort zone. Within this zone, people can use their senses to follow and understand complex systems that are currently accessible only through indirect observations. CAVERN translates nanolevel processes modeled in OPM into scenarios of human-size interacting molecules. A conceptual blueprint of CAVERN expressed via an OPM model.

References

Bolshchikov, S. 2011. Vivid OPM Evaluation. Retrieved from http://vividopmevaluation .appspot.com.

Bolshchikov, S., Renick, A., Mazor, S., Somekh, J., and Dori, D. 2011. OPM model-driven animated simulation with computation interface to Matlab. *20th IEEE International Workshops on Enabling Technologies: Infrastructure for Collaborative Enterprise* (WETICE), pp. 193–198. Paris, France.

Dori, D. 2002. *Object-Process Methodology—A Holistic Systems Paradigm.* Berlin, Heidelberg, Germany: Springer Verlag.

Dori, D. 2012. Extending the human spatiotemporal comfort zone with CAVERN—Computer-based augmented virtual environment for realizing nature. *Journal of Multidisciplinary Research*, 4(3), 23–44.

Dori, D., Linchevski, C., and Manor, R. 2010. OPCAT—A software environment for object-process methodology based conceptual modelling of complex systems. *1st International Conference on Modelling and Management of Engineering Processes*, pp. 147–151. Cambridge, United Kingdom.

Dori, D., Reinhartz-Berger, I., and Sturm, A. 2003. Developing Complex Systems with Object-Process Methodology using OPCAT. *Lecture Notes in Computer Science*, 2813, 570–572.

Erkoyuncu, J., Bolshchikov, S., Steenstra, D., Rajkumar, R., and Dori, D. 2013. Application of OPM for the healthcare sector. *Conference on Systems Engineering Research (CSER'13)*. Atlanta, GA.

Fowler, M. 1997. *Analysis Patterns: Reusable Object Models.* Boston, MA: Addison-Wesley Professional.

Fowler, M. 2004. *UML Distilled: A Brief Guide to the Standard Object Modeling Language.* Boston, MA: Addison-Wesley Professional.

Harel, D. 1987. Statecharts: A visual formalism for complex systems. *Science of Computer Programming*, 8(3), 231–274.

IBM Corporation. 2009. Systems Engineering Tutorial for Rational Rhapsody.

Marton, F. 1981. Phenomenography—Describing conceptions of the world around us. *Instructional Science*, 10, 177–200.

Mellor, S. and Balcer, M. 2002. *Executable UML: A Foundation for Model-Driven Architecture.* Boston, MA: Addison-Wesley.

No Magic, Inc. 2013a. Cameo Simulation Toolkit. Retrieved from http://www.nomagic.com/products/magicdraw-addons/cameo-simulation-toolkit.html.

No Magic, Inc. 2013b. MagicDraw. Retrieved from http://www.nomagic.com/products/magicdraw.html.

Object Management Group. 2014. UML® Resource Page. Retrieved from http://uml.org/

Oracle. 2013. The Java EE 6 Tutorial: Java Servlet Technology. Retrieved from http://docs.oracle.com/javaee/6/tutorial/doc/bnafe.html.

Peleg, M. and Dori, D. 2000. The Model Multiplicity Problem: Experimenting with Real-Time Specification Methods. *IEEE Transaction on Software Engineering*, 26(8), 742–759.

Perelman, V., Somekh, J., and Dori, D. 2011. Model Verification Framework with Application to Molecular Biology. *Symposium on Theory of Modeling and Simulation (DEVS 2011)*, Boston, MA, April 4–9, 2011.

Somekh, J., Choder, M., and Dori, D. 2012. Conceptual Model-Based Systems Biology: Mapping Knowledge and Discovering Gaps in the mRNA Transcription Cycle. *PLoS ONE*, 7(12): e51430. doi:10.1371/journal.pone.0051430, December 20, 2012.

Yaroker, Y., Perelman, V., and Dori, D. 2013. An OPM Conceptual Model-Based Executable Simulation Environment: Implementation and Evaluation. *Systems Engineering*, 16(4), 381–390.

chapter thirteen

Processes to support the quality of M&S artifacts

Jan Himmelspach and Stefan Rybacki

Contents

Introduction

It is not surprising that one of the major concerns in experimental disciplines is the quality of the derived statements that directly depend on the quality of the experiments done. Quality is a frequently used term—it is addressed in almost all areas of life and everyone has at least a feeling of the quality of objects to deal with on a daily basis.

Thereby quality itself is considered to be a multifaceted term (Garvin 1984). Garvin distinguishes between *transcendence, product centeredness, user centeredness, cost/use relation,* and the *process view.* But how do these relate to modeling and simulation artifacts?

The transcendent view assumes high standards for development and that only experts can decide on the quality of a result. An example for such a quality in modeling and simulation (M&S) is the abstraction level of models as it is typically judged by domain experts.

The product-centered view implies that quality can be measured. Balci (2001) describes a method to measure the quality of models. In addition, the quality of software code can be measured using techniques from software engineering (e.g., complexity measures, relation of documentation to code), and as computerized models are code, these measures can be applied as well.

The user centered view refers to a subjective definition of quality. Different users might have a different impression about the quality of a selection of products and make different choices among these. In M&S, this can refer to the frequently issued statement that models seem to be more credible if they ship with a nice animation. This usually has nothing to do with measurable quality but it may help to convince customers that the product delivered has a high quality.

The cost/use relation defines quality as the *best use* you can get for a given amount of money. There might be artifacts with a higher quality, but you will have to pay more for these.

Examples where this definition can be applied in M&S are the abstraction level chosen to create the model and the number of configurations and replications computed.

The process view deals with the idea that product quality depends on the product creation process. This means that a well-defined process, which is well documented when executed, leads to a product of higher quality. This notion of quality has become an industry standard (reflected in the ISO 9001 norm) and many companies producing goods or providing services are being certified based on this norm. Process models to improve the quality of M&S studies are well known as well (see also the Section "Simulation studies") (Sargent 1983; Balci 2003; Brade 2003; Law 2006; Wang and Lehmann 2008).

In M&S, all the aforementioned views on quality can be applied to judge the quality of a product (Himmelspach and Uhrmacher 2009). This applies to the creation process of M&S software as well as to conducting an experimental study using a model including all the intermediate M&S processes (and the products created therein) such as modeling, validation, fitting, simulation, and analysis.

Herein we focus on the notion of the process centered view. By reviewing process models and workflow approaches usable in the field of M&S, we try to answer the question: "How can we create 'M&S artifacts' based on processes and how can the 'production process' as well as the quality of the products be improved by process definitions?"

Based on this we discuss how the demands in M&S can be fulfilled using workflows and how workflows can be used to control M&S software.

Artifacts in M&S

The aforementioned quality dimensions always refer to a product. In the realm of M&S, products (more neutral: an artifact) are elements/results of intermediate steps that are created by a modeler, a simulation performer, or other people involved.

Artifacts in M&S are at least as follows:

1. M&S software used
2. Data/analysis/hypothesis the model is based on
3. Model(s) (conceptual, qualitative, quantitative, implemented)
4. Experiment with a model
5. Analysis plus interpretation of the results of an experiment with a model
6. Overall process of the M&S study

M&S software

The M&S software created is a product that is used to compute results in the end. To produce credible results the software needs to adhere to quality standards.

No matter what the development aim is, may it be the creation of commercial, off-the-shelf environments, highly specialized one-shot solutions, or whatever else: quality in software development is a topic addressed by *process models* (see the Section "Software development") and by concrete techniques to improve the quality (i.e., validation, verification, and code analysis). So far there is only rare evidence that these methods are consequently applied in M&S software development. But there is evidence that the product quality of M&S software could be better (Merali 2010). One of the additional chances to improve the software development process in M&S is to exploit reuse (Himmelspach 2012). If software is consequently created based on the idea of reuse it must adhere to the idea of separation of concerns, which in turn means that the M&S software is being composed from smaller elements (Himmelspach 2009).

Problem definition

The question to be answered about the system under study needs to be well formulated as it impacts the exact hypothesis building/empirical data gathering/analysis steps. Any quality issue here will have an impact on the quality of all artifacts produced in the further course of the study.

Model(s)

Model formulation is typically described as a multistep process on its own. It produces several intermediate artifacts (conceptual model, qualitative model, quantitative model, implemented/computerized model) (see also the Section "Simulation studies").

The quality of each of these intermediate artifacts plays an important role because following artifacts might be based on and verified against them.

Experiment

Well-defined and carefully executed and documented experiments are mandatory to achieve reliable results. This could be done much better as it is currently done in M&S (Pawlikowski et al. 2002). An experiment needs to be designed in a way that the answer to a specific question can be derived. Furthermore, it should produce an answer that is reliable as possible and produced with the least number of obsolete computations as possible. The execution of an experiment needs to be carefully logged, where the log describes the process of the experiment, events, and the results.

Analysis (and interpretation)

After the computations with the model have been executed, an analysis of the results considering the initial question has to be undertaken. This usually implies the application of a number of mathematical (statistical) methods on the data produced and the proper interpretation of the results produced by those methods.

Process of the M&S study

The various aforementioned artifacts can be produced in M&S studies. They are depending on each other and thus have an order (i.e., you cannot do an analysis before you have done the experiment, which cannot be done before the models have been created) and are thus to be executed one after the other. This is a process and as not all artifacts might be produced and as different techniques might be applied to produce the artifacts the process itself should be well documented.

Process definitions

That the process to create an artifact can play an essential role is anything but new as we mentioned previously (Garvin 1984). Consequently process definitions exist for software development as well as for doing simulation studies.

Process models

Software development

Process models are a common means in software engineering to define the process of developing software. Classical models similar to the water-fall model have been around for decades but still new models are being developed, such as agile software development methods. Furthermore, there is the ISO 9000-3, dedicated to software development process quality (Sommerville 2007).

Such process models guide developers through software development and support the identification of work done (deliverables). It is no secret that the software production process cannot be compared to industrial mass production as it usually means to create unique pieces using a lot of creativity, but still process descriptions support the creation processes. Detailed descriptions, for example, of test methods to be applied, can help to increase the *quality* of the software developed as they impose that these have to be executed before the deliverable is ready.

Such software development models should be used for creating M&S software as these help to improve the quality of products (Sommerville 2007).

In the process of performing a M&S study, developing software is often one of the intrinsic steps as well: Once the qualitative and quantitative models have been created they need to be transformed into a form that can be interpreted by computers. This transformation can either be done by using a dedicated language (modeling language, e.g., ML-Rules [Maus et al. 2011]), with a simulation language (which allows defining model and simulation Continuous System Simulation Language [CSSL]) or by using a general purpose programming language. Independent from the choice made, this process is intrinsically a code development process, and for this the aforementioned software development processes can be employed as well.

Simulation studies

Simulation studies comprise a multitude of steps to be executed. Comparable to software development process models, process models for M&S shall guide users through a study and show what has to be done before one can continue with the next step.

Different process models for M&S have been defined (Balci 2003; Brade 2003; Law 2006; Sargent 2008; Sawyer and Brann 2008; Wang and Lehmann 2008).

The classical process models by Balci, Sargent, and Law are sketched in Figures 13.1 through 13.3, respectively. The model from Law lacks some *restart* recommendations, for example, in case that the verification of the computer program created fails. The verification and validation (V&V) triangle for M&S (Figure 13.4) is the V-model applied to M&S (Brade 2003). It shows how subsequent artifacts depend on those being produced in the steps before. The models shown in Figures 13.1, 13.2, and 13.5 emphasize the explorative nature of many simulation studies, those in Figures 13.3 and 13.4 focus on a single pass in a study.

All these models have in common that they emphasize the model building phase incorporating the intermediate products conceptual, qualitative, quantitative, and computerized model. All model creations contain the transformational needs from the diverse intermediate models as well as V&V and the opportunity to get back to earlier phases in case of problems.

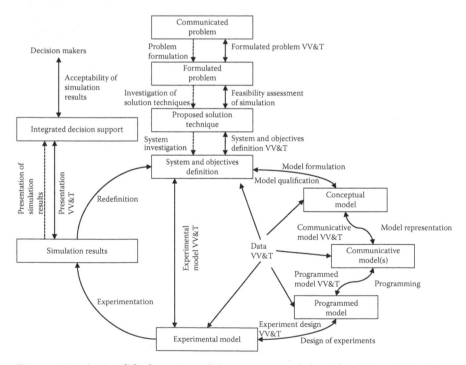

Figure 13.1 A simplified version of the process as defined by Balci (2003). This model follows the pattern that modeling and simulation shall be performed for a given problem and that this process of creating the intermediate products has a strong focus on verification, validation, and testing (VV&T).

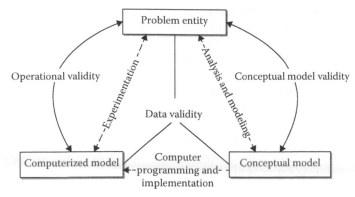

Figure 13.2 The process as defined by Sargent (1983). This lean model describes less intermediate steps and emphasizes the aspect of modeling and simulation as an experimentation process.

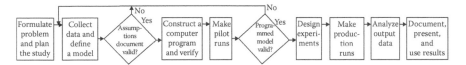

Figure 13.3 The process as defined by Law (2006). This model focuses on the basic elements and only defines jumps to previous stages.

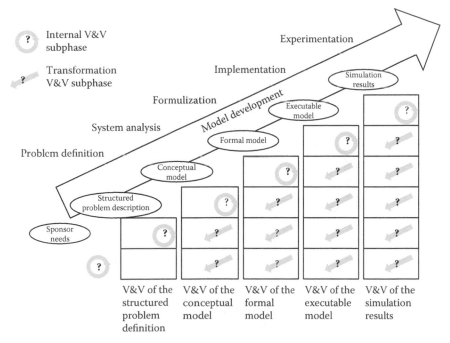

Figure 13.4 The V&V triangle (Brade 2003). It illustrates that subsequent products are always created on the preceding results creating a validity dependency that enforces the requirement to perform V&V on all intermediate results.

In the work by Sawyer and Brann (2008), the authors propose an agile software development method to create models and simulations. Thereby they point out that, especially, the frequent and early feedback rounds pay off in scenarios where customers and model/simulation developers are not the same people.

The process model shown in Figure 13.5 integrates an additional cycle into the M&S process: the hypothesis building based on experimental data. Further on, this model makes the formulation of research questions to be answered with the model/the simulation explicit. Thereby the question "B" refers to the experimental frame introduced for discrete event system models (Zeigler et al. 2000).

The shown process models describe the overall steps to be executed in the course of a simulation but they do not go into the details (subprocesses) of each of the single steps, that is, which methods are applied in the verification, validation, analysis, or experimentation steps. Consequently for a real project these need to be filled up and manually documented afterward.

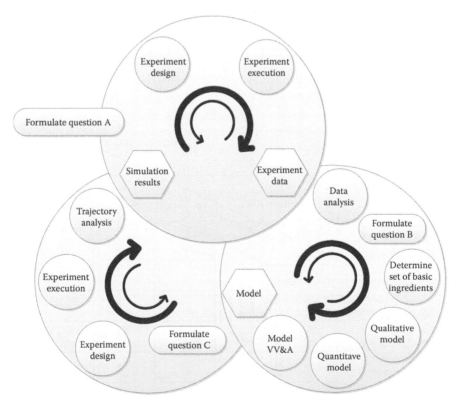

Figure 13.5 M&S study plus data collection phase. The three phases such as *empirical data, modeling,* and *simulation* are arranged in a cycle. It is always possible to step back in intermediate steps.

Scientific workflows

Scientific workflows try to fill the gap between coarse-grained process models and the detailed description and (automatic) execution of experiments. Scientific workflows are based on the idea of workflows.

Workflows

Process descriptions can be formulated as workflow descriptions. A workflow description describes how the diverse steps of a process follow each other and can be used to control a process execution (Figure 13.6). There exist different notations for specifying a workflow. Business process model and notation (BPMN) formalism introduced in 2001 can be used to describe workflows graphically. Since the recent version 2.0, its execution semantics are better defined (White and Miers 2008; Van Gorp and Dijkman 2011). Business Process Execution Language (Juric et al. 2004) and XML Process Definition Language (XPDL) (WfMC 2008) on the other hand are XML-based languages to describe workflows with well-defined execution semantics. In addition, XPDL can also be used to store the graphical representation of the described workflow, for example, it can be used to store BPMN workflow diagrams. For most of these languages conversion methods exist, however they might not work fully correctly as the languages have different expressiveness or execution paradigms (Weidlich et al. 2008). All these are used by diverse software packages and can be used to describe most of the processes appearing in M&S.

So-called workflow management systems are responsible to execute workflows—thus to move on in the processes given as soon as the current task has been manually or automatically solved.

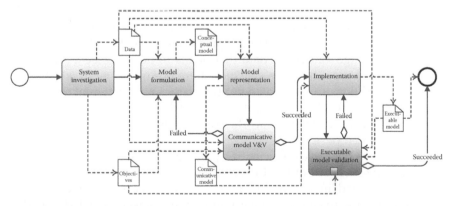

Figure 13.6 Workflow describing the creation of a model. It is described using BPMN and is inspired by the process model proposed by Balci (2003) (see Figure 13.1). (Based on Balci, O. Proceedings of the 2003 Winter Simulation Conference 1: 150–158, 2003.)

There are techniques to map workflows onto Petri nets (Hinz et al. 2005; Wynn et al. 2007; Zha et al. 2007), which make it possible to use analysis methods and software packages developed for Petri nets (Schmidt 2000; Verbeek and van der Aalst 2000).

Workflows in science

In the last decade, workflows have been considered as an interesting contribution to the scientific field and lead to the idea of *scientific workflows*. The question how to define an experiment (and how to document experiments) gained more attention.

In the early days of M&S, experiments had been defined in the used *simulation language*, for example, in CSSL (Nilsen and Karplus 1974). Most *in language* solutions lack transparency and blur the difference between an experiment and a model. In contrast, developments such as Kepler (Altintas et al. 2004), Workflows in Modeling and Simulation (WorMS) (see the Section "WorMS"), and Akaora-2/Safe (Perrone et al. 2012) separate model and the experimentation workflow.

Experiments are not special to M&S. A large community (even beyond computational sciences in general) shows significant interest in well-defined experiments. This leads to web-based platforms such as *myexperiment.org*. They allow anyone to publish their workflows that were used to obtain experimental results described elsewhere (e.g., in a publication).

Further developments, such as Taverna (Hull et al. 2006) and Trident (Barga et al. 2008), show the increased interest in using workflows to describe scientific processes.

Scientific workflows extend/modify the classical workflow descriptions given earlier in this chapter. Although business workflows are driven by the idea of controlling the order of activities within a process, which make them control-flow driven, scientific workflows focus on controlling the data flow throughout a process. Prioritizing the data perspective is more natural to the way scientific processes work. Usually, a scientist deals with a huge amount of data that needs to be processed, analyzed, and visualized according to a specific schema or question. In addition, a scientific workflow is instantiated multiple times with different data and/ or processing artifacts. This results in a number of parallel instances of the same workflow. Business workflows on the other hand are usually only instantiated once.

Secondary objectives of scientific and business workflows and their workflow systems are quite different. Users of business workflows are mainly interested in optimizing parts of the process described by the workflow to, for instance, reduce maintenance costs. Scientists on the other hand are interested in intermediate results and how results evolve over time in a scientific workflow. They like to compare different workflow executions, to determine which intermediate results occurred, what input data lead to a specific result, or what processing methods were used. Consequently, requirements for the workflow systems in regards to provenance and mining of business and scientific workflows differ.

Using workflows in M&S software to support users

As we have learned from the previously mentioned text, process models/ well-defined workflows are a widely accepted means to improve the quality of artifacts. Surprisingly, these models are usually only available on a sheet of paper. The used software does not remind you that you should do a specific step (e.g., validate your model) before you continue.

Other software solutions simply *execute* a predefined experiment step by step. Furthermore, different users might have different needs—for example, a validation phase is mandatory or not and users (or their boss) might want to specify further additional process steps (e.g., accreditation/ certification of models). Which steps are executed, using which methods provided by the software has often to be documented by the user; selections made internally by the software cannot be documented by the user although they might belong to the essential set of elements to be documented (Pawlikowski et al. 2002).

A software with internal mechanisms composed using workflows provides a high degree of freedom to its users (i.e., it does not force the user to do things that are not defined as being mandatory). The workflows can be arbitrarily created (maybe following some general, personal, or institution constraints), before the study or on-the-fly while the study is being executed, documentation about the steps undertaken can be generated automatically (being the workflow as such the very first piece of this documentation), the system can be integrated into larger workflow management systems that allow other users to observe or interact with the process, single users can be informed if others have done their job, and so forth. These aspects make workflows superior to batch scripts that have been/are traditionally used to orchestrate subsequent purely automatically executable steps.

Roles

M&S processes involve different roles (which sometimes may be impersonated by a single person). Some of these (e.g., members of the accreditation institute) should be independent from others. Workflows can guarantee a seamless process if all parties involved base their internal processes on workflows. Furthermore, different roles can get different access rights to perform duties in workflows and thus, the overall process gains a lot more transparency. In the work of Rybacki et al. (2010), we identified the following roles for workflows in M&S studies: modeling expert, simulation expert, methodology researcher, VV&A experts, workflow designer, project manager, client, and system administrator. Supporting different roles (different contributors), pre- or on-the-fly defined workflows that can be executed again if needed can be done by M&S software.

WorMS

WorMS is developed to integrate and use processes of M&S users as workflows in software products (Rybacki et al. 2010; Rybacki et al. 2011). By comprising user guidance and the orchestration of elements in the software, WorMS supports the creation of provenance for the workflow execution—every step is well defined (there has to be a workflow in which it is embedded) and every workflow execution can be recorded. In Figure 13.7, a model validation process is given. This process does not need to be hard-coded in the software—it can be adapted by end users and thus adapted to current needs without running into the need to change the code of the M&S software. The same holds true for processes deeper in M&S software such as the execution of an experiment (with configurations and replications) (Rybacki 2012a). Furthermore, using workflows allows using predefined processes for studies (e.g., following the internal norms of an institution) (see Figure 13.8). It also allows modifying predefined workflows or reusing existing workflows within the context of another (see Figure 13.9). Consequently, the elements of the software can be combined by users (either in advance or during execution); the workflows can adapt themselves according to intermediate results (e.g., a detector might realize a nonexpected value) and can introduce some further steps into the workflow, which will help to analyze the reason for the unexpected value. WorMS consists of a workflow engine that uses workflow nets (van der Aalst and van Hee 2004) to describe processes as workflows and an independent control layer allowing importing, exporting, analyzing, and administration of workflows. By separating the engine from the control layer, it is in principle possible to replace the workflow engine shipped with WorMS with another such as the one used in YAWL (Russell and ter Hofstede 2009) or Kepler (Altintas et al. 2004).

Architecture and implementation aspects of workflows for modeling and simulation

We described the following before:

1. Different views on the process of M&S exist.
2. In detail knowledge about each step is mandatory to achieve repeatable results.
3. There are different languages to express workflows.
4. Workflows can be used for user guidance as well as for the control logics of the application.

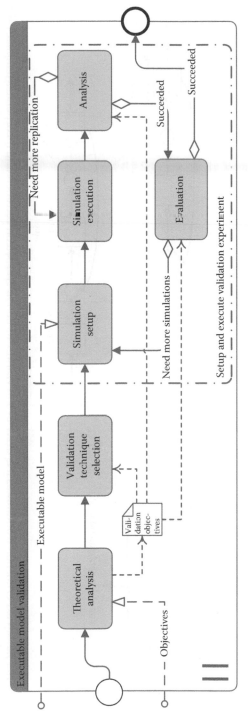

Figure 13.7 More detailed description of the "Executable Model Validation" shown in Figure 13.6.

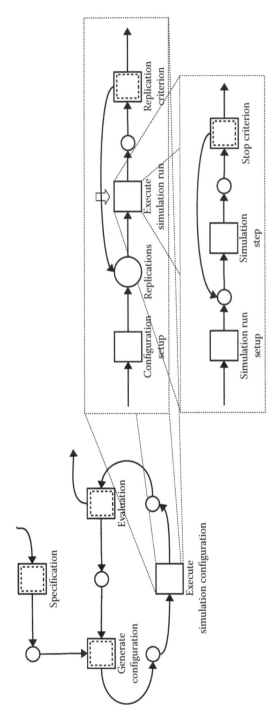

Figure 13.8 An experimental workflow as it is described in WorMS. Its notion is based on Petri nets and incorporates templates and frames (indicated by dashed lines) (Rybacki et al. 2012b) to create flexible workflow descriptions covering a wide range of scenarios.

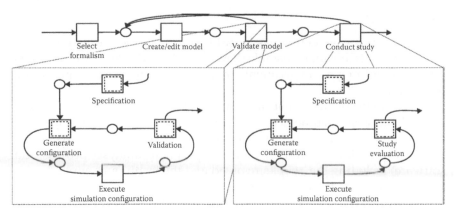

Figure 13.9 The reuse of already defined workflows in the context of another. The context represents the creation of a model including validation before conducting an actual study with it.

On the basis of these observations, we decided to create WorMS based on a double front-end/back-end architecture. WorMS thereby interacts as a middleware encapsulating its back end (workflow management system) and its front end (the interfaces toward the applications WorMS shall be used within). From the point of view of the M&S software WorMS is the back end providing the control flow for application logics (experiment control, user guidance, etc.) whereby the M&S software itself is the front end. This allows to replace the workflow management system used by WorMS without any change to the M&S software, which makes it possible to integrate the M&S software into the software landscape of a company it is used in and makes it possible to support the various workflow modeling languages around without the need to integrate all these languages natively into WorMS. Still, to extend reusability and to reduce dependency WorMS ships with an own workflow execution system that is based on workflow nets and that contains optimizations to cope with needs from M&S software (e.g., high-performance execution of parallel workflows and of single-workflow steps). This means that the control flow of the M&S is no longer hard-coded, but that it is modeled (as Petri net variant) and executed by a *simulator* on its own.

By integrating WorMS as a back end into M&S software the control logics of the application are well separated from the specialized application functions. This leads to a clean separation of concerns and, last but not least, delegates the tracking of what has been executed to the back end. This tracking is essential for repeatable experiments and can only be replaced in M&S by using concepts as aspect-oriented programming or a pervasive and expensive observation mechanism.

Discussion

In M&S literature, evidence was found that most of the results presented of M&S projects are not repeatable (Pawlikowski et al. 2002). These authors criticize that to repeat experiments mandatory information required are not given.

It is unquestionable that any empirical science is in need of well-defined and well-documented experiments. Thereby documenting simulation studies means to describe the artifacts created in the course of an M&S study (problem definition, model, experiment, analysis, process, and software) including the people involved and the techniques applied.

Over the last decades, different (computer-supported) solutions for setting up experiments have been developed but yet none of them has become *the standard*. In M&S, the first steps had been *in simulation language* solutions, as in CSSL. Later on, the separation-of-concerns paradigm supported the idea to have an explicit standalone experiment description language. However, this approach has so far not been fully exploited for the field of M&S.

Quality itself is considered to be a multifaceted and *ill-defined* term. Over the last decades, the idea of well-defined production processes (to be able to continually produce products with constant quality) gained more and more interest. This is reflected in the widely applied industry norm ISO 9001.

To support well-defined processes by computing machinery, workflow management systems have been created and deployed. Such systems can execute workflow definitions and thus allow supporting, controlling, and observing their execution.

A complete solution for M&S software could embrace such a workflow management system and experiment definitions based on workflows. One example of such a system is WorMS. With WorMS, we have shown that it is possible to base the internals of an M&S software on workflows, and that these workflow descriptions go beyond what we usually know about simulations performed: we know in greater detail how experiments have been executed (by the workflow and the execution logs) and we gain more flexibility (i.e., we can exchange experimentation control without modifying a single line of code and thus without removing the chance to repeat previous experiments). Both at the price of replacing the hardwired experimentation control with a workflow engine.

References

Altintas, I., C. Berkley, E. Jaeger, M. Jones, B. Ludäscher, and S. Mock. Kepler: An extensible system for design and execution of scientific workflows. *IEEE 16th International Conference on Scientific and Statistical Database Management*, pp. 423–424. Santorini Island, Greece, 2004.

Balci, O. A methodology for certification of modeling and simulation applications. *ACM Transactions on Modeling and Computer Simulations* (TOMACS) 11(4): 352–377, 2001.

Balci, O. Verification, validation, and certification of modeling and simulation applications. *Proceedings of the 2003 Winter Simulation Conference* 1: 150–158, 2003.

Barga, R., J. Jackson, N. Araujo, D. Guo, N. Gautam, and Y. Simmhan. The Trident Scientific Workflow Workbench. *IEEE Fourth International Conference on eScience, 2008* (eScience '08), pp. 317–318. Indianapolis, IN, 2008.

Brade, D. A Generalized Process for the Verification and Validation of Models and Simulation Results. Neubiberg, Germany: Universität der Bundeswehr, 2003.

Garvin, D. A. What does "product quality" really mean? *Sloane Management Review* 26: 25–43, 1984.

Himmelspach, J. Toward a Collection of Principles, Techniques, and Elements of Modeling and Simulation Software. *IEEE First International Conference on Advances in System Simulation, 2009* (SIMUL '09), pp. 56–61. Porto, Portugal, 2009.

Himmelspach, J. Tutorial on building M&S software based on reuse. *Proceedings of the 2012 Winter Simulation Conference.* Piscataway, NJ: Institute of Electrical and Electronics Engineers, 2012.

Himmelspach, J. and A. M. Uhrmacher. What contributes to the quality of simulation results?, pp. 125–129, edited by L. H. Lee, M. E. Kuhl, J. W. Fowler and S. Robinson. *INFORMS Simulation Society*, 2009.

Hinz, S., K. Schmidt, and C. Stahl. Transforming BPEL to Petri nets. *Proceedings of the Third International Conference on Business Process Management* (BPM 2005), pp. 220–235. Berlin, Germany: Springer-Verlag, 2005.

Hull, D. et al. Taverna: A tool for building and running workflows of services. *Nucleic Acids Research* 34: 729–732, 2006.

ISO. ISO 9001:2008: Quality management systems—Requirements. Intec, 2008.

Juric, M. B., B. Mathew, and P. Sarang. *Business Process Execution Language for Web Services.* Birmingham, UK: Packt, 2004.

Law, A. M. *Simulation Modeling and Analysis.* 4th ed. Columbus, OH: McGraw-Hill, 2006.

Maus, C., S. Rybacki, and A. M. Uhrmacher. Rule-based multi-level modeling of cell biological systems. *BMC Systems Biology* 5(1): 166, 2011.

Merali, Z. ... why scientific programming does not compute? *Nature* 467: 775–777, 2010.

Nilsen, R. N. and W. J. Karplus. Continuous-system simulation languages: A state-of-the-art survey. *Mathematics and Computers in Simulation* 16(1): 17–25, 1974.

Pawlikowski, K., H.-D. J. Jeong, and J.-S. R. Lee. On credibility of simulation studies of telecommunication networks. *IEEE Communications Magazine* 40: 132–139, 2002.

Perrone, L. F., C. S. Main, and B. C. Ward. SAFE: Simulation automation framework for experiments. *Proceedings of the 2012 Winter Simulation Conference.* Piscataway, NJ, 2012.

Russell, N. and A. H. M. ter Hofstede. *new*YAWL: Towards workflow 2.0. In *Transactions on Petri Nets and Other Models of Concurrency II*, edited by K. Jensen and W. M. P van der Aalst, pp. 79–97. Berlin, Germany: Springer, 2009.

Rybacki, S., J. Himmelspach, F. Haack, and A. M. Uhrmacher. WorMS—A framework to support workflows in M&S. *Proceedings of the 2011 Winter Simulation Conference* (WSC), pp. 716–727, edited by S. Jain, R. R. Creasey, J. Himmelspach, K. P. White, and M. Fu. Phoenix, AZ, 2011.

Rybacki, S., J. Himmelspach, E. Seib, and A. M. Uhrmacher. Using workflows in M&S software. *Proceedings of the 2010 Winter Simulation Conference* (WSC), pp. 535–545. Baltimore, MD, 2010.

Rybacki, S., J. Himmelspach, and A. M. Uhrmacher. Using workflows to control the experiment execution in M&S Software. *Proceedings of the 5th International ICST Conference on Simulation Tools and Techniques*, Desenzano del Garda, Italy, 2012a.

Rybacki, S., S. Leye, J. Himmelspach, and A. M. Uhrmacher. Template and frame based experiment workflows in modeling and simulation software with WORMS. *2012 IEEE Eighth World Congress on Services* (SERVICES), pp. 25–32. Honolulu, HI, 2012b.

Sargent, R. G. Validating simulation models. *Proceedings of the 1983 Winter Simulation Conference*, pp. 333–337. 1983.

Sargent, R. G. Verification and validation of simulation models. *Proceedings of 2008 Winter Simulation Conference*, pp. 157–169. 2008.

Sawyer, J. T. and D. M. Brann. How to build better models: applying agile techniques to simulation. *Proceedings of the 2008 Winter Simulation Conference*, pp. 655–662. Baltimore, MD, 2008.

Schmidt, K. *LoLA: A Low Level Analyser*, Vol. 1825, pp. 465–474. Berlin, Germany: Springer, 2000.

Sommerville, I. *Software Engineering*. Ed. 8. Boston, MA: Addison-Wesley, 2007.

van der Aalst, W. M. P. and K. M. van Hee. *Workflow Management: Models, Methods, and Systems*. Vol. 1., Cambridge, MA: The MIT Press, 2004.

Van Gorp, P. and R. Dijkman. BPMN 2.0 Execution Semantics Formalized as Graph Rewrite Rules: extended version. 2011.

Verbeek, E. and van der Aalst, W. M. P. Woflan 2.0: A Petri-Net-Based Workflow Diagnosis Tool. Berlin, Heidelberg, Germany: Springer, pp. 475–484. 2000.

Wang, Z. and A. Lehmann. Expanding the V-Modell® XT for verification and validation of modelling and simulation applications. *7th International Conference on System Simulation and Scientific Computing, 2008 (ICSC 2008) Asia Simulation Conference*, pp. 404–410. Beijing, China, 2008.

Weidlich, M., G. Decker, A. Großkopf, and M. Weske. *BPEL to BPMN: The Myth of a Straight-Forward Mapping*, Vol. 5331, pp. 265–282. Berlin, Germany: Springer, 2008.

WfMC. Workflow Management Coalition Workflow Standard Process Definition Interface—XML Process Definition Language. 2008.

White, S. A. and D. Miers. *BPMN Modeling and Reference Guide*. Lighthouse Point, FL: Future Strategies Inc., 2008.

Wynn, M. T., H. M. W. Verbeek, W. M. P. van der Aalst, A. H. M. ter Hofstede, and D. Edmond. Business process verification—finally a reality! *Business Process Management Journal* (Emerald) 15(1): 74–92, 2007.

Zeigler, B. P., H. Prähofer, and T. G. Kim. *Theory of Modeling and Simulation*. San Diego, CA: Academic Press, 2000.

Zha, H., Y. Yang, J. Wang, and L. Wen. Transforming XPDL to Petri nets. *Proceedings of the 2007 international conference on Business process management*, pp. 197–207. Berlin, Germany: Springer, 2007.

chapter fourteen

Formal validation methods in model-based spacecraft systems engineering

Joost-Pieter Katoen, Viet Yen Nguyen, and Thomas Noll

Contents

Introduction

Verification and validation (V&V) are key processes in the engineering of safety-critical hardware and software systems. They check whether the system under construction or its artifacts meet their requirements and its intended functions. The current industry practices for conducting V&V are rather labor intensive. There are severe concerns on scaling these practices to deal with the ever-growing complexity of systems and in particular of software. The trend is to incorporate the use of *formal methods*. In particular, automated verification techniques are attractive for supporting more rigorous V&V. Formal methods, however, tend to require a high degree of expertise and specialized know-how. These incur substantial investments before their cost and efficiency benefits can be reaped.

This chapter presents a recently developed system validation methodology. It exploits formal method tools and techniques, while substantially reducing the engineer's learning curve by automating and hiding technical intricacies that usually come with formal methods. It has been developed within several European Space Agency (ESA) projects aimed at advancing methods for spacecraft system and software co-engineering. Our methodology features a dialect of the Architecture Analysis and Design Language (AADL), which is standardized by the Society of Automotive Engineers (SAE) (2004). This component-oriented modeling language enables system engineers to describe the software and hardware aspects together with their interaction and reliability aspects. Our methodology features a comprehensive toolset, called the COMPASS toolset (http://compass.informatik.rwth-aachen .de/). COMPASS is an acronym for correctness, modeling, and performance of aerospace systems. It analyses models with respect to a wide range of V&V objectives, such as functional correctness, safety and dependability, and performance using state-of-the-art model checking techniques and probabilistic variants thereof developed by the formal methods community. Key and novel in this is that the engineer only has to specify a single system model in our AADL dialect. This is in contrast to industrial practice. It is currently common to construct separate and loosely related models each targeted for a certain V&V objective, like a dedicated functional model, a dedicated reliability model, a dedicated performance model, and so forth. Our AADL dialect is equipped with a formal semantics that is cornerstone to the integration of manifold analyses by several model checking tools (Baeier and Katoen 2008). in a semantically consistent manner. This benefits engineers by hiding the intricacies of each model checker, and the combination of them, without losing consistency and accuracy.

Our methodology was initially evaluated by subsystem validation case studies. This has led to subsequent studies of a more comprehensive nature. There we investigated the methodology from both a technical and process perspective. The outcome and the lessons learned from these studies are reported in this chapter. Guided by a larger case study, a satellite platform, we show how to embed such formal system validation activities in the context of the overall systems engineering life cycle. We argue that our model-based approach particularly suits the early phases of system development, that is, those dealing with the identification of requirements and, in particular, the preliminary design. Employing our formal modeling and analysis techniques at this stage enables an improved understanding of the system and its expected behavior. The improved understanding supports informed design decisions. This is crucial as choices during the design have a significant impact on the subsequent engineering phases.

Formal methods

Formal methods in computer science are mathematically based techniques for the specification, development, and verification of software and hardware systems. Their foundations are in theoretical computer science, in particular automata theory, logic, and program semantics.

The starting point is a description of the possible behaviors of the system under study. This is the *model*. It is typically described in a modeling language such as Unified Modeling Language (UML) state charts, MATLAB/Simulink models, or AADL (SAE 2004). Modeling languages often come with a well-defined syntax in which the model can be described. Their semantics are often, however, left to one's interpretation making it subjective and possibly ambiguous. We therefore advocate a formal description of not only the syntax but the semantics as well. A *formal semantics* gives a mathematically precise meaning to all language constructs. Thus, a model description is uniquely mapped onto a mathematical object such as an automaton, a Markov chain, a set of sequences, or the like. This has several advantages. First, as the semantics is uniquely defined, it avoids any discussion about the meaning of model specifications. There is no room for ambiguities. Perhaps more important though is the fact that a formal semantics allows for rigorous tool development. The interpretation is not determined by a single tool, but acts as a blueprint for all tool implementations. Formal semantics therefore increase interoperability: different analyses of the same model with distinct tools yield consistent results. In the absence of such semantics, this is not guaranteed, and interpretations are tool-vendor dependent.

The second ingredient for the analysis is a *precise description of the model properties* that are to be analyzed. As before, a mathematically precise property description leaves no room for multiple interpretations. In addition, the mathematical logics that are typically used allow specifying a broad range of properties in a rather succinct way. The ability to nest formulas, and to equip formulas with features referring to timing aspects or randomness in the systems, enable an enormous plethora of system properties. This range goes beyond the imaginations of most system engineers. The real crux is that a single analysis algorithm suffices to check them all. This is the enormous strength of model checking and also the reason for being a successful branch of formal methods. A single algorithm that is capable of checking all system properties, being it absence of deadlock, progress properties, or bounded reachability (can a situation occur within the next bound dumber of computation steps?), does the job.

Formal methods are often too rigorous to be cost-effective for designing every piece of software or hardware. This is, however, not the case for safety-critical systems. Aerospace systems and high-volume consumer products are characterized by the tremendous impact failures can have. They could be destructive, life threatening, or costly to be repaired. Standard techniques for their V&V such as peer review, simulation, or testing are tough and labor-intensive for such systems tend to exhibit concurrency, nondeterminism, and randomness. Although simulations can (and preferably should) be based on mathematical models, they can only show the presence of anomalous behavior, but not prove their absence. Simulation is nevertheless extremely useful for a first sanity check of system models. To obtain hard guarantees about the absence of flaws, such as deadlock situations in which concurrent tasks mutually wait for each other, formal verification techniques are indispensable. For these reasons, formal methods are one of the *highly recommended* verification techniques for software development of safety-critical systems according to, for example, the best practices standard of the International Electrotechnical Commission (IEC) 61508, EN 50128 (trains), ISO 26262 (cars), and IEC 62304 (medical). The resulting report of an investigation by the Federal Aviation Authority and NASA (National Aeronautics and Space Administration) about the use of formal methods concludes that "Formal methods should be part of the education of every computer scientist and software engineer, just as the appropriate branch of applied mathematics is a necessary part of the education of all other engineers" (Rushby 1995).

Spacecraft engineering life cycle

The spacecraft is the key technical system for fulfilling a space mission. Its life cycle, therefore, strongly correlates with the space mission itself. An example space mission project life cycle is depicted in Figure 14.1. It is the life cycle according to ECSS M-ST-10C, ESA's standard on space mission project planning and implementation (ECSS 2009b). It bears many similarities with other codified practices, such as those by NASA (2007). The life cycle is split in phases 0 to F. In each phase, activities are performed, such as mission definition and requirements engineering. A continuous V&V process is in place throughout the life cycle to manage various risks, like budget overruns, procurement difficulties, political instabilities, and technical failures. We generalize the latter as technical risks and they are our prime interest. Typical V&V activities are reviews, inspections, tests, and analyses. They are conducted to increase the current understanding of the system (under design), and in particular its (current) weaknesses. Verification checks the system (or a part of it) against the requirements it was built from. During verification, it is assumed that the requirements are those intended. Validation on the other hand does reflect on this assumption. Both verification and validation strive to improve quality by increasing the understanding of the system. As such, the term V&V is often coined in one stretch for this aim. The methodology in this chapter therefore is applicable for both verification and validation.

Current practices employ a wide range of modeling tools and techniques for V&V. However, all of them investigate particular system aspects such as functional correctness, safety and dependability, and performance in relative isolation, holding strong mutual assumptions. For systems in which these aspects strongly interweave, such as spacecraft and their fault management systems, a deeper and more coherent and integrated perspective is desired. We developed a formal-methods-supported model-based approach that enables this view.

Our methodology specifically supports V&V in the early design phases. For example, according to ESA's prescribed practices (ECSS 2009b; ECSS 2009c; ECSS 2010), this is in late phase A to early phase C in Figure 14.1. For NASA's prescribed practices (NASA 2007; NASA 2009), this is in prephase A to early phase C. In these early design phases, the system's possible architectures in terms of subsystems, modes, components, functions, and component interactions (i.e., behavior) are being investigated and evaluated by V&V. Eventually, a final system architecture is determined, which is refined into an implementation in the subsequent engineering phases. As such, our

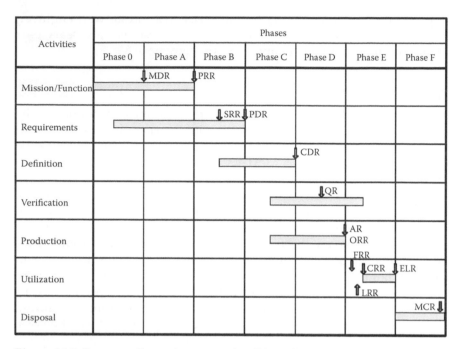

Activities	Phases						
	Phase 0	Phase A	Phase B	Phase C	Phase D	Phase E	Phase F
Mission/Function	↓MDR ↓PRR						
Requirements			↓SRR ↓PDR				
Definition				↓CDR			
Verification					↓QR		
Production					↓AR ORR FRR		
Utilization					↓	↓CRR ↓ELR ↑LRR	
Disposal							MCR↓

Figure 14.1 European Space Agency project life cycle.

methodology provides for a modeling language that gracefully captures the system architecture design, and provides a wide range of analyses from formal methods in a software toolset that are the automated counterparts of V&V analyses that are currently conducted manually. The benefit is that variations of system architectures can be evaluated faster and more effectively.

Modeling language: An AADL dialect

The AADL (SAE 2006) is an industry standard for modeling safety-critical system architectures. It was developed and is governed by the SAE. AADL provides a cohesive and uniform approach for modeling heterogeneous systems, consisting of software (e.g., processes and threads) and hardware (e.g., processors and buses) components, and their interactions. It allows analysis of system designs prior to development and supports a model-based and model-driven development approach throughout the system life cycle.

Our dialect of AADL was designed to meet the needs of the European space industry. The original language mainly focused on the architectural organization of a system *under nominal and degraded modes of operation*. The nominal modes indicate that the system is operating normally, whereas degraded modes typically indicate that the system's functions are (partially) impaired due to issues. Our goal was to extend beyond AADL's focus of expressing architecture by also analyzing its dynamic behavior, namely both its nominal and degraded modes of operation and their interweaving.

In particular, *quantitative aspects* such as the timing of operations and the likelihood of faults should be covered. To this end, we built on a core fragment of AADL version 1 and extended on it, essentially by supporting the following features:

- Modeling both the system's *nominal and faulty behavior.* To this aim, primitives are provided to describe software and hardware faults, error propagation (i.e., turning fault occurrences into failure events), sporadic (transient) and permanent faults, and degraded modes of operation (by mapping failures from architectural to service level).
- Modeling *(partial) observability* and the associated *observability requirements.* These notions are essential to analyze the effectiveness of fault management systems. These subsystems, being part of the overall system, monitor it, identifying when a fault has occurred, pinpointing the type of fault and its location, and finally recovering from it by, for example, switching to a backup system configuration.
- Specifying *timed and hybrid* behavior. In particular, to analyze physical systems with nondiscrete behavior, such as mechanics and hydraulics, the modeling language supports continuous real-valued variables with (linear) time-dependent dynamics.
- Modeling *probabilistic aspects,* such as random faults, repairs, and stochastic timing.

A complete system specification consists of three parts, namely a description of the nominal behavior, a description of the error behavior, and a fault injection specification that describes how the error behavior influences the nominal behavior. This separation approach is different from the one taken in AADL and its Error Model Annex (SAE 2006), which interacts through an explicit specification of mangling error and nominal events. In contrast, our dialect provides an automated mechanism called model extension that enables engineers to keep the nominal model completely separate from the error model. In the following, we give a comprehensive presentation using a running example.

Nominal behavior

The system model is hierarchically organized into components, distinguished into software (processes, threads, data), hardware (processors, memories, devices, buses), and composite components (called systems). Components are defined by their type (specifying the functional interfaces as seen by the environment) and their implementation (representing the internal structure). An example of a component's type and implementation is shown in Model 14.1, which represents a simple battery device.

The component type specifies the ports through which the component communicates with its environment. There are two kinds of ports, namely event and data ports. Event ports enable components to synchronize their state on each other whereas data ports are used to expose component variables to neighboring components. For example, the battery interface features two ports, namely an outgoing event port *empty*, which indicates that the battery is about to become discharged and an outgoing data port *voltage*, which makes its current voltage level accessible to the environment.

A component implementation describes the internal structure of a component through the definition of its subcomponents, their interaction through (event and data) port connections, the (physical) bindings at runtime, and the behavior via modes and transitions between them. This behavior, which basically is a finite state automaton, describes how the component evolves from mode to mode while being triggered by events, or by spontaneously triggering events at the ports. On a transition, data components (like integers, reals, and Boolean variables) may change values due to transition assignments. Modes can be further annotated with invariants on the value of data components (continuous or clock variables), restricting, for example, residence time. They furthermore may contain trajectory equations, specifying how continuous variables evolve while

Model 14.1

```
device Battery
  features
    empty: out event port;
    voltage: out data port real default 6.0;
end Battery;

device implementation Battery. Imp
  subcomponents
    energy: data continuous default 1.0;
  modes
    charged: activation mode while energy' = -0.02 and
      energy > = 0.2;
    depleted: mode while energy' = -0.03 and energy > = 0.0;
  transitions
    charged - [then voltage : = 2.0 * energy + 4.0] ->
      charged;
    charged - [empty when energy = 0.2] -> depleted;
    depleted - [then voltage : = 2.0 * energy +
      4.0] ->depleted;
end Battery.Imp;
```

residing in a mode. This is akin to timed and hybrid automata, standard models in formal methods.

For example, the implementation of Model 14.1 specifies the battery to be in the *charged* mode whenever activated, with an energy level of 100% as indicated by the default value of 1. This level is continuously decreased by 2% (of the initial amount) per time unit (*energy* denotes the first derivative of *energy*) until a threshold value of 20% is reached, on which the battery changes to the *depleted* mode. This mode transition triggers the *empty* output event, and the loss rate of energy is increased to 3%. Moreover, the voltage value is regularly computed from the energy level (ranging between 6 and 4 [volts]) and made accessible to the environment via the corresponding outgoing data port.

Mode transitions may give rise to modifications of a component's configuration: subcomponents can become (de)activated and port connections can be (de)established. This depends on the *in modes* clause, which can be declared along with port connections and subcomponents. This is demonstrated by the specification of a redundant power system as given in Model 14.2. It employs two instances of the battery device, namely *batt1* and *batt2*, being respectively active in the *primary* and the *backup* mode. The mode switch that initiates reconfiguration is triggered by an *empty* event arriving from the battery that is currently active. It also reroutes the data flow of the *voltage* information.

Model 14.2

```
system Power
  features
    voltage: out data port real;
end Power;

system implementation Power. Imp
  subcomponents
    batt1: device Battery. Imp in modes (primary);
    batt2: device Battery. Imp in modes (backup);
  connections
    data port batt1.voltage -> voltage in modes
      (primary);
    data port batt2.voltage -> voltage in modes
      (backup);
  modes
    primary: activation mode;
    backup: mode;
  transitions
    primary -[batt1.empty]-> backup;
    backup -[batt2.empty]-> primary;
end Power.Imp;
```

Error behavior

Error models are an extension to the specification of nominal models and are used to conduct safety and dependability and performability analyses. For modularity, they are defined separately from nominal specifications. Akin to nominal models, an error model is defined by its type and its associated implementation.

An error model *type* defines an interface in terms of error states and (incoming and outgoing) error propagations. Error states are employed to represent the current configuration of the component with respect to the occurrence of errors. Error propagations are used to exchange error information between components.

An error model *implementation* provides the structural details of the error model. It defines a (probabilistic) machine over the error states declared in the error model type. Transitions between states can be triggered by error events, reset events, and error propagations. Error events are internal to the component; they reflect changes of the error state caused by local faults and repair operations, and they can be annotated with occurrence distributions to express probabilistic error behavior. Moreover, reset events can be sent from the nominal model to the error model of the same component, trying to repair a fault that has occurred. Whether or not such a reset operation is successful has to be modeled in the error implementation by defining (or, respectively, omitting) corresponding state transitions. Outgoing error propagations report an error state to other components. If their error states are affected, the other components will have a corresponding incoming propagation.

Model 14.3 presents a basic error model for the battery device. It defines two probabilistic error events. The *die* event occurs once every 1000 time units on average. The *glitch* event occurs once every 10 time units on average. The battery can fail and enter the *dead* state immediately, or by glitching first by entering the *glitched* state. Once the battery is in the *dead* state, the battery failure is signaled to the environment by means of the outgoing error propagation *batteryDied*. Moreover, the battery is enabled to receive a *reset* event from the nominal model to which the error behavior is attached. It causes a transition to the *ok* state.

Fault injections

As error models bear no relation with nominal models, an error model does not influence the nominal model unless they are linked through a *fault injection*.

A fault injection describes the effect of the occurrence of an error on the nominal behavior of the system. More concretely, it specifies the value update that a data element of a component implementation undergoes when its associated error model enters a certain error state.

Model 14.3

```
error model Battery Failure
  features
    ok: activation state;
    dead: error state;
    glitched: error state;
    resetting: error state;
    batteryDied: out error propagation;
end BatteryFailure;

error model implementation BatteryFailure. Imp
  events
    die: error event occurrence poisson 0.001;
    glitch: error event occurrence poisson 0.1;
  transitions
    ok -[die]-> dead;
    ok -[glitch]-> glitched;
    glitched -[die]-> dead;
    dead -[batteryDied]-> dead;
    dead -[reset]-> ok;
end BatteryFailure.Imp;
```

To this aim, each fault injection has to be provided by specifying three parts: a state *s* in the error model, an outgoing data port or subcomponent *d* in the nominal model, and the fault effect given by the expression *a*. Multiple fault injections between error models and nominal models are possible. For example, when the battery error model enters the error state *dead*, the voltage variable in the nominal model becomes 0, expressing that the battery is flat when it dies.

The automatic procedure that integrates both models and the given fault injections, the so-called *model extension*, works as follows. The principal idea is that the nominal and error models are running concurrently. That is, the state space of the extended model consists of pairs of nominal modes and error states, and each transition in the extended model is due to a nominal mode transition, an error state transition, or a combination of both (in case of a reset operation). The previous fault injection becomes enabled whenever the error model enters state *s*. In this case, the data element *d* becomes mangled and is assigned the expression *a*, that is, the data subcomponent is assigned with the fault effect. This error effect is maintained as long as the error model stays in state *s*, overriding possible assignments to *d* in the nominal model. When *s* is left, the fault injection is disabled (but possibly another one is enabled).

Formal semantics

Cornerstone to our methodology is a coherent and consistent *formal semantics*. It describes how each component is mapped onto its formal counterpart, a so-called event-data automaton. A system, comprising interacting components, is formally mapped to a network of event-data automata. The formal semantics moreover describe, in the form of an operational semantics, how a network of event-data automata behaves in terms of transition systems. This formal underpinning is fundamental for mapping AADL models to the theoretical formalisms, like finite state machines, Büchi automata, timed automata, and Kripke structures, which are reasoned over by the employed model checkers. As we apply several different model checkers, including probabilistic ones, the formal semantics function as a tool-independent layer for ensuring that analysis results are consistent regardless of the implementation. We refer the interested reader to Bozzano et al. (2011) for further information.

Graphical modeling

The AADL models can be either expressed using a textual format through a conventional text editor or using graphical notation through our click'n'connect visual editor. The visual representation of AADL models is attractive to engineers accustomed to modern engineering tools. The component hierarchy is shown as a nesting of boxes along with the port connections. Each box represents a component. See also Figure 14.2. Its zooming capabilities enable users to focus on a particular level of detail in the component hierarchy, thereby visualizing directly the scope of interest. The visual models can be exported to their textual representation for formal analyses.

V&V analyses with formal methods

Models expressed in our dialect of AADL can be analyzed using the COMPASS toolset (see Figure 14.3). This graphical-driven toolset supports the following V&V aspects:

- *Functional correctness*: Whether the system's behavior fulfills its specified function
- *Effectiveness of fault management*: Specialized correctness, safety, and dependability analyses that investigate the observables and alarms of the fault management system
- *Safety and dependability*: Whether the system provides sufficient readiness and continuity of its service (i.e., availability and reliability), whether it can undergo modifications and repairs (i.e., maintainability) or does not lead to catastrophic consequences to the user(s) and its environment (i.e., safety)
- *Performability*: Whether the system provides sufficient performance under dependability constraints, like the presence of (possibly multiple) faults

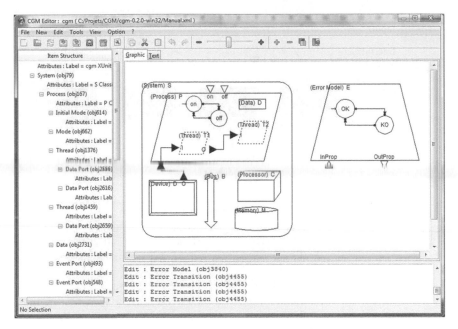

Figure 14.2 Click'n'connect graphical editor for the Architecture Analysis and Design Language.

The power of this toolset comes from the formal semantics. It is equipped with our AADL dialect and provides a consistent and automatic mapping to the input languages of several model checkers, and furthermore, maps the results back again in symbols only occurring in the AADL dialect. The user is therefore not exposed to the underlying formal methods and no thorough understanding of it is required to conduct the system validation activities.

Properties

Analysis begins first by stating hypotheses or requirements on the expected system behavior. In our methodology, these are properties that are formally given in a temporal logic. This ensures a high-quality and unambiguous specification. The use of formal logic makes subtle questions explicit that otherwise might be hidden in the ambiguity of natural language. We use temporal logics like Linear Temporal Logic (LTL) or Computational Tree Logic (CTL) for expressing qualitative properties. For quantitative properties, we use Continuous Stochastic Logic (CSL).

To ease the specification by nonexpert users, we provide *property patterns*. On the basis of statistical studies (Dwyer et al. 1999; Grunske 2008)

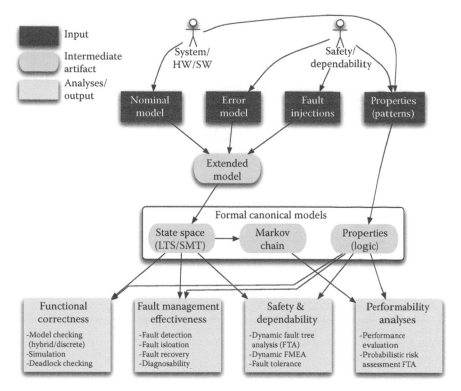

Figure 14.3 Overview of the COMPASS toolset.

from requirements engineering, it has been determined that 92% of the requirements can be captured using eight qualitative patterns and eight quantitative patterns. An example is the response pattern, stating that whenever X holds, this is eventually responded to by Y. The variables X and Y are placeholders that have to be filled by the user. The patterns cover cases such as the absence and existence of events, responses, and precedences between two events, or system invariants. Each pattern has a direct mapping to LTL, CTL, or CSL. A table of six example qualitative patterns is shown in Figure 14.4.

For example, we can use the global response pattern to check whether the battery system switches to the second battery on battery failures. This is expressed by taking φ to be *voltage = 0* and ψ to be *mode = mode:backup*. In general, placeholders are Boolean expressions that may contain operators such as greater than, less than, and equal to. These operators take data ports, data subcomponents, modes and, states as their left- and right-hand sides. The full syntax and semantics of the placeholders can be found in the *COMPASS User's Manual* (COMPASS Consortium 2012).

Pattern	Description	Logic (CTL/LTL)
Global absence	The atomic proposition $\boxed{\phi}$ never holds.	$\forall\Box\neg\phi$ $\Box\neg\phi$
Global existence	The atomic proposition $\boxed{\phi}$ shall eventually hold.	$\forall\Diamond\phi$ $\Diamond\phi$
Global universal	The atomic proposition $\boxed{\phi}$ globally holds.	$\forall\Box\phi$ $\Box\phi$
Global precedence	The atomic proposition $\boxed{\phi}$ globally precedes $\boxed{\psi}$.	$\neg\exists(\neg\phi\bigcup(\psi\wedge\neg\phi))$ $(\Box\neg\psi)\vee(\neg\phi\bigcup\psi)$
Global response	Whenever the atomic proposition $\boxed{\phi}$ holds, this is eventually responded with $\boxed{\psi}$.	$\forall\Box(\phi\implies\forall\Diamond\psi)$ $(\Box\phi\implies\Diamond\psi)$
Exists response	Whenever atomic proposition $\boxed{\phi}$ holds, it may be eventually responded with $\boxed{\psi}$.	$\forall\Box(\phi\implies\exists\Diamond\psi)$

Figure 14.4 Six qualitative property patterns and their underlying logical formats.

Functional correctness

The COMPASS toolset supports both traditional methods for analysis of functional correctness, such as simulation, and automated techniques for property verification, based on model checking.

Model-based simulation can be performed in two variants: random simulation and guided simulation. Guided simulation comes in two different flavors:

1. Guided-by-transitions simulation: The user can execute a step-by-step-simulation by choosing the next transition to be taken, among the enabled ones.
2. Guided-by-values simulation: The user can perform a step-by-step simulation by choosing a target value for one or more variables, and let the toolset choose a transition that drives the system into a corresponding target state, if any exists.

The generated traces can be inspected using a trace manager that displays the values of the model variables of interest (filtering is possible) for each step, in a human-readable form; in case of timed and hybrid systems, timed transitions are highlighted.

Property verification is carried out by *model checking* (Baier and Katoen 2008). Model checking is an automated technique that verifies whether a property holds for a given model. Symbolic techniques are used to tackle the combinatorial explosion of the amount of possible states, that is, the state space explosion. The COMPASS platform integrates the model checking capabilities provided by the NuSMV model checker (Cimatti et al. 2002). In case of finite-state systems, it can encode the model as a Binary

Decision Diagram (BDD) and uses BDD-operations such as conjunction, disjunction, and existential quantification for spawning the accompanying state space or use propositional formulae for encoding the model and its state space. For the latter, Boolean satisfiability solvers can be used to check whether properties hold. Models and state spaces of timed and hybrid systems are encoded as a subset of first-order logics, and properties over them can be checked using Satisfiability Modulo Theories solvers. The encoding techniques are further described in the work of Biere et al. (1999) and Audemard et al. (2005). On refutation of a property, the model checker generates a *counterexample*. This counterexample is an execution trace of the model violating the property and can be inspected using the trace manager.

In addition to property model checking, it is also possible to run deadlock checking, to pinpoint deadlocks (i.e., states with no outgoing transitions) in the model, if there are any.

Safety and dependability assessment

Model-based safety assessment of AADL models aims at reducing the effort involved in safety assessment and at increasing the quality of the results by automatically generating safety and dependability artifacts from the system specification, rather than constructing them manually. Our toolset is capable of dealing with two such artifacts that are predominantly used, namely *Failure Modes and Effect Analysis* (FMEA) tables (ECSS 2009a). and *Fault Trees* (ECSS 2008).

FMEA is an inductive (bottom-up) technique that starts by identifying a set of failure modes. Then, using forward reasoning, it assesses their impact on a set of events (system properties). FMEA requires as input a set of failure modes and a set of (failure) events of interest. FMEA typically considers single faults. Fault configurations involving several faults can be investigated as well by increasing the cardinality of the FMEA table. This means that sets of failure modes of cardinality at most k are considered and analyzed. This comes on the expense of increasing the computational complexity of the FMEA algorithm. The analysis results are summarized in an FMEA table, which links the given fault configurations with their effect(s) on the given events. An example of that is shown in Figure 14.5. It is also possible to generate dynamic FMEA tables. The dynamic option enforces an order of occurrence between failure modes within a fault configuration.

Fault Tree Analysis is a deductive (top-down) technique, which, given a top-level event (TLE) (i.e., the specification of an undesired condition), constructs all possible combinations of one of more basic faults that contribute to the occurrence of the top level event. These causal dependencies are then depicted in a fault tree. COMPASS also supports the generation

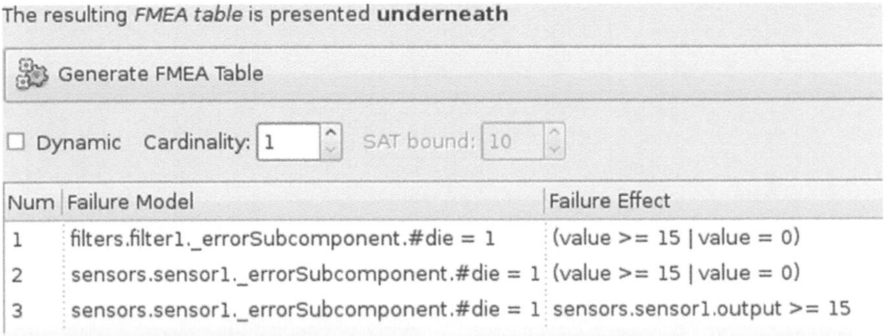

The resulting *FMEA table* is presented **underneath**

Num	Failure Model	Failure Effect
1	filters.filter1._errorSubcomponent.#die = 1	(value >= 15 | value = 0)
2	sensors.sensor1._errorSubcomponent.#die = 1	(value >= 15 | value = 0)
3	sensors.sensor1._errorSubcomponent.#die = 1	sensors.sensor1.output >= 15

Figure 14.5 An example Failure Modes and Effect Analysis table of cardinality 1 generated by the COMPASS toolset.

of dynamic fault trees. In the dynamic case, it analyzes for each combination of basic faults in which order they have to occur to trigger the TLE. Pictorially, such ordering constraints are represented using a priority AND (PAND) logical gate in the generated fault tree. An example is shown in Figure 14.6.

Failure detection, isolation, and recovery effectiveness analysis

The COMPASS toolset also supports *diagnosability analysis* and *fault management effectiveness analysis*. These analyses work under the hypothesis of partial observability. Rarely systems are fully observable: parts of their state are hidden, and sensors are used to expose (partial) information about otherwise unobservable aspects. Diagnosis starts from the observed runtime behavior of a system, and tries to provide an explanation of observations (in terms of hidden states). Variables and ports in AADL models can be declared to be observable. These observables indicate which variables or states are visible during execution. Diagnosis amounts to identifying the set of possible causes of a specific unexpected or faulty behavior. Several techniques are available in COMPASS to analyze the suitability of the observables.

Diagnosability analysis is typically carried out before a fault management subsystem is available. It investigates the possibility for an ideal diagnosis system to infer accurate and sufficient run-time information on the behavior of the observed system. Given a diagnosability property, diagnosability depends on the observed system and the available observations. The COMPASS toolset uses a TwinPlant framework to determine this (Cimatti et al. 2003). In this framework, the violation of a diagnosability condition (i.e., the impossibility of detecting and isolating faults) is reduced to the search of critical pairs. A critical pair is a pair of paths in the state space that are indistinguishable (i.e., they share the same inputs

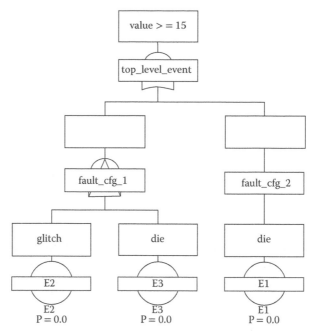

Figure 14.6 An example dynamic fault tree generated by the COMPASS toolset. The gate of fault_cfg_1 is a priority AND gate whereas the top-level event is an OR gate. This fault tree describes that the top-level event, value > = 15, is triggered by a failure of the battery (e.g., the *die* event, see E3) followed by *glitch* (event E2), or just by the battery dying (event E1).

and observations), but hide conditions that should be distinguished (i.e., the fault is triggered only in one of the two executions). It is a proof of undiagnosability and can be used to refine the model accordingly.

There are additionally three analyses that specifically assess the effectiveness of the fault management system by analyzing its fault detection capabilities, its fault isolation capabilities, and its fault recovery capabilities (failure detection, isolation, and recovery [FDIR]).

Fault detection analysis is concerned with detecting whether a given system is malfunctioning. It checks whether a candidate observable signal can be considered as a fault detection means for a given fault, that is, whether every occurrence of the fault will eventually cause the observable to become true. The COMPASS toolset reports all such observables as possible detection means.

Fault isolation analysis is concerned with identifying the specific cause of malfunctioning. It generates a fault tree that contains the minimal explanation for triggering an observable port. In case of perfect isolation, the fault tree contains a single fault configuration consisting of one fault, indicating that the fault has been identified as the single cause of the malfunctioning. A fault tree with multiple fault configurations indicates that there may be several explanations for the same observable malfunctioning.

Fault recovery analysis is used to check whether a system is able to recover from a fault (e.g., by reconfiguring the system using backup components), according to a user-specified recoverability property.

Performability

To guarantee the required performance under dependability constraints, that is, performability, an AADL model can be evaluated using *probabilistic model checking* techniques (Baier et al. 2010). The COMPASS toolset supports model checking of properties expressed in CSL, which allows for the formal specification of steady-state and transient probabilities, and more intricate performance measures such as combinations thereof (Baier et al. 2003). Typical performance requirements of interest are "the probability that the first battery dies within 100 hours" or "the probability that both batteries die within the mission duration." The COMPASS toolset automatically generates the underlying continuous-time Markov chain from the AADL model and then checks it against the logical representation of the performance requirement (in CSL). The result is a graph showing the cumulative distribution function over the time horizon specified in the performance requirement. See also Figure 14.7.

The same probabilistic model checking techniques are used for fault tree evaluation. Fault tree evaluation computes the probability of the TLE in fault trees by transforming it into its underlying Markov chain (Boudali et al. 2010).

Industrial case studies

Our methodology has been progressively assessed by several industrial case studies, steadily increasing their size and scope. Thales Alenia Space conducted the first evaluation. They developed two case studies of their satellite subsystems and analyzed them using the COMPASS toolset. These subsystem case studies demonstrated the potential of understanding the subtleties between the system, software, and the fault management system. It furthermore raised follow-up questions: how would models with a greater level of detail be handled? In which phases of the systems engineering life cycle is our methodology particularly suitable? Thus afterwards, the ESA conducted a laboratory project to model a full satellite platform using our methodology. This was performed at phase B of the space systems engineering life cycle, the preliminary system design. A subsequent laboratory project was initiated afterwards to model a full satellite platform at phase C, the detailed system design. The outcomes of the studies are discussed as follows.

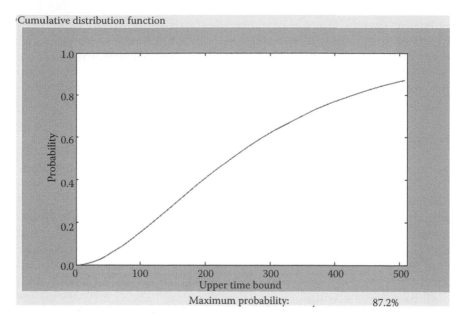

Figure 14.7 A reliability curve plotted as the cumulative distribution function generated by the COMPASS toolset.

Satellite subsystem case studies

The first subsystem case study relates to the definition of the satellite mode management and its associated fault management strategy.

Satellite mode management case study

This case study features the satellite mode management, the Attitude and Orbit Control System (AOCS) mode management, the (re-)configuration sequences, and an abstract model of the AOCS equipment and other functional subsystems. The (re-)configuration sequences represent sequences of commands that are sent to the AOCS units; they are used in case of a detected failure.

The satellite mode management consists of three global modes and seven elementary transitions. The AOCS mode management has six modes and about 20 transitions. The model of the equipment is based on a six-state automaton that determines whether it is on, off, operational, and controllable (and combinations thereof). Injected faults are transmission errors and data inconsistencies. The system-level alarm represents several hardware-detected faults, such as loss of pointing to Earth or Sun. The model has been checked for deadlocks, and simulations were then used to generate traces of the model. Both random and guided-by-transitions simulations have been used. Traces have been generated with and without fault injections. Then, the following three items were verified:

- Identification of all the failures leading to a given level of fault management
- Identification of those failures which can lead to the activation of a reconfiguration sequence
- Impact of a reconfiguration sequence on the satellite mode and AOCS mode

The first two items have been verified by generating fault trees. For each level of fault management, a fault tree was generated. Each leaf represents a fault that can lead to this level. The third item was verified using model checking. We wanted to check that the AOCS unit would be in the required status, depending on the AOCS mode. All results have been validated manually, and were as expected.

Thermal regulation case study

The second subsystem case study models a thermal subsystem that regulates the satellite's temperature. It is a codesigned system and performs both active and passive regulation. Passive regulation is achieved by optical solar reflectors, shields, and heat pipes. They have no behavior and are therefore not modeled. Active regulation relies on sensors and heaters at dedicated positions in the satellite. Their position is not taken into account as no requirement uses that information. Two subcases have been derived at functional and behavioral levels. The model at the functional level covers the whole perimeter of the thermal regulation function. It consists of five subfunctions and 12 elements (of which five are passive and seven active). Fault trees and FMEA tables have been generated to identify the critical path in the function tree. For further analysis, this model was refined into a more detailed logical model, taking into account the behavior of the thermal lines. The thermal subsystem consists of thermal lines, heaters, and sensors. A thermal line consists of two heater lines (nominal and redundant), two safety switches, and three thermistors in hot redundancy. If a failure occurs, the software switches the nominal heater line off and enables the redundant one to resume temperature regulation. The behavior of the line depends on the computation of the median of the three thermistors' measurements. On the basis of this value, it switches the heaters on or off, according to the thresholds of the measured temperature.

Two error states have been considered for each unit of the model: for a temperature sensor, being stuck at the minimal value and being stuck at the maximal value; for a heater, no heating and always heating; for a switch, stuck at ON and stuck at OFF. An environment model has been used to describe the evolution of temperature. Several reconfiguration sequences have been modeled to reconfigure switches and heaters. Three models, describing alternative system designs, have been developed.

All the models were analyzed for correctness, reliability, safety, and diagnosability capabilities. Model checking was used to verify that the temperature is regulated between the lower and upper bounds of the safe range under nominal operation. Several simulations have been performed for sanity checking the model. See Figure 14.8. Fault trees and FMEA tables have been generated. All the fault configurations have at least cardinality two. Fault detection analysis has been used to verify that there exists a means for detecting each TLE. Fault isolation analysis has provided for each observable the list of faults to which the observable is sensitive. The two analyses are sufficient to validate the observability of the modeled system. The collected data has been used to improve the dependability of the thermal regulation line from a hardware and software point of view. Finally, the models have been checked using fault recovery analysis. All results conformed to the expectations.

Lessons learned

Both the satellite mode management and the thermal subsystem case studies have resulted in several learned lessons that apply generally the following:

- Our dialect of AADL allows for *incremental modeling* by starting with a functional representation of the subsystem and by refining it into a more detailed one without breaking the structure of the model. The modeling language facilitates this by supporting component

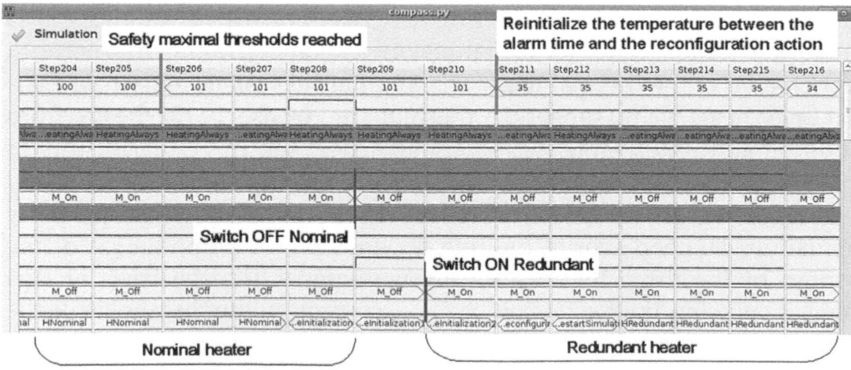

Figure 14.8 An example simulation of the thermal regulation system with a fault injected annotated with the interpretations.

hierarchies and the separation of component interface and implementation. In addition, this separation also eases comparison of alternative designs. Different component implementations, representing design alternatives, can be easily exchanged while preserving the same component type. The resulting overall model can then be automatically validated using the toolset.

- Furthermore, our modeling language has a *proper level of detail* suitable for the design phase. As it expresses behavior in an automata-like way, it encourages the modeler not to represent overly many details by descending to a code-like level.
- From a modeling process perspective, it is important to predefine the perimeter of the system model. Otherwise, the resulting model diverges easily to cover irrelevant or low-criticality details.
- The COMPASS toolset allows for running analyses with fault injections on and off through a simple option. This separation and ease of integration of concerns allow the modeler to see directly the effects of faults on the system with respect to its nominal behavior.
- It is difficult to determine the right property for functional correctness verification using model checking. The predefined set of property patterns aids significantly to this, by providing structure on the most common property specifications. The correct understanding of the property patterns without a priori knowledge of its underlying temporal logics comes through use and experience with them.
- It is important to critically interpret the generated results. A particular output, either expected or unexpected, or desired or undesired, depends on the model, the used properties, and the employed analysis. All three inputs have to be critically interpreted with respect to the generated outcome.

Altogether, the evaluation demonstrated that the basic functionalities for safety and dependability analysis are mature enough for adoption by commercial space industry. The overall methodology and technology is promising, and subsequent investigation was required to integrate it further in a (space) systems engineering life cycle. This investigation has been performed in the satellite platform case study, which is detailed in the subsequent section.

Satellite platform preliminary design case study

The subsystems case study in the previous section raised follow-up questions. The most important were the following: How well does the methodology scale? How should it fit in the (space) systems engineering life cycle? We decided to model and analyze a full satellite in its preliminary design phase. The total duration of the pilot was six months including the initial learning phase.

Satellite architecture

At the highest conceptual level, the satellite is composed of the payload and the platform. The payload comprises mission-specific subsystems and the platform contains all subsystems needed to keep the satellite orbiting in space. The payload is usually designed and tailored from scratch, whereas for the platform lots of design heritage applies. For this reason, our case study focuses on an AADL model of the platform, as this might benefit future projects too. The platform itself is decomposed into several other subsystems, including propulsion, AOCS, command and data handling system, power system, and telemetry, tracking, and control system. See Figure 14.9.

These subsystems are designed with certain degrees of fault tolerance, depending on the respective criticality. Redundancies with reconfigurations, voting algorithms, correcting codes, and compensation procedures are part of comprehensive strategies for achieving fault tolerance. In the extreme case, the satellite should survive a particular number of days without ground intervention assuming no additional failure occurs. As faults could occur at any level in the system's hierarchy (system, subsystem, equipment), the fault management system obeys a cross-cutting design according to the FDIR paradigm. This paradigm separates fault management into three functions. The function of fault *detection* continuously monitors the system and in case of anomalous values,

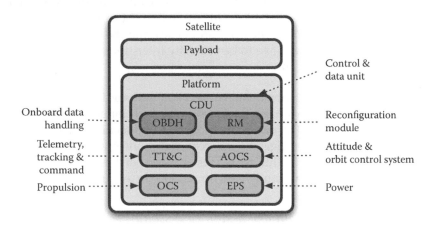

Figure 14.9 Decomposition of the case study's satellite. Orbit control system (OCS) consists of a series of controllable thrusters for orbital corrections. Attitude and Orbit Control System is a control system consisting of several kinds of sensors for acquiring and maintaining a correct attitude and orbit. Control and data unit (CDU) is the main computer. External power supply (EPS) consists of solar arrays and batteries for powering both the platform and the payload. Telemetry, tracking and command (TT&C) is the radio communication interface to ground control on Earth.

emits appropriate events to react on them. Monitoring is decentralized and performed at all levels of the system's hierarchy. After emitting fault detection events, fault *isolation* kicks in. This function is responsible for identifying the affected system's scope by determining the cause of the fault events. The function of fault *recovery* then takes appropriate actions to mitigate the fault events and, if possible, return to a nominal state. As faults can occur at all levels, and as their effects can propagate throughout the system horizontally (within level) and vertically (across levels), the system is partitioned into five levels depending on the complexity of the respective FDIR functions:

- *Level 0*: Failures are associated to a single unit and recovery can be performed by the unit itself.
- *Level 1*: Failures are associated to a single subsystem, and an external subsystem, the on-board software, is responsible for its mitigation.
- *Level 2*: Failures are associated to multiple subsystems, and an external subsystem, the on-board software, is responsible for their mitigation.
- *Level 3*: Failures are occurring in the on-board software or in the processor modules. Dedicated reconfiguration modules are responsible for their mitigation.
- *Level 4*: Failures that are not covered by lower levels and that are completely managed by hardware.

Failures are mitigated at their appropriate level. As with most Earth orbiting satellites, this satellite is required to be single-fault tolerant. If a fault is detected, all fault monitoring is ignored and the isolation and recovery of the detected fault is prioritized. As such, it is designed to handle one fault at a time.

The overall composite system is described by two important modes of satellite operation: nominal and safe. The nominal mode describes a set of satellite configurations in which the system functions within nominal conditions. On the detection of faults, recoveries might be attempted for resuming nominal operation. Otherwise a transition is made into safe mode for which the system reconfigures itself for survival until the ground control station performs an intervention. This important transition has system-level effects and hence is critical. We shall call this important event TLE-1 (*first top-level event*).

Modeling process
During modeling, we focused on one subsystem/equipment at a time. This followed the system decomposition structure, as each subsystem/equipment is typically specified in its own (section of the) design and requirement document. We progressively increased coverage by adding

more detailed subsystems to the overall model, while keeping high-level abstractions or stubs for the remaining subsystems.

The satellite design documents often express various aspects in terms of ranges. We discretized them to avoid a combinatorial explosion of the state space. *Real-time* functionality, like a recovery procedure, was modeled with their tight deadlines. We used mode invariants to ensure that the system cannot stay in a mode forever (i.e., to avoid time divergence), while we used transition guards to ensure time passes in the mode (i.e., to guarantee progress). We accidentally modeled Zeno paths (Herbreteau and Srivathsan 2011) several times. On a Zeno path, infinitely many actions take place in a finite amount of time. Such paths obviously cannot occur in the implementation. They are due to the abstraction of timed behavior of actions. As algorithmic detection of such paths is still heavily studied, manual inspection of the model is currently the only detection method. *Hybrid* aspects, like the evolution of temperature or fluid pressure over time, typically involve dynamics. As the COMPASS toolset only supports constant linear differential equations (due to computational tractability), the involved dynamics had to be severely abstracted to simpler equations. Errors and fault injections were obtained from the preliminary FMEA (which was constructed manually). The FMEA document lists the possible detectable failures as events and relates them to their effect on the system. This mapping is nearly equal to the fault injections. It also provides the information for constructing the error models. We found that in all cases, the probabilistic behavior was either shaped as a single step from an error-free state to an error state (so-called permanent errors), or that they follow a fault-repair loop structure on the error-free/error states (so-called transient errors). The FMEA is also the source for failure rates. They are expressed as failures in time, which indicates the expected number of failures within 10^9 hours.

As our case study was run in parallel with satellite development, the design documents were in a preliminary state and susceptible to change. Every few months, new versions of the design documents, which included detailed change-logs, were distributed. To keep track of the changes, we maintained a traceability spreadsheet that maps each AADL entity to the corresponding points in the design documents. On a new revision, we simply traversed the change-logs, pinpointed the affected parts in the AADL model, and updated them to reflect the change accordingly.

Furthermore, because we are not experts in all of the satellite's subsystems, a major part of the satellite's design was unfamiliar to us. The amount of information is so overwhelming that it is inconceivable to comprehend the system all at once. Information might be perceived as incomplete, unclear, or wrong due to this, and this delays the modeling process. We developed a practice of quickly continuing modeling using assumptive modeling decisions:

- Abstraction: Describes how an AADL element abstracts a part of the system
- Assumption: Describes how an AADL element captures an assumptive understanding of the system
- Direct conversion: Describes how an AADL element maps directly to a part of the system
- Underspecified: Describes why parts of the system design documentation were insufficient for formal representation
- Explained: Are assumptions that have been clarified during review meetings

During review meetings, we had the opportunity to check assumptions, and once these were clarified, the assumption was resolved (i.e., explained).

The resulting model is one of the largest formal models made of a satellite platform design. It consists of 86 components, 937 ports, 244 modes, 20 error models, and 16 recovery procedures. Its nominal state space approximately comprises *48 million states*. After fault injections, the state space is multiplied and the degree of multiplication depends on the kind of injections. For equipment-level failures, we have measured a triple increase of the nominal state space and for system-level failures we have observed up to *213653-fold multiplication* of the nominal state space. See Figure 14.10.

From requirements to formal properties

Requirements documents are developed for all parts of the satellite at all levels, such as at the system level, subsystem level, equipment level, and so on. Furthermore, a particular set of requirements (e.g., system-level requirements) could function as a baseline for a set of lower-level requirements (e.g., subsystem requirements). Not all requirements were amenable to formal analysis. This has several reasons.

High-level requirements typically function as an umbrella for more detailed requirements. They typically lack the detail needed for formal validation. This can be seen in their form, which is typically prose-like, for example, "fault management must be active in all AOCS modes."

A significant part of the requirements do not only cover behaviors but also reflect the system's structural organization. They state which components should be present, and they state how a component is structured in subcomponents, and which components may communicate with each other. Thus they give rise to the system hierarchy. Because we focus on the analysis of dynamic behavior, the validation of these static elements are out of scope in our activities.

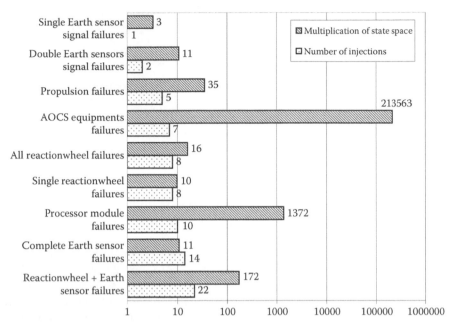

Figure 14.10 Degrees of state space increase with respect to nominal state space size when injecting particular kinds of failures. The scale is logarithmic.

In the early phases of a design, requirements are often specified about the existence of a behavior, without describing how to realize this behavior. An example is that the fault management system should be present to detect, isolate, and mitigate faults, but it is left unspecified how to achieve this. These kinds of requirements are typically subject to refinement in detailed design, where decisions are made on the exact expected behavior. Thus, they generally represent intended underspecification.

Also, unintended underspecification of requirements might occur. These are not trivially to spot. The requirement specifications we used do not explicitly mark requirements as intentionally underspecified. Without an extensive broad background in satellite engineering, and as such a clear picture of unintended underspecification, it is difficult to detect them.

In other cases, requirements can address system aspects that are out of scope of our objectives. Such requirements could, for example, cover behaviors that are intentionally abstracted away in the model. The behavior of the AOCS control loop is an example. It is typically expressed in terms of transformations of sensor data (e.g., sunlight angles, Earth sensor intensities). Such computations are usually abstracted to the status of the behavior, for example, whether it is nominal, degraded, or failed. Also, requirements could refer to the payload, and this was intentionally left out of the scope of our objectives.

From all the requirements we obtained, we found 24 meeting the above criteria and used them for verifying our model. They were each mapped to a specification pattern. In many cases, clarifications were needed during the mapping. For example, which analysis (and why) is suitable for validation? What are the affected modeled components? What constitutes the atomic propositions (e.g., what is exactly a fault recovery function)? Hence for the analysis and mapping of requirements, we additionally maintained an assumptions spreadsheet and a traceability spreadsheet, just like we did for modeling.

Validation activities

Modeling is highly intertwined with analysis, because the output from analysis provides valuable information for possible refinements of the model. The most widely used analysis method during modeling is model simulation, as inspection of traces is a fast sanity check before running a resource-consuming analysis.

During all analyses, particular sets of fault injections were disabled or enabled depending on the aim. This was needed for this case study as we observed that fault injections lead to a significant increase of the state space. This is not surprising. A fault injection basically yields the cross product of the subsystem to which the error is injected, and the error model. As explained before, there is no direct correlation between the amount of fault injections and state space size, although there is a relation between the *kind* of fault injections and the state space size. Fault injections that have system-level impact (e.g., processor module failures) add more (recovery and degraded) system behavior than fault injections with lower-level impact (e.g., Earth sensor failures) as the former affect a larger fragment of the state space. This is depicted in Figure 14.10.

Functional validation. As the COMPASS toolset employs different techniques for discrete and real-time/hybrid models, we modeled a strictly discrete part and a strictly real-time/hybrid part.

On the discrete part of the satellite model, which constitutes the majority, we validated 16 properties. Noteworthy here is that the COMPASS toolset does validation in the absence of any fairness constraints. This means there are no scheduling assumptions, and thus all interleavings are considered. If the model heavily abstracts from the scheduling behavior of components, there can be traces in which a component never performs an action even though the action is enabled because it is outrun by other actions. This means that the component suffers from starvation. Starvation, and other fairness issues, can be avoided by explicitly modeling a scheduling strategy of the system under design. This typically increases the size of the model.

The real-time/hybrid parts of the model are relatively small and were joined together into a single hybrid model. This alternative model was developed to check a requirement stating that the redundant heater is only active in degraded operations. As real-time/hybrid systems cannot be generally verified exhaustively, we used the bounded-model checking as a pragmatic alternative analysis technique. This method is inherently incomplete, as the user-defined bound limits the perimeter of the searched state space (Biere et al. 1999). We experimented with increasing bounds to measure the limit for our model. On our model, we observed that the time needed to validate properties grew exponentially and the memory grew linearly. We stopped our measurements at bound 79, as this exhausted our machine on this model.

Safety and dependability. The platform's most critical event that affects safety and dependability is TLE-1, that is, the transition to safe mode. The transition is triggered on the occurrence of severe failures. In the design documents, a (static) fault tree of 66 nodes is provided relating TLE-1 with the failures. Using our toolset, we could produce the same fault tree from our AADL model in a *fully automated* manner. We also generated a fault tree for the setting of the fail-operational flag. This flag indicates that the satellite's payload services might be impaired due to platform failures. The dynamic variant of fault tree analysis delivered similar results. A FMEA table was generated for mapping the sensor failures with three system effects: detection of failures, the setting of the fail-operational flag, and TLE-1. It often suffices to explore single-fault configurations, although the COMPASS toolset also offers the possibility of inspecting multiple-fault configurations by increasing the fault configuration cardinality. One can then investigate various degrees of fault tolerance and their effects, albeit at the expense of significantly increased computation time.

Fault management effectiveness. For fault detection, we checked which observables are triggered when the transition to safe mode is made. This affects 129 observables. Subsequently, fault isolation was performed on all 129 observables. No properties are used for this, because the observables themselves are the only required inputs. Diagnosability analysis was performed to see whether a double Earth sensor failure is diagnosable (for the satellite operator) when TLE-1 occurs. Because of the size of the model and the time complexity of the algorithms used to check for diagnosability, we had to stop the analysis after seven days without any conclusive result.

Performability.   Reliability requirements are usually defined as a cumulative distribution function and furthermore state that the foreseen reliability must meet or exceed that baseline. Its probabilistic nature fits performability analysis. On our model, we wanted to determine the reliability of the satellite in the presence of a sensor failure. Performability analysis in the COMPASS toolset involves the use of a state space reduction algorithm, which given the large state space of the model, ran out of allocatable memory.

Another approach to verifying the reliability requirements is by computing the probabilities of the TLE-1 fault tree, which was generated during safety and dependability analysis. This is called *fault tree evaluation*. The computation is undertaken almost instantaneously.

Note that though both approaches can be used for this requirement, there are substantial differences. Fault trees are essentially abstract state spaces where the relations between the TLEs and the failures are conservatively over-approximated by AND, OR, and PAND gates. With performability on the other hand, these relations are precisely preserved, because the employed state space reduction algorithm keeps the branching structure of all traces related to the failure intact. For large state spaces, however, this can be resource consuming and in that case the use of fault tree evaluation is the more pragmatic solution.

Lessons learned

- Preliminary assessment of the model perimeter can be derived from analyzing the available requirements. They should be filtered regarding temporal behavior. The criticality of having the respective requirement covered by the model under development should also be determined.
- When a system consists of many concurrently active components, and when the scheduling scheme that manages this concurrency is overly abstracted, starvation might occur. *Fairness constraints* can be added by modeling the scheduler explicitly.
- One has to be attentive of possible *Zeno behaviors* (Herbreteau and Srivathsan 2011). They may occur from modeling real-time and/or hybrid systems. Time divergent traces should also be avoided.
- Our AADL dialect supports *depth-first modeling*. This is especially useful if the modeler has no deep and broad extensive experience of all subsystems under design. By focusing on familiar subsystems first, and keeping the remaining subsystems as stubs, early validation results can be obtained.
- Related to the degree of experience of the modeler, it is wise to keep track of modeling assumptions, applied abstractions, directly formalizable design content, and intended underspecification. This can

be used to assess the fidelity of the model, as well as the maturity of the design of the system.

- Design is in a volatile state during the preliminary design phases of a system. Thus *traceability aspects* of the formal model should be maintained such that design refinements and modifications can quickly be reflected in the formal model.
- Our case study was run on the preliminary design of a system in development. By performing both activities in parallel, we were faced with the same design questions and issues as the engineering team would. This clarified the use of our methodology with respect to the overall engineering life cycle much more than if we had applied our methodology to a fully crystallized and matured system design.

Related work

Several works are known in literature that deal with high-level specification languages like AADL and implementations such as the COMPASS toolset that support quantitative analyses in the context of component- and hierarchy-based models. These approaches are briefly discussed in the following sections.

High-level specification languages

Another important high-level approach is the UML (OMG 2014b). Although initially coping with software models, UML has been extended so as to model application structure, behavior, and architecture, and also business processes and data structures. Recently, UML has been customized to fit the system engineering domain in the form of SysML (OMG 2014a). It uses seven out of the 13 types of diagrams in UML 2.0 and adds two new types (Requirements and Parametric diagrams). Unfortunately, SysML also inherits particular weaknesses from UML, of which most important in this context is the loose formalism concerning nominal and error behavior modeling.

Closer to our approach is AUTOFOCUS (Hölzl and Feilkas 2010), a formalism for component-based modeling of reactive, distributed systems that provides validation and verification mechanisms. In the associated modeling language, called Focus, a system is represented as a set of communicating components, each having its own behavior specification. Components exchange pieces of data in the form of typed messages via channels and are synchronized by a global clock, assuming a discrete-time model. Focus models can be formally verified by a translation to Isabelle/HOL. It is also possible to generate C code. The COMPASS toolset does not provide the latter functionality, but instead adds a safety and dependability dimension to the modeling and analysis of nominal system software.

The Behavior Interaction Priority (BIP) framework (Chkouri et al. 2008) allows the construction of complex systems by coordinating the behavior of a set of atomic components. Coordination in BIP is based on connectors, to specify possible interaction patterns between components, and priorities, to select amongst possible interactions. In contrast to our approach, failure modeling is not (explicitly) supported.

AADL tools

Several tools have been developed to analyze AADL specifications. However, these tools either focus on functional correctness or safety and dependability analysis, whereas our methodology allows for both using the same model.

The AADL2BIP tool (Chkouri et al. 2008) translates an AADL specification into a BIP component specification. This translation enables the simulation of the specification and the application of formal verification techniques developed for BIP. At the moment, these tools allow one to perform only simulation and deadlock detection.

Within the TOPCASED project (Berthomieu et al. 2008), an intermediate modeling language called Fiacre was developed to ease the translation from higher-level specification languages like Specification and Description Language (SDL) and AADL to the lower-level formalisms used by model checkers. Specifically, a toolchain was created that translates AADL and its Behavior Annex to Fiacre, and from Fiacre to Timed Petri nets. The latter are then model checked using the TINA toolbox. Another backend for Fiacre targets LOTOS-NT, a process-algebraic modeling language, which is supported by the CADP toolbox. Both Fiacre backends cover support of AADL, its Behavior Annex and real-time properties. Contrary to COMPASS, however, they do not integrate erroneous behavior.

The ontology-based transformation from AADL to Altarica (Mokos et al. 2010) takes advantage of the ontology languages built-in reasoning capabilities to bridge the semantic gap between these two languages. Moreover, it allows one to detect the lack of model elements and semantically inconsistent parts of the system design.

Case studies

In the past, other case studies have applied formal methods within the space domain. We highlight a few that bear some similarity with the ones described in this chapter.

In the work by Rugina et al. (2012), the validation of a satellite's design is supported by simulation of several linked models comprising parts of the spacecraft and its FDIR system. The result is an integrated executable model of high fidelity on which functional correctness properties can be validated. Our case studies do not reuse and link existing models, and the fidelity of our model is lower so as to mitigate the state space problem. We, however, validate beyond functional correctness, covering safety and dependability and performance aspects.

In the work by Iliasov et al. (2013), a formal approach is described for the specification of an AOCS. Central to that approach is that correctness properties are preserved by each refinement of the specification. The COMPASS approach allows for incremental specification by deepening the system hierarchy, but verifies correctness properties on a whole system architecture at once. Our case studies furthermore cover the full satellite platform, which include the AOCS.

In the work by Brat et al. (2004), NASA studied the use of formal analysis techniques for finding bugs in the Mars Rover software. They focused on program verification of spacecraft software implementation. Our case study focuses on the spacecraft system software design, and analyses it from a correctness, safety, and dependability and performance perspective.

Conclusion

Our methodology is a model-based approach supporting a wide range of V&V analyses for safety-critical systems, in particular spacecraft. It leverages decades of theoretical and tool advances from the formal methods community. Several model checkers and probabilistic variants thereof are joined together in a consistent and coherent way by a formal semantics underlying our AADL modeling dialect. The resulting toolset is designed to be friendly to engineers without prior knowledge of model checking or formal methods in general. We demonstrated the toolset's effectiveness for spacecraft design validation using industrial case studies. These case studies, described in this chapter, were performed on system-level and subsystem-level satellite designs of past and ongoing space missions. They pinpointed strengths of our methodology as well as areas for improvements, both theoretical and practical. More information about the modeling language and the toolset is available from Bozzano et al. (2011). Distribution of the COMPASS toolset is restricted to ESA member states. The exact license, along with manuals, tutorials, and presentations are available on the COMPASS website, http://compass.informatik.rwth-aachen.de/.

Bibliographic notes

The majority of this chapter is based on the work published by Esteve et al. (2012) and Bozzano et al. (2011).

Spacecraft system engineering builds on a wide range of various natural sciences such as physics, chemistry, mathematics, and computer science. Each spacefaring nation has its own distinct engineering tradition in this field. The American and European traditions are nevertheless quite similar. A comprehensive overview of this subject matter from an American/European perspective can be found in the work by Fortescue et al. (2011).

In this chapter, we have emphasized a model-based approach toward the technical aspects of the engineering process. Specifically, we capture the system (e.g., spacecraft) as a model in the AADL. AADL is standardized by the SAE and a comprehensive introduction can be found in the work of Feiler and Gluch (2013).

We furthermore stress the value of formal methods for model-based engineering of safety-critical systems. Formal methods provide for rigor and automation. There is a wide range of formal techniques and associated tools out there. Our work fits in the model checking branch, which has been highly successful in industry. The seminal work by Baier and Katoen (2008) provides both an in-depth and broad overview of model checking concepts, techniques, and algorithms.

Acknowledgments

This work has been supported by ESA/ESTEC (contracts no. 4200021171 and 4000100798) and Thales Alenia Space (contract no. 1520014509/01).

References

Audemard G., Bozzano M., Cimatti A., and Sebastiani R. "Verifying Industrial Hybrid Systems with MathSAT." In: *Proceedings of the 2nd International Workshop on Bounded Model Checking (BMC)*. Ed. by Biere A. and Strichman O. *Electronic Notes in Theoretical Computer Science* 2. Elsevier, 119: 17–32 (2005).

Baier C., Haverkort B. R., Hermanns H., and Katoen J. P. "Model-Checking Algorithms for Continuous-Time Markov Chains." In: *IEEE Transactions on Software Engineering* 29(6): 524–541 (2003).

Baier C., Haverkort B. R., Hermanns H., and Katoen J. P. "Performance Evaluation and Model Checking Join Forces." *Communications of the ACM* 53(9): 76–85 (2010).

Baier C. and Katoen J. P. *Principles of Model Checking*. MIT Press, Cambridge, MA, 2008.

Berthomieu B., Bodeveix J., Farail P., Filali M., Garavel H., Gaufillet P., Lang F., and Vernadat F. "Fiacre: An Intermediate Language for Model Verification in the TOPCASED Environment." In: *Proceedings of Embedded Real-Time Software and Systems (ERTSS)*, Toulouse, France, 2008.

Biere A., Cimatti A., Clarke E. M., and Zhu Y. "Symbolic Model Checking without BDDs." In: *Proceedings of the 5th International Conference on Tools and Algorithms for Construction and Analysis of Systems (TACAS)*. Ed. by R. Cleaveland. *Lecture Notes in Computer Science*. Springer-Verlag, Heidelberg, Germany, 1579: 193–207 (1999).

Boudali H., Crouzen P., and Stoelinga M. "A Rigorous, Compositional, and Extensible Framework for Dynamic Fault Tree Analysis." *IEEE Transactions on Dependable Secure Computing* 7(2): 128–143 (2010).

Bozzano M., Cimatti A., Katoen J. P., Nguyen V. Y., Noll T., and Roveri M. "Safety, Dependability and Performance Analysis of Extended AADL Models." *The Computer Journal* 54(5): 754–775 (2011).

Brat G., Drusinsky D., Giannakopoulou D., Goldberg A., Havelund K., Lowry M., Pasareanu C., Venet A., Visser W., and Washington R. "Experimental Evaluation of Verification and Validation Tools on Martian Rover Software." *Formal Methods in System Design* 25(2–3): 167–198 (2004).

Chkouri M. Y., Robert A., Bozga M., and Sifakis J. "Translating AADL into BIP - Application to the Verification of Real-Time Systems." In: *Proceedings of the 1st International Workshop on Model-Based Architecting and Construction of Embedded Systems*, Toulouse, France, 2008.

Cimatti A., Clarke E., Giunchiglia E., Giunchiglia F., Pistore M., Roveri M., Sebastiani R., and Tacchella A. "NuSMV 2: An Open- Source Tool for Symbolic Model Checking." In: *Proceedings of the 14th International Conference on Computer Aided Verification (CAV)*. Ed. by E. Brinksma and K. G. Larsen. *Lecture Notes in Computer Science*, Springer-Verlag, Heidelberg, Germany, 2404: 359–364 (2002).

Cimatti A., Pecheur C. and Cavada R. "Formal Verification of Diagnosability via Symbolic Model Checking." In: *Proceedings of the 18th International Joint Conference on Artificial Intelligence (IJCAI)*. Ed. by G. Gottlob and T. Walsh. Morgan Kaufmann, San Francisco, CA, pp. 363–369 (2003).

COMPASS Consortium. *COMPASS Toolset User Manual*, 2012.

Dwyer M. B., Avrunin G. S., and Corbett J. C. "Patterns in Property Specifications for Finite-State Verification." In: *Proceedings of the 21st International Conference on Software Engineering (ICSE)*. Ed. by Boehm B. W., Garlan D. and Kramer J. ACM, 1999.

ECSS. Failure Modes, Effects (and Criticality) Analysis. ECSS- Q-ST-30-02C. ESA Requirements and Standards Division, Noordwijk, Netherlands, Mar. 2009a.

ECSS. Fault Tree Analysis. ECSS-Q-ST-40-12C. ESA Requirements and Standards Division, Noordwijk, Netherlands, Jul. 2008.

ECSS. Project Planning and Implementation. ECSS-M-ST-10C Rev. 1. ESA Requirements and Standards Division, Noordwijk, Netherlands, Mar. 2009b.

ECSS. Verification. ECSS-E-ST-10-02C. ESA Requirements and Standards Division, Noordwijk, Netherlands, Mar. 2009c.

ECSS. Verification Guidelines. ECSS-E-HB-10-02A. ESA Requirements and Standards Division, Noordwijk, Netherlands, Dec. 2010.

Esteve M. A., Katoen J. P., Nguyen V. Y., Postma B., and Yushtein Y. "Formal correctness, safety, dependability, and performance analysis of a satellite." In: *Proceedings of the 2012 International Conference on Software Engineering (ICSE)*. 2012.

Feiler P. H. and Gluch D. "Model-Based Engineering with AADL." Addison-Wesley, Boston, MA, 2013.

Fortescue P., Swinerd G., and Stark J. *Spacecraft Systems Engineering*. John Wiley & Sons, West Sussex, United Kingdom, 2011.

Grunske L. "Specification Patterns for Probabilistic Quality Properties." In: *Proceedings of the 30th International Conference on Software Engineering (ICSE)*. Ed. by Schäfer W., Dwyer M. B., and Gruhn V. ACM, New York, 2008.

Herbreteau F. and Srivathsan B. "Coarse Abstractions Make Zeno Behaviours Difficult to Detect." In: *Proceedings of the 22nd International Conference on Concurrency Theory (CONCUR)*. Ed. by Katoen J. P. and König B. *Lecture Notes in Computer Science*, Springer-Verlag, Heidelberg, Germany, 6901: 92–107 (2011).

Hölzl F. and Feilkas M. "AutoFocus 3: A scientific tool prototype for model-based development of component-based, reactive, distributed systems." In: *Proceedings of the 2007 International Dagstuhl conference on Model-based engineering of embedded real-time systems*, pp. 317–322 (2010).

Iliasov A., Troubitsyna E., Laibinis L., Romanovsky A., Varpaaniemi K., Ilic D., and Latvala T. "Developing Mode-Rich Satellite Software by Refinement in Event-B." *Science of Computer Programming* 78(7): 884–905 (2013).

Mokos K., Meditskos G., Katsaros P., Bassiliades N., and Vasiliades V. "Ontology-Based Model Driven Engineering for Safety Verification." In: *Software Engineering and Advanced Applications (SEAA), 2010 36th EUROMICRO Conference on*, Lille, France, pp. 47–54 (2010).

NASA. NASA Software Engineering Requirements, NPR 7150.2A, Nov. 2009.

NASA. *NASA Systems Engineering Handbook*. NASA/SP-2007- 6105 Rev. 1. NASA Center for AeroSpace Information, Hanover, MD, Dec. 2007.

OMG. Systems Modeling Language. Available at http://www.omgsysml.org/. Last accessed 10 Sept. 2014a.

OMG. Unified Modeling Language. Available at http://www.um.org/. Last accessed 10 Sept. 2014b.

Rugina A., Leorato C., and Tremolizzo E. "Advanced Validation of Overall Spacecraft Behaviour Concept Using a Collaborative Modelling and Simulation Approach." In: *21st International Workshop on Enabling Technologies: Infrastructure for Collaborative Enterprises (WETICE)*, Toulouse, France, pp. 262–267, Jun. 2012.

Rushby J. Formal Methods and Their Role in Digital Systems Validation for Airborne Systems, NASA Contractor Report 4673, Aug. 1995. Available at http://ntrs.nasa.gov/archive/nasa/casi.ntrs.nasa.gov/19960008816.pdf.

SAE. Architecture Analysis and Design Language (AADL) Annex Volume 1. AS5506/1. Society of Automotive Engineers, Jun. 2006.

SAE. Architecture, Analysis and Design Language. AS5506. Society of Automotive Engineers, Nov. 2004.

chapter fifteen

Modeling and simulation framework for systems engineering

Saikou Diallo, Andreas Tolk, Ross Gore, and Jose Padilla

Contents

Introduction

In many application domains (military, homeland security, health care, and business), modeling and simulation (M&S) has become a standard approach for systematically developing, testing, and acquiring systems. The use of M&S usually involves three major steps: (1) identifying user requirements to capture stakeholders' needs, (2) conceptual modeling to capture system's parts and relationships, and (3) verification and validation (V&V) to evaluate the accuracy, correctness, and credibility of the simulation (Balci 1994; Balci 1998). Although these activities are supported successfully in systems engineering (SE) through standard processes, a formal (in the mathematical sense) process that ties in all SE activities has not been formulated. Further, SE projects that involve both humans and computer systems are very difficult to formulate, track, and validate because computers can only support simplified aspects of a real system

(Zeigler 1976; Balci 1998; Dill 1998; Pace 2004). Tolk et al. (2013) addressed this gap by proposing an M&S system development framework (MS-SDF) that unifies SE and M&S processes through model theory. The MS-SDF is a framework for modeling real-world complex system often called *problem situations*. A *problem situation* occurs when there are multiple, possibly competing, perspectives on how to deal with a problem or represent a system. The MS-SDF allows for multiple viewpoints of a system to be captured and for a single representation of the system to be specified and tracked. The result is an end-to-end framework that ensures each requirement specified is satisfied in the conceptual model and the system, and that the requirements, conceptual model, and system are all consistent with one another. Thus, by applying the framework, we can prove that each requirement is always satisfied, exactly met, and most importantly that there is no emergence in the system. The balance of the chapter is organized as follows: the Section "MS-SDF" introduces the MS-SDF and discusses how it can be applied to develop systems; the Section "Application example: Building an experimental system to support decision making before sea level rise flooding" presents a real-world problem and describes how the MS-SDF is applied to develop a system. The Section "Conclusion" presents recommendations and future work. To avoid confusion, in the balance of this chapter, the term *system* refers to a system that we develop (the purposeful simplification of a real system) and the term *real system* refers to the actual system we are attempting to represent.

MS-SDF

The goal of the MS-SDF is to capture a complex system explicitly through an iterative process. The foundation of MS-SDF is model theory, a branch of mathematics focused on the study of mathematical structures (Hodges 1993; Ebbinghaus and Flum 2005). The basic tenet of this approach is that truth is relative, and therefore a system is a reflection of the stakeholders' perspective and understanding of the real system. In other words, although the real system stands on its own, each stakeholder has a representation of that system based on how they interact with it, what they know about it, and, as importantly, what they assume about it. Therefore, the process of developing a system should reflect the subjective nature of the stakeholders' understanding of the system along with its more objective functions. The MS-SDF process shown in Figure 15.1 aims at capturing and tracing all aspects of the system development process to better understand what was included in a system, what was left out and to provide a context in which V&V can meaningfully occur.

The MS-SDF processes are described in the Section "Reference modeling." We purposefully skip the mathematical formalism that justifies the MS-SDF and focus on its practical implications instead.

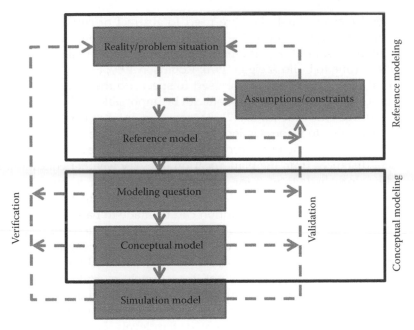

Figure 15.1 MS-SDF. (From Tolk, A. et al., *Journal of Simulation*, 7(2), 69–82, 2013. With permission.)

In the work by Tolk et. al (2013), we motivate the need for a framework such as the MS-SDF, but do not explicitly elaborate on the different types of V&V that have to be performed to ensure consistency between the different products of the framework.

Reference modeling

The MS-SDF starts with the construction of a *reference model*. The goal during the *reference modeling* phase is to establish what is known and assumed about the system (Tolk et al. 2013). It captures the knowledge of every stakeholder's understanding of the real system. This includes capturing what each stakeholder knows or believes to be true (or false) about the system. The reference model includes knowledge about inputs, outputs, actors, components, behaviors, and relationships. Recall that our main concern at this moment is to arrive at a consensual description of the system. It is very likely that contradictions will emerge between the different views expressed by stakeholders. As our goal is to be as complete as possible, it is essential not to try to resolve any contradictions within the reference model. Instead contradictions are noted and stakeholders are made aware of their existence.

In addition, the MS-SDF requires all assumptions about the system to be explicitly captured in the reference model. Assumptions are made for several reasons. First, they are used to abstract or *simplify* information considered irrelevant to the system being developed. Without the ability to make simplifying assumptions, a system would take an infinite amount of time to build (Zeigler 1976; Robinson 1997; Spiegel et al. 2005; Robinson 2007). They can also be used to speed up the system development process in situations where data can be impossible to obtain because of (1) an unrealistic time frame, (2) placement in an unreachable location, or (3) danger due to its possession. For example, in representing a power plant, it might be difficult to account for the radiation levels of a leak at different distances from the plant location in every weather condition. It would be unrealistic to wait for an actual leak to occur to measure the radiation levels and also be too dangerous to intentionally start a leak to collect data. Therefore, an assumption can be made to base the radiation levels of the leak on data obtained from other radiation leaks or disasters in the past. Finally, assumptions can be used to construct systems where specific data are theoretical or unknown. This is useful for experimental systems where the purpose is not to accurately reproduce a real system but to investigate new systems or test hypotheses. King and Turnitsa (2008) also note the use of assumptions in experimental systems to clarify the interactions between the different system levels and to establish the relationship between the system elements.

The list of assumptions needs to come from every stakeholder involved in the system development process. This allows for different views of the system to be represented based on the viewpoints of each stakeholder. The result is a reference model that contains a list of *statements* that not only describes the system but also bounds it around all of its assumptions.

Once we have captured what we know and assume about the system, we need to conduct three steps of *model level validation*, which are shown in the green boxes in Figure 15.2:

- *Model assumptions validation* (Process 1): In this process, we consult with the stakeholders to ensure that all assumptions are *reasonable*. The definition of what constitutes a *reasonable assumption* is context sensitive and rests solely on the appraisal of the stakeholder. The goal is to ensure that stakeholders are aware of each other's

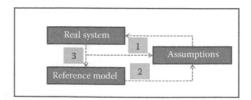

Figure 15.2 Reference model construction.

assumptions about the system and that they acknowledge the different viewpoints involved. Assumptions can be changed or modified if a stakeholder wishes to amend his/her position or viewpoint. This process is iterative, and we recommend starting it early and continuing it throughout the system development process.

- *Contradiction and equivalence validation* (Process 2): In this process, we ensure that all assumptions are recorded in the reference model and we identify all contradictions and equivalences that might emerge among all combinations of assumptions and knowledge. The contradictive and equivalent statements are not removed; instead they are presented to the stakeholders for discussions, which might lead to changes to the initial set of assumptions. The process needs to be iterated with the assumption validation process.

- *Emergence validation* (Process 3): When we combine two or more assumptions or when we combine assumptions with knowledge, additional statements about the system emerge. A simple example is the set of sentences that emerge when two or more sentences are connected logically (*a* AND *b*, *an* OR *b*, *a* IMPLY *b*). These statements need to be validated against the real system either directly or through the stakeholders.

These three processes are lengthy, cumbersome, error prone, and difficult to track for nontrivial system. For these reasons and others, many researchers recommend using formal means to capture the reference mode (Milne 1993; Clarke and Wing 1996; Tolk et. al 2013). For instance, the static portion (entities, actors, and relationships) can be captured as an ontology whereas the dynamic portion (functions, behaviors, algorithms) can be captured in a specification language suitable for model checking, such as the Alloy Specification Language (Jackson 2002) or the Prototype Verification System (Owre et al. 1995). The use of a formal means allows us to automate processes 2 and 3 and semiautomate process 1.

The created reference model acts as a *surrogate* for the real system. This implies that validation of the resulting system is done against the reference model and not the real system. Recall that at this moment, the reference model contains what is known and assumed about the system including contradictive, equivalent, and emergent statements. We can now proceed to building a conceptualization of the system. It is important to emphasize that when we are dealing with a problem situation, there is a strong disagreement on what the problem is and how to describe it. The reference modeling phase is an attempt to identify and capture all problem descriptions. The conceptual modeling phase on the other hand is an attempt to arrive at a consistent problem description based on the reference model.

Conceptual modeling

The reference model acts as a representation of the real system. After we are confident that all stakeholders agree that their understanding of the system has been captured, we move to capture a set of *modeling questions* that they would like to have the system answer. From an M&S perspective, the *modeling question* is the reason a simulation of the system is being built. From an SE perspective, the *modeling question* is the goal of the system. If the system is built for experimentation, the goal is formulated as a set of questions similarly to an M&S project. If the goal is to replicate the real system, the goal is formulated as a set of functional requirements for the system. Either way, a modeling question is a statement that a stakeholder would like to know the answer to, meaning it is not already known or assumed. Formally, it is a statement for which there is no truth value. It turns out that these types of statements are hard to come by for a complex system. Often formulating the modeling question is the hardest part of the project because each stakeholder can have multiple and possibly competing questions. The goal of this phase is to collect all of the modeling questions from every stakeholder and most importantly ensure that the modeling question is answerable using the specified reference model. In most cases, the formulation of a modeling question might lead to the introduction of new assumptions or knowledge, which might lead us to validate the reference model anew. This is another reason for using formal means to conduct the system development process—we can automatically detect new contradictive, emergent, or equivalent statements.

Once we have a stable set of modeling questions and relevant information on the questions added to the reference model, we can construct a *conceptual model*. For the purpose of the MS-SDF, the *conceptual model* is the subset of the reference model that can answer the modeling questions. This means that every statement in the conceptual model is also a statement of the reference model. In addition to being a subset of the reference model, we require that the conceptual model be *consistent*. Here, *consistent* means the conceptual model is free of contradictions. This is very important because if the conceptual model is inconsistent the system will also be inconsistent. At best an inconsistent system is undesired and at worst it is dangerous (Zeigler 1976; Jackson 2002; Tolk et al. 2013). Consequently, we have to resolve contradictions in the reference model between stakeholders for the purpose of answering the modeling questions. Note that, the resolution is bounded by the modeling questions the stakeholders wish to answer. This forces them to compromise on what is necessary and sufficient to answer the question as opposed to the system developer deciding what is best to include and exclude.

We introduce the notion of *constraints* to capture additional assumptions that have to be made to resolve contradictions and arrive at a consistent conceptual model. *Constraints* are the type of assumptions made to remove irrelevant knowledge and assumptions from the reference model. Once the conceptual model is specified and constrained, the system can be built. The system must be a consistent subset of the conceptual model. This means that the system itself might need to be additionally constrained. Here, constraints are usually introduced by the (1) environment in which the system is built (computational resources, money, manpower, etc.), (2) restrictions in terms of time and data availability, and (3) other system requirements (speed, reliability, usability, etc.) (Brooks 1995; Verner et al. 1999). We advocate, capturing constraints with the same formal means as the reference model to inherit the benefits of automation in tracking and validating the system against the conceptual model and the reference model. Figure 15.3 shows all of the validation paths that have taken place to ensure that the system is an acceptable representation of the reference model. More precisely, we need to ensure that the system is a valid representation of the reference model such that it can answer the modeling questions consistently.

We call this level of validation *system level validation* with the following steps:

- *Modeling question validation* (Process 4): In this process, we ensure that the modeling questions can be answered by the existing reference model. If not, we have to introduce additional knowledge and/ or assumptions and go through the entire system validation process.
- *Conceptual model validation* (Process 5): In this process, we ensure that the constrained conceptual model is both consistent and a subset of the reference model. We have to be careful not to introduce new knowledge or assumptions directly in the conceptual model. This means that every statement in the conceptual model can be obtained directly from the reference model or it can be obtained by combining a sentence in the reference model with a set of assumptions and constraints.
- *System validation* (Process 6): In this process, we ensure that the system, as constructed, is a consistent subset of the conceptual model and we have to be careful not to introduce new knowledge or assumptions in the system. Every statement in the system must also

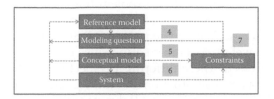

Figure 15.3 Validation path from system to reference model.

be a statement of the conceptual model or be obtainable from a combination of statements with constraints.

- *Constraints validation* (Process 7): In this process, we ensure that the constraints are *reasonable*. Again, as with the assumptions, the nature of what is *reasonable* is subject matter expert dependent. We recommend asking the stakeholders to validate constraints and at a minimum agree that they are needed. Constraints can also be competing and contradictive. This has to be solved by stakeholders and engineers to ensure that we have a *consistent* set of constraints. Again by *consistent* we mean not contradictive in the formal sense.

All of these processes are iterative and the labeling of processes in terms of numbers is to uniquely identify them and not to indicate a sense of order. In practice, we recommend, validating constraints as they are identified and using formal means to automate the deductive process of identifying contradictions and emergence. In the Section "Application example: Building an experimental system to support decision making before sea level rise flooding," we present a practical application of the MS-SDF.

Application example: Building an experimental system to support decision making before sea level rise flooding

The destruction of coastal regions of the United States is becoming a growing concern as the level of the sea is slowly increasing along the coasts and in some areas the level of the earth is slowly sinking as well. The U.S. Environmental Protection Agency (EPA) states that sea level rise may cause some ecosystems and communities, especially those that fall in low-lying areas, to become permanently lost (Titus 1999). Scientific data show that the city of Norfolk, Virginia, is sinking while the sea level is rising, and sea level rise by the year 2100 is expected to approach or exceed 1 m (Weiss et al. 2011). Current policies for handling sea level rise have ranged from relocating people or resources to creating dikes or dredging the beaches to slow the effects of sea level rise. Other approaches have included abstracting the sea level rise to a problem that will not have any effect on coastal regions in the future. This viewpoint is taken by the state of North Carolina whose state senate passed Bill 819 (2012) that bans the use of projected sea level rise data in creating rules for governing how coastal regions are addressed. These differing perspectives on the effects of sea level rise make it a problem situation and thus, an ideal candidate for application of the MS-SDF. In the Section "Reference model of sea level rise," we discuss how to construct a reference model of the sea level rise situation.

Reference model of sea level rise

To create a system that can be used to study the effects of sea level rise on an area, a consensus must first be reached on what the problem is and what questions we want the system to answer. The very first step is to collect information about the effect of sea level rise on an area as well as possible mitigation factors. In this case, we mined the web and identified authoritative data sources to extract information. Authoritative data sources include scientific literature on sea level rise, subject matter experts, government agencies, and websites. In accordance with the recommendation to use formal means, we organize the knowledge gained from these sources into an ontology that we formally capture in Protégé (Fensel 2001; Protégé-Ontology 2007).

To identify critical entities susceptible of being damaged by sea level rise in a region, we use the infrastructure taxonomy created by the U.S. Department of Homeland Security (DHS, 2009). This taxonomy contains all the physical and organizational aspects that create value in an area separated into 18 different factors shown in Table 15.1.

This taxonomy forms the basis of our reference model. Our next step is to start interacting with the stakeholders at the local level. Stakeholders include city managers, health-care professionals, and engineers (transportation, civil, etc.).

Table 15.1 Physical Assets in an Area

Assets
Agriculture and food
Banking and finance
Chemical and hazardous materials industry
Defense industrial base
Energy
Emergency services
Information technology
Telecommunications
Postal and shipping
Health care and public health
Transportation
Water
National monuments and icons
Commercial facilities
Government facilities
Dams
Nuclear facilities
Manufacturing

After discussions with the health-care stakeholders, we added socio-economic and political factors that cover aspects such as vulnerable populations and political will. All of these factors are entered into Protégé where they are organized into a hierarchical structure and their relationships are also explicitly defined in various contexts. An example diagram from Protégé is shown in Figure 15.4. This diagram shows the possible mitigation factors that were collected for addressing sea level rise, including water channeling, raising elevations, and shoreline armoring. These mitigation factors are then further refined into area-specific factors such as raising dock heights at amphibious bases within the area.

The discussions also lead to the identification of three perspectives on how to deal with sea level rise:

- *Option 1: Maintain*: Maintaining an area means that the area will continue to receive budgeted funds to protect or reduce the effects of sea level rise in the area.
- *Option 2: Vacate*: Vacating an area means that budgeted funds will be reallocated to other areas.
- *Option 3: Invest*: Investing in the area means that additional funds will be procured to mitigate the effects of sea level rise in the area.

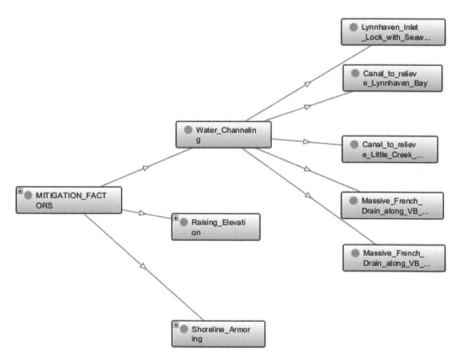

Figure 15.4 Mitigation factors of sea level rise.

However, the determination on whether to vacate, maintain, or invest in an area is based on how important the area is from the viewpoint of the stakeholder. Therefore, it became important to define the *value* of an area. This is difficult because of the subjective nature of the term, *value*. The reference model captures value objectively by incorporating the traditional and quantitative metrics, for example, GDP, economic activity, population, commodity production, and consumption, and subjectively by incorporating social welfare factors such as health care, cultural and historical landmarks, and so on. Stakeholders want these intangibles to be evaluated side by side with the more concrete causes. Thus, the focus of our reference model is to identify assets in a region and a way to *value* them as well as mitigation factors that a stakeholder might want to implement. Figure 15.5 shows a snapshot of the sea level rise ontology. To illustrate, let's focus on the water ontology.

The major classes of the water ontology are water system, water type, and consumer. Other water consumption points such as agriculture and food and commercial facilities (e.g., manufacturing) are not included in the water ontology because they are annotated in their own factor ontology. The factor of water encompasses the subclasses *Raw Water Storage, Raw Water Supply, Raw Water Transmission, Treated Water Storage, Treated Water Distribution Control Centers, Treated Water Distribution Systems, Treated Water*

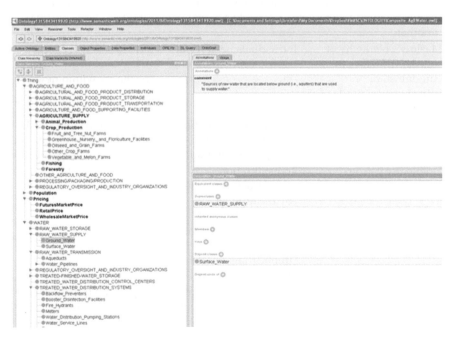

Figure 15.5 A snapshot of the sea level rise ontology in Protégé.

Monitoring Systems, Wastewater Facilities, and *Water Treatment Facilities.* For example, the subclass *Treated Water Distribution Control Centers* is related to *Treated Water Distribution Systems* by the necessary and sufficient condition otherwise known as a defined class that the *"Treated Water Distribution Control Centers* monitors and operates only *Treated Water Distribution Systems."* In this case, any instance, that is, an actual named and functioning facility, that is categorized as monitoring and operating only a treated water distribution system will also be categorized as being assigned the class of *Treated Water Distribution Control Center* and vice versa.

The reference model also captures key assumptions that we made when talking to the stakeholders. Our very first assumption was that sea levels are rising and we made sure that every stakeholder who disagrees assumes that it is true otherwise they would not have anything to contribute to the project. Other key assumptions are (1) mechanisms by which sea levels are rising are not important, (2) the exact numbers in terms of how many millimeters of sea level rise causes a flood are not important, and (3) specific events (hurricane, storm, etc.) are not important to be described in minute details. Table 15.2 shows a list of assumptions about

Table 15.2 Assumption Related to the Affected Areas

Assertion: Population	Meaning	Implementation
People have the ability to move from one area to another.	A single person is not confined to a single area.	Each person has the ability to leave an area.
Maslow's hierarchy of needs—provides the fundamental physiological needs of individuals.	Individuals need food, water, sleep, breathing, etc.	Individuals will check the availability of basic resources within new areas.
People use resources in the area where they live.	Each person uses some amount of water, food, energy, etc., annually.	Number of people affects the total water usage, total energy usage, etc.
People use various transportation methods.	Transportation infrastructure is affected by the number of people utilizing it (utilization of private vehicles as well as public transportation).	People contribute to traffic congestion and increase the need for public transit.
People need water.	Higher populations increase the amount of water that needs to be distributed across an area.	Water distribution expansion (pipelines) is affected by the total number of people.

the population that we collected and included in the reference model. Some of these assumptions came directly from stakeholders and others were generated iteratively after we went through the reference model. We also captured assumptions about the area affected by sea level rise and the decision makers involved in defining and implementing policy.

The reference model was iterated over several times, and we constructed several small prototype systems to help stakeholders visualize their assertions. We put the model through the validation process and identified several contradictions. For instance, an area cannot be both *invested in* and *vacated*. However, an area can be both *maintained* and *invested in* or both *maintained* and *vacated*. This leads us to propose the introduction of a threshold for maintaining, *investing*, and *vacating* an area. Each stakeholder can set the thresholds based on what they believe the value of an area to be. Once we have a reference model, we can construct a conceptual model.

Conceptual model of the effects of sea level rise on an area

To construct a conceptual model, we must first identify one or more modeling questions. After talking to the stakeholders, the modeling questions in Table 15.3 were identified and validated as being relevant to the problem at hand. Note that not all of the questions are directly relevant to the effect of sea level rise.

On the basis of the level of granularity of these questions (person, area, resources), we decided to construct a conceptual model that can help us answer these questions. We used the Unified Modeling Language (UML) to capture the conceptual model (D'Souza and Wills 1998). As the conceptual model is what we implement, we decided to use well-known

Table 15.3 Sample of Modeling questions

Actor	Modeling question
Decision maker	What is the best combination of factors that would make an *area* as attractive as possible for as long as possible?
Person	What is the best *area* (area that meets most of the needs of most people) for a person to live in? How about businesses?
Area	What types of areas are most/least attractive; under what conditions?
Critical infrastructure	What is the combination of critical infrastructure that should be protected for an *area* to survive or thrive?
Mitigation factors	What combination of mitigation factors is most/least useful in protecting an area?
Resource	What is the best use of resources for an *area* to thrive?

paradigms to facilitate the transition between conceptual modeling and implementation. We used agent-based modeling (ABM) (Bonabeau 2002) to create autonomous entities and to define rules for the entities within the system and system dynamics (SD) modeling (Karnopp et. al 1976) to represent the behavior of the areas (i.e., cities) within the environment. ABM allows for the beliefs, desires, and intentions of each of the entities to be set and then allows for the behavior of the system to be observed with respect to the actions of the agents. SD is better suited to represent the complex behaviors of areas and helps in understanding the long-term effects of a complex system over time. Discrete event simulation (DES) (Kelton et. al 2002) allows us to track the flow of people and businesses within and outside of an area. The agents of interest for the simulation are the *people* in the system and the *investments* in the system.

The *people* agents represent the general population that is moving into or out of an area whereas the *investment* agents represent businesses or government investments moving into areas. The agents contain pre-defined values for how they interact with the area while they are in it and follow specific sets of rules for determining if they want to remain in their current area or if they want to leave and find a new area. Each rule set is captured in the UML and validated with the stakeholders.

The *people* agents were given a worldview to follow within the system. The worldview represents an initial set of conditions that must be met in order for a person to move into a new area. The *people* agents move through the simulation environment looking for new areas and when they arrive at the area they check their requirements (different for each *people* agent) against the conditions of the area. If these conditions are not met then the *people* agent leaves the area that it is currently at and looks for a new area. If the conditions are met, then the *people* agents check a more specific set on conditions called their impact attributes against the values of the area. The impact attributes represent the specific factors within the area that the *people* agent will have the greatest interaction with while in the area. The impact attribute values per *people* agent are assigned as a random distribution so that each of the agents has a different set of values. This is designed to represent the different types of people that move into an area, such as a banker or a farmer. These conditions are different for the *people* agents and the *investment* agents.

Figure 15.6 shows the high-level use case diagram for the simulation system.

The *people* agent first looks to see if the area can meet its basic need requirements as defined by *Maslow's hierarchy of needs*. This introduced theory, which must be captured in the reference model, needs to be validated anew with the stakeholders. The area must be able to provide food, water, and shelter for all agents within the area, and the *people* agents have a threshold that the area must be greater than in order for the agent to

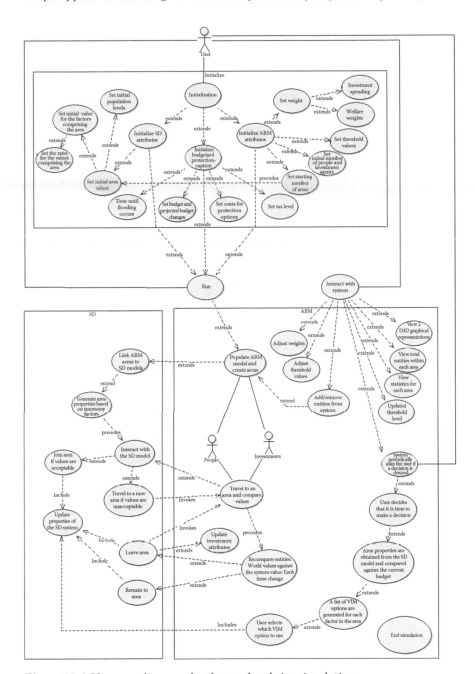

Figure 15.6 Use case diagram for the sea level rise simulation.

consider joining the area (i.e., the area must be able to provide a minimum of X gallons of water per person per day based on the value required by each agent). If these conditions pass, then the *people* agent compares its impact attribute values against the values of the area. The impact attribute is the main component in deciding if the *people* agent finds the area acceptable to live in because it represents the factors that the agent finds most valuable to itself. For example, if the *people* agent's impact attributes want an area with a high agricultural base then that agent will refuse to stay in an area with a low agricultural presence even if every other component in the area is prospering. The *people* agent continues to recheck its worldview and impact attribute values while within the area to make sure that the area remains an acceptable living environment. Figure 15.7 shows

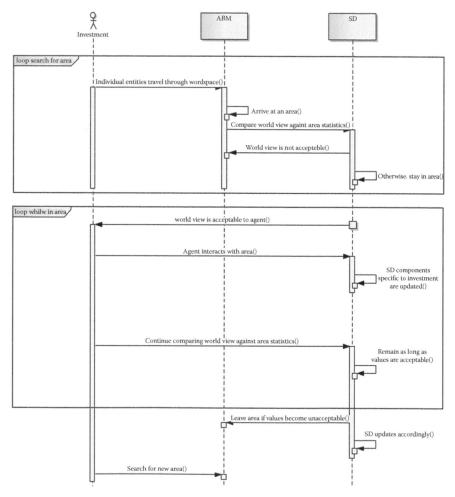

Figure 15.7 Sequence of interactions in the sea level rise model.

a high-level view of the interactions between the investment agents and their environment.

The *investment* agent also checks a worldview value and an impact attribute value against the areas, but these conditions are created differently for the *investment* agents. The *investment* agent represents a wide variety of businesses and government expenditure values based on the values that are randomly assigned to the entities. In terms of the worldview for the investment, the area must be able to financially support the addition of new businesses. In terms of the impact attribute, the area must have the necessary infrastructure to justify the new investment joining the area. For example, if the *investment* agent represents a government expenditure on dam repair then there has to be at least one dam in need of repair in the area for the investment agent to join the area.

SD modeling is used to handle the changing dynamics within the areas of interest. An SD model is created for every area that is being represented within the environment. This model represents the 18 factors that are of greatest interest within the area. The SD model is designed to represent how each of these factors is related to each of the other factors. These factors are also affected by the number of *people* agents and *investment* agents that have entered the area.

By relating all of the factors of the area to each other, changes in one of the factors can create drastic observable changes in an area. This allows for the area to be tested under different conditions to observe how the dynamics of the area respond. For example, if an area has a large energy infrastructure and the sea level rise is threatening to cause serious damage to that particular infrastructure then the derivative or cascading effects can be observed in the system. The system can also be configured to view what happens if money is invested in protecting the energy infrastructure from the potential damage. This can then allow for the costs of saving the energy infrastructure to be compared against the cost of not saving the infrastructure to provide an estimate of how the system will be affected either way. This same test can be run for all of the factors within the area to help understand how the system will respond under different conditions.

Changing the dynamics of the system also serves to make the area more or less attractive to the *people* agents that are living in the areas. If the conditions in the area become unbearable to the *people* agent then the agent will leave the system. As more agents leave the system, thus removing their financial contributions to their areas (i.e., disposable income), the area may become even less desirable and an even greater number of people will start to leave until the economic value of the area completely collapses. This affect can also be observed while testing changes in the dynamics of the system to provide a check on the emergent behavior of the system. Protecting one aspect of the area completely while completely ignoring other parts of the system may show the best results for the area financially, but the ABM portion of the model helps to determine if the change will cause a reaction forcing all of the population to vacate the area and cause the area to die in the future. At the same time, it could also show that making certain changes will make the area more desirable to live in and increase the number of *people* and *investment* agents into the area.

We also identified constraints captured in Table 15.4 and added them to the reference model. In this case, the constraints are mainly present to reduce the size of the problem to an implementable subset and are validated with the stakeholders before they are implemented.

Once we have a conceptual model, we can implement it using tools that can best help us maintain consistency between the implementation and the conceptualization.

Table 15.4 Constraints on Decision Maker Entity

Decision makers	Meaning	Implementation
Decision/policy makers have several options for dealing with sea level rise.	Build dikes, raise land, relocate people/assets, dredge coastlines, etc.	Decisions are selected by the user.
Each decision option has a cost.	Monetary, environmental, political, etc.	Only monetary costs are considered.
Decision makers can have different goals for a given area.	A decision maker for the Department of Transportation is probably more concerned with transportation infrastructure than with dams.	Decisions are selected by the user and reflect the goals of the user.
Decision makers have different terms of employment.	If sea level rise is not expected to affect an area during a decision makers term in office, then sea level rise may be overlooked.	Decision maker (the user) can decide to do nothing in response to sea level rise.
Decision makers do not have unlimited funds to allocate for sea level rise.	Decision makers can allocate funding from their budgets or try to borrow money.	Decision makers can only allocate money that they have or that they can potentially collect.

System implementation

We implemented this system in Anylogic 6.6 because of its ability to support ABM, SD, and DES in a single environment (Borshchev and Filippov 2004). The implementation of the system resulted in a tool that can be used by decision makers to observe the effect of policy choices on an area. An area in this case can be of any size such as a state, city, region, or school district. This tool helps determine what factors within the area should be invested in, maintained, or vacated depending on the value of the factor (i.e., transportation or medical infrastructures) with respect to the area to protect the essential parts of the area. In addition, the system

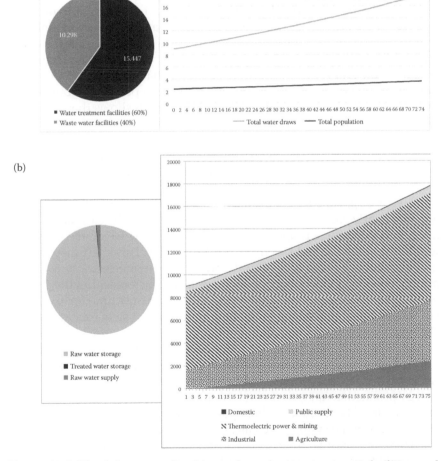

Figure 15.8 Water factor results: (a) number of water treatment facilities and (b) water storage and usage.

is designed to account for the movement of people and resources into and out of the areas.

M&S allows for potentially high-cost solutions to be tested without having to physically implement the solution such as actually relocating people or building a dike. The decision makers would then be able to make a more informed decision about how to protect their areas by gaining a better understanding of the effects, benefits, and costs associated with implementing different mitigation policies in their areas.

Different policies can be put in place to deal with rising sea levels, and the system is designed to allow for different strategies to be tested. Policies can range from evacuating areas to doing nothing to establishing criteria to mitigate the effects of sea level rise.

The two graphs in Figure 15.8a display the total number of water treatment facilities in an area including both wastewater and non-wastewater facilities. The pie chart displays the percentage of water treatment facilities versus the number of waste water facilities at the current model time within the system. The time plot (right) displays the total water draws per day in billions of gallons. The upper line represents the water draw and the lower line represents the number of people that need this water in millions. This amount is reliant on the total number of people and businesses that need supplied water in the area, and the graph shows that the water draw increases as the number of people increases.

Figure 15.8b displays information on water storage and water usage. The pie chart (left) displays the amount of water being stored in the area in the forms of raw water storage, treated water storage, and also the supply of raw water in the area per day. The time chart (right) displays the amount of water in millions of gallons that is used by various other factors within the system on a yearly basis. The amount of water that is used by each other factor increases or decreases proportionally over time as the factor itself expands within the system or reduces in its need for water. This chart displays the usage amounts for the agricultural, industrial, thermoelectric power and mining industry, public supply, and domestic use factors within the system.

Figure 15.9 displays the total economic output over time for six different factors within the system over the course of the system execution. This allows for the increases and decreases in the economic output levels of the factors to be observed visually by the user while the system is running. This graph displays amounts resulting from bank asset amounts, chemical industry production rates, information technology production rates, energy production rates, commercial expansion rates, and agriculture production rates. These amounts fluctuate during the system as the values of other factors in the system, such as population level, increase or decrease.

These are sample graphs. Any group of variables can be visualized or exported for analysis. It is noted that the graphs provide a means of analyzing causal effects between factors (Figure 15.9).

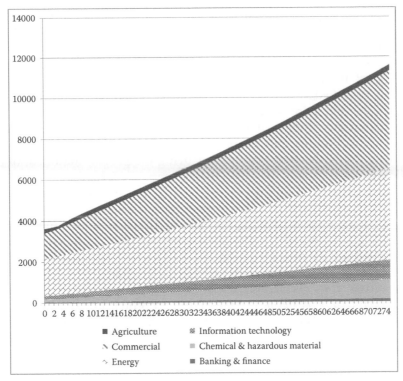

Figure 15.9 Stack chart of six factors.

Conclusion

The MS-SDF is a framework that allows us to build and track complex systems. Methodologically, the framework allows for designing, implementing, and testing systems that are considered problem situations. The idea of using ontology as a reference model from which conceptual models can be derived is essential because it (1) facilitates the design of the system and (2) provides the means of formal validation using model checking. We have shown how to apply it to develop an experimental system to study the effects of sea level rise on an area. We have also applied the MS-SDF to the development of a simulation of the Ballistic Missile Defense System (Lynch et al. 2013).

The practical feasibility of this approach has recently been shown in two PhD dissertations, although they are not directly using the model theoretic foundations described in this chapter. However, both dissertations utilize common conceptualization captured in a formal way to derive consistent and executable simulation systems from these specifications. Cetinkaya (2013) describes the model-driven development for M&S framework as an integrated approach to bridge the gaps between the different steps of a simulation study, as they are identified in our chapter as well. She is using metamodeling and model mappings. These metamodels are languages as identified in this chapter, and the mappings are the required filters. Her work is therefore closely related to the insight and practical applications described above. Similarly, Huang (2013) approached the problem from the challenge to better model increasingly larger and more complex systems. Using a comparable method, her dissertation shows how to define domain-specific model components and use them as building blocks for model generation of complex systems applying graph theoretic concepts ensuring the proper alignment of all building blocks. Again, the graph theory maps the formal representation that must guide the building process. Huang applied her work practically to light-rail transportation in the Netherlands on a significant scale. These are just two examples showing the general direction of the utilization of model theoretic concepts, although they are not always immediately recognizable from the direct application.

There is a long road to comprehensively evaluate the validity of systems across domains. It requires both an empirical and an axiomatic evaluation that is particularly challenging for representations of problem situations. For instances, the generation of an axiomatic structure poses many questions: what do we keep in the model, what do we leave out, do we have theories/data, and are they the appropriate theories/data? Explicitly, modeling problem situations using the MS-SDF will facilitate an objective evaluation, and the use of formal approaches should facilitate traceability, reuse, and interoperability over time.

References

Balci, O. (1994). Validation, verification, and testing techniques throughout the life cycle of a simulation study. *Annals of Operations Research*, 53(1), 121–173.

Balci, O. (1998). Verification, Validation, and Testing. In *The Handbook of Simulation*, J. Banks (Ed.). pp. 335–393. New York: Wiley.

Bonabeau, E. (2002). Agent-based modeling: Methods and techniques for simulating human systems. In *Proceedings of the National Academy of Sciences of the United States of America*, 99(Suppl 3), 7280–7287.

Bill 819 (2012). House Bill 819, General Assembly of North Carolina.

Borshchev, A. and Filippov, A. (2004). Anylogic—Multi-paradigm simulation for business, engineering and research. In *The 6th IIE Annual Simulation Solutions Conference*, Vol. 150, Orlando, FL.

Brooks, F. P. (1995). *The Mythical Man-Month: Essays on Software Engineering*, Anniversary Edition. New York: Pearson Education.

Cetinkaya, D. (2013). Model Driven Development of Simulation Models: Defining and Tranforming Conceptual Models into Simulation Models by Using Metamodels and Model Transformations. PhD Dissertation, Delft University of Technology, the Netherlands.

Clarke, E. M. and Wing, J. M. (1996). Formal methods: State of the art and future directions, *ACM Computing Surveys*, 28(4), 626–643.

DHS. (2009). National Infrastructure Protection Plan: Executive Summary. Retrieved from http://www.dhs.gov/xlibrary/assets/nipp_executive_summary_2009 .pdf; accessed on July 28, 2013.

Dill, D. L. (1998). What's between simulation and formal verification? In *Proceedings of Design Automation Conference*, pp. 328–329. Piscataway, NJ: IEEE.

D'Souza, D. and Wills, A. C. (1998). *Catalysis: Objects, Components, and Frameworks with UML*. Vol. 223. Object Technology Series. United Kingdom: Addison-Wesley.

Ebbinghaus, H. and Flum, J. (2005). *Finite Model Theory*. New York: Springer-Verlag.

Fensel, D. (2001). *Ontologies*, pp. 11–18. Berlin, Heidelberg, Germany: Springer.

Hodges, W. (1993). *Model Theory*, Vol. 42. Cambridge, MA: Cambridge University Press.

Huang, Y. (2013). Automated Simulation Model Generation. PhD Dissertation, Delft University of Technology, the Netherlands.

Jackson, D. (2002). Alloy: A lightweight object modelling notation. *ACM Transactions on Software Engineering and Methodology* (TOSEM), 11(2), 256–290.

Karnopp, D., Rosenberg, R., and Perelson, A. S. (1976). System dynamics: A unified approach. *IEEE Transactions on Systems, Man and Cybernetics* 6(10), 724–724.

Kelton, W. D., Sadowski, R. P., and Sadowski, D. A. (2002). *Simulation with ARENA*, Vol. 3. New York: McGraw-Hill.

King, R. D. and Turnitsa, C. D. (2008). The landscape of assumptions. In *Proceedings of the 2008 Spring Simulation Multiconference*, pp. 81–88. San Diego, CA: Society for Computer Simulation International.

Lynch, C. J., Diallo, S. Y., and Tolk, A. (2013). Representing the ballistic missile defense system using agent-based modeling. In *Proceedings of the Military Modeling & Simulation Symposium*, p. 3. San Diego, CA: Society for Computer Simulation International.

Milne, G. J. (1993). *Formal Specification and Verification of Digital Systems*. New York: McGraw-Hill.

Owre, S., Rushby, J., and Shankar, N. (1995). Formal verification for fault-tolerant architectures: Prolegomena to the design of PVS. *IEEE Transactions on Software Engineering*, 21(2), 107–125.

Pace, D. K. (2004). Modeling and simulation verification and validation challenges. *Johns Hopkins University Applied Physics Laboratory Technical Digest*, 25, 163–172.

Protégé-Ontology. (2007). Knowledge Acquisition System. http://protege.stanford .edu.

Robinson, S. (1997). Simulation model verification and validation: increasing the users' confidence. In *Proceedings of 1997 Winter Simulation Conference*, pp. 53–59. Piscataway, NJ: IEEE Computer Society.

Robinson, S. (2007). Conceptual modelling for simulation. Part I: Definition and requirements. *Journal of the Operational Research Society*, 59(3), 278–290.

Spiegel, M., Reynolds, P. F., and Brogan, D. C. (2005). A case study of model context for simulation composability and reusability. In *Proceedings of the 2005 Winter Simulation Conference*, pp. 8–18. Piscataway, NJ: IEEE.

Titus, J. G. (1999). Rising seas, coastal erosion, and the takings clause: How to save wetlands and beaches without hurting property owners. *Coastlines: Information About Estuaries and Near Coastal Waters*, 9(6), 3–6.

Tolk, A., Diallo S. Y., Padilla, J., and Herencia-Zapana. H. (2013). Reference modelling in support of M&S—Foundations and applications. *Journal of Simulation,* 7(2), 69–82.

Verner, J. M., Overmyer, S. P., and McCain, K. W. (1999). In the 25 years since The Mythical Man-Month what have we learned about project management? *Information and Software Technology,* 41(14), 1021–1026.

Weiss, J. L., Overpeck, J. T., and Strauss, B. (2011). Implications of recent sea level rise science for low-elevation areas in coastal cities of the conterminous U.S.A. *Climate Change,* 105(3/4), 635–645. Doi:10.1007/s10584-011-0024-x

Zeigler, B. (1976). *Theory of Modeling and Simulation,* Vol. 19. New York: Wiley.

chapter sixteen

Liquid business process model collections

Wil M. P. van der Aalst, Marcello La Rosa, Arthur H. M. ter Hofstede, and Moe T. Wynn

Contents

Introduction

Business process management (BPM) has become an established discipline (Dumas et al. 2013b; van der Aalst 2013a) dedicated to the way an organization identifies, captures, analyzes, improves, implements, and monitors its business processes. Through the management of the *process life cycle*, BPM influences the effectiveness and efficiency of a corporation and is a significant contributor to its overall performance and competitiveness. Business processes are thus seen as strategic corporate assets and, in the case of comprehensive supply chains, complex call center operations, or advanced distribution networks, can represent *multimillion-dollar assets* (Gotts 2010). As processes determine how an organization operates, what activities need to be fulfilled, and what data and resources are required for their successful execution, they are crucial to a plethora of key performance indicators. This importance of processes has motivated organizations to *significantly invest* in methods, tools, and techniques facilitating process life cycle management. For example, Wolf and Harmon (2012) report that 37% of organizations surveyed spend more than US $500,000, and 4% spend more than US $10 million, on investments in business process analysis, management, monitoring, redesign and improvement, and similar amounts for related software acquisitions.

At the core of the process, life cycles are conceptual models of processes, which help involved stakeholders gain a shared understanding of their processes. Such visual depictions are called *process models*. A process model is a directed graph describing the triggers, activities, data, and resources of a business process. Such models can be used for documentation, discussion, enactment (e.g., to configure a BPM system), and various types of analysis. *Simulation* is a classical model-based analysis technique used for *what-if* analysis. Given an *as-is* model describing the current situation, multiple *to-be* models can be constructed to explore different alternatives while measuring key performance indicators such as flow times, utilization, response times, faults, risks, and costs. The value of such analyses stands or falls on the quality of the *as-is* model: Is the simulation model able to capture reality?

As the ultimate source of process information, a process model *informs critical decisions* such as investments related to process-aware information systems, complex organizational redesigns or crucial compliance assessments (Davies et al. 2006). These requirements demand that process models accurately reflect the corresponding real-world processes, that is, the actual *organizational behavior*.

As a consequence, a substantial research branch of BPM, namely *process mining* (van der Aalst 2011, IEEE Task Force on Process Mining 2012), has been dedicated to techniques for extracting organizational behavior from *event logs* and using this information to enhance existing process models (Fahland and van der Aalst 2012) or discover new ones (van Dongen et al. 2009). Event logs are recorded by a variety of IT systems commonly available within organizations, such as Enterprise Resource Planning systems, Content Management Systems, Customer Relationship Management systems, Database Management Systems, or e-mail servers. Process mining is fueled by the incredible growth of event data (Hilbert and Lopez 2011).

In the practical deployment of process modeling, however, a new challenge emerges, that is, scaling up or *process modeling in the large* (Raduescu et al. 2006; Rosemann 2006; van der Aalst 2013a). For large organizations, it is common to maintain collections of thousands of complex process models, which all need to be managed appropriately to cater for the various demands of their stakeholders. For instance, Suncorp, the largest Australian insurer, manages a collection of 3000+ process models with models ranging from 25 to 500 activities (La Rosa et al. 2013).

These large organizations employ some form of process model repository management, for example, Suncorp uses ARIS (Software AG). However, such forms of repository management are often not adequate as the models in these repositories tend:

Issue 1: To be based on *subjective* perceptions about the real processes rather than actual *objective* process data

Issue 2: To be *out of sync* with organizational behavior, as the frequency of real-world process changes is such that cost-effective process model updates are not possible

Issue 3: To be designed for the most *common purposes* of the stakeholders instead of catering for specific demands of individual stakeholders

These three issues lead to *severe limitations* of process model collections, dramatically compromising the quality of business decisions that are made based on these artifacts. Current process mining techniques are not adequate to address these issues, as they focus on *single* process models as opposed to collections thereof. Concomitantly, recently emerged research techniques for managing large process model collections (Dijkman et al. 2012) are concerned with challenges such as how to identify similar models in a repository (Dijkman et al. 2011a), how to merge these models (La Rosa et al. 2013), and how to modularize them (Reijers et al. 2011), but have never considered organizational behavior despite the wide availability of event logs in today's organizations.

This chapter proposes the new notion of a *liquid business process model/ log collection*, that is, a collection of process models that does the following:

1. Is *aligned* with the organizational behavior, as recorded in event logs
2. Can *self-adapt* to evolving organizational behavior, thereby consistently remaining current and relevant
3. Incorporates relevant *execution data* (e.g., process performance and resource allocations) extracted from the logs, thereby allowing insightful reports to be produced from factual organizational data

Such collections provide a rich source of input for various types of analysis, including process mining and simulation.

The remainder of the chapter is organized as follows. The Section "BPM use cases" provides an overview of BPM by discussing selected BPM use cases. The process mining spectrum and the importance of aligning observed and modeled behavior are discussed in the Section "Process mining," whereas the management techniques for process model collections are discussed in the Section "Management of large process model collections." The Section "Research innovations" elaborates on the main innovations needed to make business process model collections *liquid* using event data. The Section "Realization" discusses the different challenges in terms of five research streams. Initial tool support realized through ProM and Apromore is described in the Section "Supporting software." The Section "Conclusion" concludes the chapter.

BPM use cases

In this chapter, we propose a *liquid* business process model collection where modeled and observed behaviors are aligned and multiple evolving processes are considered. To position such liquid business process model and event data collections, we first provide an overview of classical BPM approaches using typical use cases.

In the work by van der Aalst (2013a), 20 use cases are used to structure the BPM discipline and to show *how, where, and when* BPM techniques can be used. These are summarized in Figure 16.1. Models are depicted as pentagons marked with the letter M. A model may be descriptive (D), normative (N), and/or executable (E). A "D|N|E" tag inside a pentagon means that the corresponding model is descriptive, normative, or executable. Configurable models are depicted as pentagons marked with CM. Event data (e.g., an event log) are denoted by a disk symbol (cylinder shape) marked with the letter E. Information systems used to support processes at run time are depicted as squares with rounded corners and marked with the letter S. Diagnostic information is denoted by a star shape marked with the letter D. We distinguish between conformance-related diagnostics (star shape marked with CD) and performance-related diagnostics (star shape marked with PD). The 20 atomic use cases can be chained together in so-called composite use cases. These composite cases can be used to describe realistic BPM scenarios.

In the work by van der Aalst (2013a), the BPM literature is analyzed to see trends in terms of the 20 use cases, for example, topics that are getting more and more attention. Here we only mention the use cases most related to process mining.

- Use case *Log Event Data* (LogED) refers to the recording of event data, often referred to as event logs. Such event logs are used as input for various process mining techniques. Extensible event stream (XES), the successor of mining XML format (MXML), is a standard format for storing event logs (www.xes-standard.org).
- Use case *Discover Model from Event Data* (DiscM) refers to the automated generation of a process model using process mining techniques. Examples of discovery techniques are the alpha algorithm, language-based regions, and state-based regions. Note that classical synthesis approaches need to be adapted as the event log only contains examples.
- Use case *Check Conformance Using Event Data* (ConfED) refers to all kinds of analysis aiming at uncovering discrepancies between modeled and observed behavior. Conformance checking may be done for auditing purposes, for example, to uncover fraud or malpractices. Token-based (Rozinat and van der Aalst 2008) and alignment-based (van der Aalst et al. 2012) techniques replay the event log to identify nonconformance (De Weerdt et al. 2012).

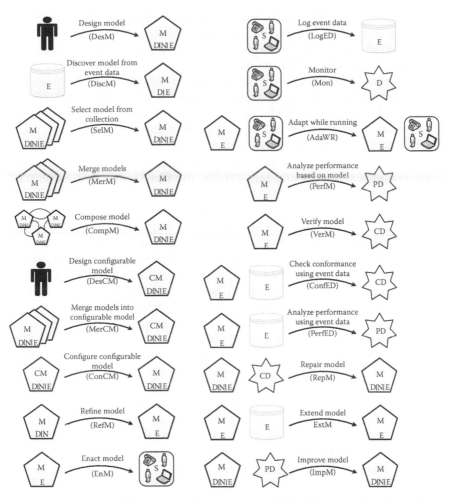

Figure 16.1 Twenty BPM use cases (van der Aalst, 2013a). Use cases Log Event Data (LogED), Discover Model from Event Data (DiscM), Check Conformance Using Event Data (ConfED), Analyze Performance Using Event Data (PerfED), Repair Model (RepM), Extend Model (ExtM), Improve Model (ImpM) are most related to process mining. Use case Analyze Performance Based on Model (PerfM) includes traditional forms of simulation. (Derived from van der Aalst, W. M. P. Business process management: A comprehensive survey. *ISRN Software Engineering 2013*: 1–37, 2013a. doi:10.1155/2013/507984.)

- Use case *Analyze Performance Using Event Data* (PerfED) refers to the combined use of models and timed event data. By replaying an event log with timestamps on a model, one can measure delays, for example, the time in-between two subsequent activities. The results of timed replay can be used to highlight bottlenecks. Moreover, the gathered timing information can be used for simulation or prediction techniques (De Weerdt et al. 2012).

- Use case *Repair Model* (RepM) uses the diagnostics provided by use case ConfED to adapt the model such that it better matches reality. On the one hand, a process model should correspond to the observed behavior. On the other hand, there may be other forces influencing the desired target model, for example, a reference model, desired normative behavior, and domain knowledge.

- Event logs refer to activities being executed and events may be annotated with additional information such as the person/resource executing or initiating the activity, the timestamp of the event, or data elements recorded with the event. Use case *Extend Model* (ExtM) refers to the use of such additional information to enrich the process model. For example, timestamps of events may be used to add delay distributions to the model. Data elements may be used to infer decision rules that can be added to the model. Resource information can be used to attach roles to activities in the model (Rozinat et al. 2009).

- Use case *Improve Model* (ImpM) uses the performance-related diagnostics obtained through use case PerfED. ImpM is used to generate alternative process models aiming at process improvements, for example, to reduce costs or response times. These models can be used to do *what-if* analysis. Note that unlike RepM, the focus of ImpM is on improving the process itself.

These use cases illustrate the different ways in which models and event data can be used. Use case *Analyze Performance Based on Model* (PerfM) includes traditional forms of simulation not directly driven by event data. See van der Aalst (2010) for a discussion on the relation between simulation, BPM, and process mining.

Process mining

Over the last decade, process mining emerged as a new scientific discipline on the interface between process models and event data (van der Aalst 2011). On the one hand, conventional BPM and workflow management (WfM) approaches and tools are mostly model driven with little consideration for event data. On the other hand, data mining (DM), business intelligence (BI), and machine learning (ML) focus on data without considering end-to-end process models. Process mining aims to bridge the gap between BPM and WfM on the one hand and DM, BI, and ML on the other hand.

The starting point for process mining is not just any data, but *event data* (IEEE Task Force on Process Mining 2012). Data should refer to discrete events that happened in reality. A collection of related events is referred to as an event log. Each event in such a log refers to an activity (i.e., a well-defined step in some process) and is related to a particular case (i.e., a process instance). The events belonging to a case are ordered and can be seen as one *run* of the process. It is important to note that an event log contains only example behavior, that is, we cannot assume that all possible runs have been observed. In fact, an event log often contains only a fraction of the possible behavior (van der Aalst 2011). Frequently, event logs store additional information about events and these additional data attributes may be used during analysis. For example, many process mining techniques use extra information such as the resource (i.e., person or device) executing or initiating the activity, the timestamp of the event, or data elements recorded with the event (e.g., the size of an order).

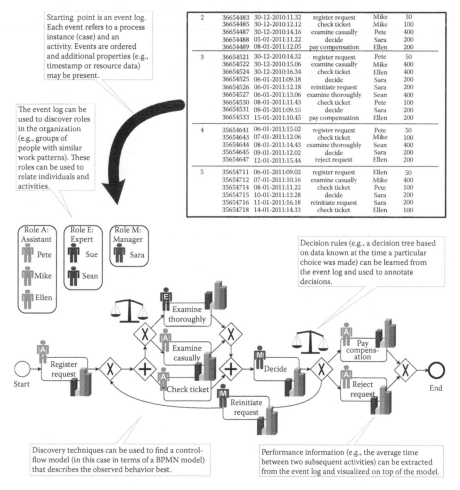

Figure 16.2 Process mining techniques extract knowledge from event logs to discover, monitor, and improve processes.

Event logs can be used to conduct various types of process mining, as is illustrated in Figure 16.2. Here we only mention the three main forms of process mining. The first type of process mining is *discovery*. A discovery technique takes an event log and produces a model without using any a priori information. Process discovery is the most prominent process mining technique. For many organizations, it is surprising to see that existing techniques are indeed able to discover real processes merely based on example executions in event logs. The second type of process mining is *conformance*. Here, an existing process model is compared with an event log of the same process. Conformance checking can be used to check if reality, as recorded in the log, conforms to the model and vice versa. The third type of process mining is *enhancement*. Here, the idea is to extend or improve an existing process model using information about the actual process recorded in some event log. Whereas conformance checking measures the alignment between model and reality, this third type of process mining aims at changing or extending the a priori model. For instance, by using timestamps in the event log, one can extend the model to show bottlenecks, service levels, throughput times, and frequencies. Note that the three main forms of process mining correspond to some of the use cases mentioned earlier.

One of the key contributions of process mining is its ability to relate observed and modeled behavior at the event level, that is, traces observed in reality (process instances in event log) are aligned with traces allowed by the model (complete runs of the model). As shown in Figure 16.3, it is useful to align both even when model and reality disagree (van der Aalst et al. 2012). First of all, it is useful to highlight where and why there are discrepancies between observed and modeled behavior. Second, deviating

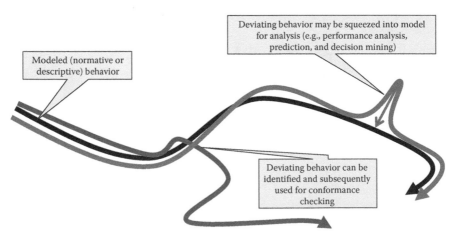

Modeled (normative or descriptive) behavior

Deviating behavior may be squeezed into model for analysis (e.g., performance analysis, prediction, and decision mining)

Deviating behavior can be identified and subsequently used for conformance checking

Figure 16.3 Process mining aligns observed and modeled behavior: "moves" seen in reality are related to "moves" in the model (if possible).

traces need to be *squeezed* into the model for subsequent analysis, for example, performance analysis or predicting remaining flow times. The latter is essential in case of nonconformance (van der Aalst et al. 2012). Without aligning model and event log, subsequent analysis is impossible or biased toward conforming cases.

Management of large process model collections

As organizations start to develop and maintain large collections of process models, there is an increasing need for continuous and efficient management of these process repositories. To reduce redundancy and improve maintainability of process model collections, efficient techniques to manage multiple process models in relation to each other as well as management of different versions of a single model are required. In the work by Dijkman et al. (2012), an overview of state-of-the-art management techniques for process model collections is provided. A pictorial representation of such techniques is provided in Figure 16.4.

As a collection of process models evolves over time, it may start to display unnecessary internal complexity. A common example is redundancy in the form of exact or approximate clones. Such clones are typically the result of copy/paste activities and they adversely affect the maintainability of process model collections, besides leading to unwanted inconsistencies in the repository, if they are modified independently of each other. Clones can manifest themselves both at the level of entire process models as well as fragments thereof. Researchers have proposed various techniques for

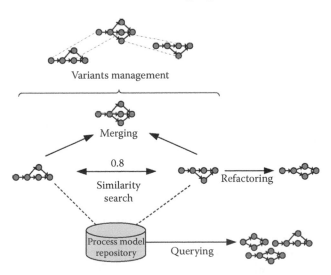

Figure 16.4 Overview of techniques for the management of process model collections. (Adapted from Dijkman, R.M. et al., *Computers in Industry*, 63(2), 91–7, 2012.)

detecting such clones within process model repositories (Guo and Zou 2008; Ekanayake et al. 2012). *Refactoring* techniques, inspired from software engineering, have been explored to improve the maintainability and readability of process model collections. Examples of such refactoring techniques are extracting the identified clones and storing them as reusable subprocesses (Dumas et al. 2013a), standardizing approximate clones (Ekanayake et al. 2012), and modularizing process models into different levels of abstraction (Dijkman et al. 2011b; Weber et al. 2011).

Another set of management techniques relate to the notion of *similarity search* (Becker and Laue 2012). Given a collection of process models and a search process model, similarity search techniques identify and return those models from the collection that are deemed similar (e.g., potentially inexact matches) to the search model. A potential use case of similarity search is for an organization to identify which of its own processes are similar to a standardized (reference) process model. Research in this area approaches this challenge from two angles: (1) the definition and implementation of similarity measures that return a similarity rating (e.g., between 0 and 1) for two process models (Dijkman et al. 2011a), and (2) the implementation of indexing techniques for improving the retrieval of similar process models (Yan et al. 2010; Kunze and Weske 2011) or of models containing the query model as a subgraph (Jin et al. 2013). More in general, such indexing techniques can be used as the backbone for efficient *querying* of process model repositories. The query could be expressed as a model (Awad and Sakr 2012; Jin et al. 2013) or in textual form (Jin et al. 2011).

Process model merging is concerned with merging a collection of process variants into one consolidated process model, which can be very useful in the context of organizational mergers, restructurings, and rationalizations. This can lead to a collection of reduced size that has been standardized and optimized for the current business context, which in turn can significantly improve the maintainability of the collection as a whole. Some merging techniques enforce behavior preservation such that the merged model maintains the behavior of all individual models allowing one to replay the behavior of each input variant on the merged model (La Rosa et al. 2013). Some of the merging techniques take into account notions of label similarities when merging so that it is possible to merge activities from different models that have similar but not identical labels (Gottschalk et al. 2008; La Rosa et al. 2013) whereas others only merge activities with identical labels (Mendling and Simon 2006; Sun et al. 2006; Reijers et al. 2009).

Given a collection of process models, *variants management* mechanisms (Pascalau et al. 2011; Ekanayake et al. 2011; Weidlich et al. 2011) are required to keep track of the organization of the collection such that users can browse it and view its evolution as changes are made. These mechanisms are based on various types of relations between the process models of the repository. For example, Ekanayake et al. (2011) exploited information on shared clones across process models and versions thereof to provide change propagation and access control features. Other common relations that can be used to manage variants are aggregation and generalization relations (Kurniawan et al. 2012). An aggregation relation exists between a business process model and it parts (usually subprocesses) whereas a generalization relation exists between a more general process model and a more specific one. Aggregation and generalization are typically used to develop a hierarchical classification of process models, enabling users to navigate a collection of process models by traversing the hierarchy.

Although there is a plethora of techniques for managing large process model collections, a collection remains *static* in most cases until a process improvement initiative is carried out. This is an area where considerable progress can still be made toward *self-healing* or *self-adapting* process model collections.

Research innovations

We propose to integrate process mining and the management of process model collections, leading to a solution for the management of liquid process model collections. This results in the following innovations:

- **Innovation 1**: *Confidence → Evidence*. To address Issue 1 mentioned in the Section Introduction (process model collections based on subjective perceptions), we propose to create a new entry point to the process lifecycle. Traditionally, business processes are first designed and then executed on the basis of these designs, which are informed by domain expertise (*confidence-based process design*). This has proven to be time-consuming and error-prone as organizational behavior is inherently difficult to capture formally. This problem is exacerbated in the context of large firms executing many complex and disparate business processes. We propose to leverage off the organizational knowledge stored in event logs to provide an alternative starting point to process model design (*evidence-based process design*). By developing semiautomated techniques, this design activity will be less *expensive* and results in *more current* process models.

- **Innovation 2**: *Static → Dynamic*. In response to Issue 2 (process model collections out of synch with organizational behavior), we challenge the *static* nature of process model collections, manually built once, and updated from time to time. Rather, we propose *liquid* process model collections as a *dynamic* artifact that can self-adapt to evolving business operations, as recorded in logs, thereby consistently remaining current and relevant. This *continuous realignment* of process model collections with organizational behavior can virtually eliminate the risk of obsolete process models and concomitantly increase the value of BPM initiatives. As opposed to traditional process monitoring and controlling, it will be possible to adapt process model collections based on event logs, even if their constituent process models are not automatically enacted by a BPM system.

- **Innovation 3**: *Generic → Demand-driven.* Today's process model collections are *common ground*: they are inspired by *generic* business needs and designed before specific stakeholder demands are articulated. As such, they are problem-independent and user-independent. With the availability of liquid process model collections, which *incorporate execution data* inferred from event logs, such as actual process performance metrics or resource allocations, it will be possible to generate insightful reports, the result of which will be new, *demand-driven* process models tailored to the needs of specific stakeholders. This will address Issue 3 (process model collections designed for the most common purposes).

These innovations result in the so-called *liquid* business process model/log collection mentioned in the Section Introduction.

Realization

To realize the above innovations, we propose a research agenda consisting of five interrelated research streams (RS1–RS5), which are illustrated in Figure 16.5. Each stream aims to realize one or more of the innovations identified in the Section Research innovations. Details for each stream are provided below.

- **Research Stream 1 (RS1)**: Fundamentals of liquid process model collections: This research stream focusses on the foundational notions behind *liquid process model collections* and techniques for operationalizing them. As such, RS1 is the enabler for the realization of innovations 1–3 through the other research streams. The first priority in this stream should be the extension of the concept of *alignment* between a log and a *single* process model (Adriansyah et al. 2011; van der Aalst et al. 2012) to the realm of process model collections to determine an *overall alignment score* between logs and process model collections. This notion, illustrated in Figure 16.6, will need to consider different ways of partitioning the log, for example, along entire

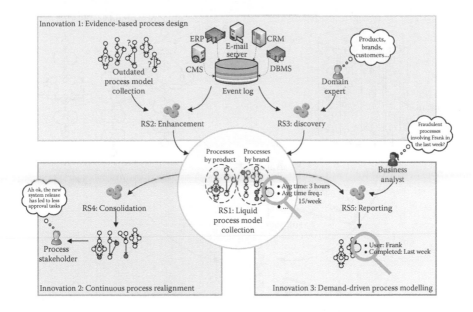

Figure 16.5 The proposed research agenda consists of five research streams (RS1-RS5), each mapped to one or more innovations.

traces or portions thereof (i.e., subtraces) to identify the sequence of events that best matches an individual process model. Further, for each model in the collection, its execution occurrence frequency (i.e., number of cases) needs to be considered, as inferred from the logs, to weigh individual alignments.

Existing process discovery (van der Aalst 2011) and graph matching (Bunke 1997) algorithms may serve as the basis for a novel algorithm to compute the overall alignment score. The process mining environment ProM can be extended to incorporate these algorithms.

RS1 should also identify suitable *execution properties* inferred from the logs (e.g., performance indicators, bottlenecks, and resource utilization) that can be linked to different elements of the process model collection to enrich it (e.g., the average duration of single tasks or the frequency of a whole process). The types of relations should be captured in the form of a *conceptual model*. We need to formalize a *data structure* to persist both the alignment and these properties for subsequent business analysis (proposed to be realized as part of RS5). Finally, a *customizable dashboard* needs to be designed and implemented, which uses this data structure to visualize relevant execution properties at various levels of abstraction

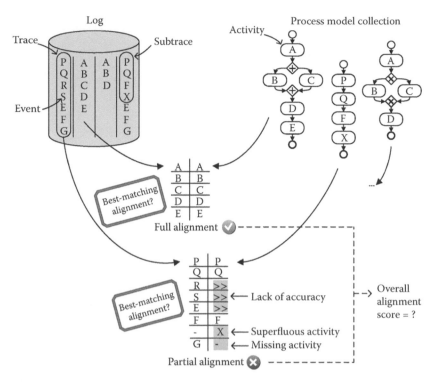

Figure 16.6 The backbone of liquid process model collections is a notion of over-all alignment score between logs and process model collections.

and allows one to navigate across these levels. Such a dashboard can be implemented on top of the Apromore repository.

- **Research Stream 2 (RS2)**: Enhancement of static process model col-lections: This stream aims to make existing, static process models liquid, thus contributing to the realization of innovation 1 (evidence-based process design). This involves devising and implementing techniques for applying appropriate changes to an existing process model collection to improve its alignment (as defined in RS1) with the organizational behavior recorded in the log. Genetic algorithms such as simulated annealing (Suman 2004) can be leveraged for this purpose to apply perturbations to the models in the collection that keep change to a minimum.

In this stream, we also develop a method for enriching an aligned process model collection with log-inferred execution properties, as identi-fied in RS1. Once again, the techniques envisaged in this research stream can be realized on top of ProM.

- **Research Stream 3 (RS3)**: Domain-driven discovery of liquid process model collections: This stream, complementary to RS2, aims to develop a set of parametric algorithms for the semiautomatic discovery of liquid process model collections from scratch. Therefore, it contributes to the realization of innovation 1 (evidence-based process design).

The novelty of these algorithms is that they will operate over process model collections, rather than single models, and be *driven by domain knowledge*, which can be provided by subject-matter experts via parameters. In fact, although a plethora of process discovery algorithms is available (van der Aalst 2011), none of them allows domain knowledge to influence the discovery. Different algorithms should be developed based on the possible types of input. Examples of different types of input are company-specific dimensions (e.g., products, brands, customer types) as well as quality metrics mandated by the company (e.g., a threshold for process model size). Findings from multidatabase mining (Wu et al. 2005) can potentially inform the envisaged algorithms to discover process model collections from heterogeneous logs, that is, logs generated from various systems. Finally, mechanisms should be developed so that the discovered collections can be enriched with execution properties relevant to the domain knowledge used as input. These mechanisms should be based on the techniques defined in RS2.

This stream also aims to develop a technique for inferring relationships between the discovered process models (e.g., abstraction and order relationships) and use these to compose a process architecture (Eid-Sabbagh et al. 2012) to accompany the collection. Similar to RS2, the discovery algorithms and the technique for inferring process architectures can be implemented in ProM.

- **Research Stream 4 (RS4)**: Consolidation of liquid process model collections: This stream aims to design and implement an approach for retaining model alignment with organizational behavior, as the latter evolves over time. As such, this stream contributes to the realization of innovation 2 (continuous process realignment).

A challenge to achieve this continuous realignment will be how to distinguish transient changes in organizational behavior from those of a more permanent nature, also known as *concept drifts* (Bose et al. 2011), that need to be reflected onto the process model collection. Techniques for process model change required to realign a collection should be informed by RS2 and extended to deal with updates in the associated execution properties. Conversely, a mechanism should be developed to recognize changes intentionally made to the process model collection that may have not yet been observed in the log. This situation is illustrated in Figure 16.7.

Another component of this research stream involves the design of a console that notifies process stakeholders about potential changes to the process models of the collection that concern them, and which can visualize these changes as delta differences on top of the existing models. Such a console will assist the stakeholders with change assessment, that is, deciding whether and if so, to what degree to commit changes. For example, one may decide to only incorporate those changes that have been observed relatively frequently as part of a new organizational behavior. This console should also allow stakeholders to provide feedback on those changes that are considered undesirable. This may provide input to targeted analyses, which is the scope of RS5.

Further, in this stream process model version, control mechanisms (Ekanayake et al. 2011) should be extended so that one can keep track of all versions of process model collections and their execution properties by storing delta differences. This may provide input to targeted analyses in RS5, for example, by comparing changes that have occurred over given time frames.

The consolidation approaches and console can be implemented in Apromore.

- **Research Stream 5 (RS5)**: Reporting for liquid process model collections: This stream aims to develop techniques to generate *sophisticated reports* on a liquid process model collection. Such reports should cover historical organizational behavior (descriptive nature) or forecasts of future organizational behavior (predictive nature).

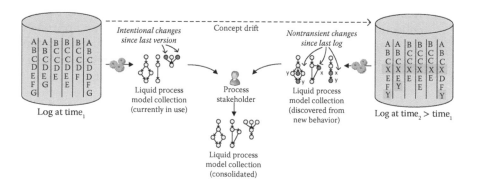

Figure 16.7 Consolidating a process model collection to cope with evolutions of organizational behavior (e.g., due to new laws).

These reports are likely to provide deep insights into different organizational aspects, ranging from performance issues to compliance violations and fraudulent activities. For example, it will be possible to generate all process models in which the employee Frank was involved, in the last month, that lasted 10 days; or all process models affected by a fraud through a change of account in the last year; or all process models that are likely to be impacted by the bankruptcy of an important client in the next month.

The results of these reports are presented in the form of new process models, driven by specific stakeholders' demands, and serve as the basis for more informed organizational decisions. As such, this stream aims to realize innovation 3 (demand-driven process modeling).

Process mining techniques are able to reveal the reasons for a good or bad performance and may even provide predictions and recommendations. However, for *what-if* analysis, we will often need to resort to simulation. See van der Aalst (2010) for a discussion on the interplay between simulation and process mining.

This stream also involves the design and development of a query language exploiting the data structure defined in RS1. This language is to be supported by a dashboard that business analysts can use to submit their queries.

Indexing techniques from graph databases (Jin et al. 2013) will need to be extended to index-specific elements of a liquid process model collection (e.g., execution properties) to efficiently execute queries over large, property-rich collections, whereas statistical techniques (e.g., regression analysis [Freedman 2005]) and data mining algorithms (e.g., decision tree building algorithms [Quinlan 1993]) can be used as a basis for generating process model forecasts. The liquidity property of the collection will guarantee that the results of the queries will always be relevant.

The envisioned dashboard and the underlying reporting techniques can be implemented in Apromore.

Supporting software

To realize the intended innovations, we build on ProM and Apromore and tightly integrate both. ProM is a generic open-source framework for implementing process mining tools in a standard environment. It can be downloaded from www.processmining.org. The ProM framework can load event logs in standard formats such as XES and MXML. The ProM toolkit has been around for about a decade. During this period, the ProM framework has matured to a professional level. Dozens of developers in different countries contributed to ProM in the form of plug-ins. In the current version, more than 600 plug-ins are available distributed over 100 packages that can be loaded separately. Through these plug-ins ProM supports the following entire process mining spectrum:

- Online and offline process mining: Event data can be partitioned into *premortem* and *postmortem* event logs. The term *postmortem* event data refer to information about cases that have completed, that is, these data can be used for process improvement and auditing, but not for influencing the cases they refer to. The term *premortem* event data refer to cases that have not yet completed. If a case is still running, that is, the case is still *alive* (premortem), then it may be possible that information in the event log about this case (i.e., current data) can be exploited to ensure the correct or efficient handling of this case. *Postmortem* event data are most relevant for offline process mining, for example, discovering the control flow of a process based on 1 year of event data. For online process mining, mixtures of *premortem* (current) and *postmortem* (historic) data are needed. For example, historic information can be used to learn a predictive model. Subsequently, information about a running case is combined with the predictive model to provide an estimate for the remaining flow time of the case.
- Different model types: Two types of models can be identified: *de jure models* and *de facto models*. A de jure model is normative, that is, it specifies how things should be done or handled. For example, a process model used to configure a BPM system is normative and forces people to work in a particular way. A de facto model is descriptive and its goal is not to steer or control reality. Instead, de facto models aim to capture reality. Both de jure and de facto models may cover multiple perspectives including the control-flow perspective (*How?*), the organizational perspective (*Who?*), and the case perspective (*What?*). These are supported by ProM. The control-flow perspective describes the ordering of activities. The organizational perspective describes resources (worker, machines, customers, services, etc.) and organizational entities (roles, departments, positions, etc.). The case perspective describes data and rules. Process mining can be used to determine the degree to which organizational aspects, as modeled in the above process perspectives, conform to observed behavior.
- Different types of process mining: As discussed before, process mining includes discovery, conformance checking, and enhancement. However, also more advanced forms of process mining such as prediction, recommendation, and concept-drift analysis are supported by ProM.

Currently, ProM does not support the management of collections of models and logs.

Apromore is an open and extensible repository to store and disclose business process models of a variety of languages, such as business process model and notation formalism (BPMN), Event-Driven Process Chains (eEPCs),

Yet Another Workflow Language (YAWL), and workflow nets. Apromore provides state-of-the-art features to facilitate the management of large process model collections.

The backbone of Apromore is a fragment-based version control mechanism (Ekanayake et al. 2011). Accordingly, each single-entry single-exit fragment of each process model is independently versioned, and can be associated with an owner. This mechanism allows automatic detection of cloned fragments both between different versions of the same process model as well as between different process models. Cloned fragments are only stored once (thus allowing *vertical sharing* between different versions of the same process model and *horizontal sharing* between different process models).

Sharing fragments vertically allows Apromore to easily track differences between versions of the same model. Sharing fragments horizontally enables change propagation features. When one makes a change to a process model fragment, the owners of all process models in the repository that will be impacted by this change (because these models contain clones to the fragment being changed) will be notified. Depending on the change propagation policy, one can decide to commit the change, or not (in the latter case, a separate version of that fragment will be created).

This fragment-based version control mechanism also allows access control at the level of single fragments, and concurrency control: multiple users can simultaneously work even on the same process model, provided they edit unrelated fragments.

Moreover, like in software code, process model versions are organized in branches, where one branch may be created by *branching out* from a version in an existing branch.

Besides this fragment-based version control mechanism, Apromore provides a wealth of features to manage large process model models, such as, searching for similar models, merging similar models into a consolidated process model, identifying clones within and across process models, and cluster them.

Apromore relies on an internal format called *canonical process format* to support the previously mentioned features. Whether one is after finding the similarity between your process models, or detecting clones, Apromore performs all these operations on the canonical format of the process models stored in its repository. This way one can, for example, compare process models defined in different languages such as EPCs and BPMN, or merge them into a process model and then decide the target language for this new model.

The canonical process format provides a common, unambiguous representation of business processes captured in different languages and/or at different abstraction levels, such that all process models can be treated alike.

Apromore is available as a Software as a Service at www.apromore .org. It relies on a plug-in framework based on OSGi. This way, new features can be added in the form of OSGi plug-ins on-the-fly, and similarly, existing features can be uninstalled without the need to restart Apromore.

Presently, Apromore is tailored toward the management of process models rather than event logs. Thus, the tool does not provide any feature to align modeled and observed behavior. To address the previously mentioned challenges, a tight integration between Apromore and ProM is needed. Moreover, we also envision the integration of other tools. For example, from ProM we can call the simulation software CPN Tools (www.cpntools.org) for large-scale simulation experiments.

Conclusion

This chapter discussed the need to *relate modeled and observed behavior for large collections of processes*. After introducing the main BPM use cases, we discussed the state-of-the-art in process mining and managing large process model collections. The *disconnect* between process mining research and the management of large model collections is severely limiting the application of BPM technologies. Therefore, we suggest three major research innovations and five different research streams to realize these innovations. The aim is to create *liquid* business process model collections, that is, collections of process models that are synchronized with the organizational reality and continuously adapt to evolving circumstances. This is the only way to breathe life into business process model collections. Without it, model collections will be static, and thus of limited value; they will soon become outdated unless they are manually updated, which is often an expensive operation.

In this chapter, we invite the research community to contribute to the research streams identified in this chapter, and hope we made a good case that Apromore and ProM provide a good starting point for realizing the ideas presented. As described in the five research streams, the challenges that need to be dealt with are manifold, ranging from the continuous alignment of models and event logs to refactoring of process model collections with event data and advanced reporting on such collections.

In the work by Ekanayake et al. (2013), we did some initial work in the direction of bridging this gap between process mining and management of large process model collections. Specifically, we developed a technique on top of Apromore and ProM, called *Slice, Mine, and Dice*, that can mine a hierarchical collection of process models from an event log, where each model in the collection has a bounded complexity (e.g., on the model size) that can be set by the user. Further, in the work by van der Aalst (2013c), we introduced the *process cube* notion. The process cube structures event data using different dimensions (type of process instance, type of event and time window) to discover multiple interrelated processes or check the conformance thereof. Moreover, by precisely aligning event data and process models, we enable new types of simulation (van der Aalst 2010, 2013b) where real and simulated behaviors are combined. These ideas illustrate that *liquid* business process model collections are likely to trigger new forms of process management.

References

Adriansyah, A., van Dongen, B., and van der Aalst, W.M.P. Conformance checking using cost-based fitness analysis, *Proceedings of the 15th IEEE International Enterprise Distributed Object Computing Conference,* EDOC 2011, August 29–September 2, Helsinki, Finland, 2011.

Awad, A. and Sakr, S. On efficient processing of BPMN-Q queries. *Computers in Industry* 63(9): 867–88, 2012.

Becker, M. and Laue, R. A comparative survey of business process similarity measures. *Computers in Industry* 63(2): 148–67, 2012.

Bose, R.P.J.C., van der Aalst, W.M.P., Zliobaite, I., and Pechenizkiy, M. Handling Concept Drift in Process Mining. *Proceedings of CAiSE,* LNCS 6741: 391–405, 2011.

Bunke, H. On a relation between graph edit distance and maximum common subgraph. *Pattern Recognition Letters* 18(8): 689–94, 1997.

Davies, I., Green, P., Rosemann, M., Indulska, M., and Gallo, S. How do practitioners use conceptual modeling in practice? *Data & Knowledge Engineering* 58(3): 358–380, 2006.

De Weerdt, J., De Backer, M., Vanthienen, J., Baesens, B. A multi-dimensional quality assessment of state-of-the-art process discovery algorithms using real-life event logs. *Information Systems* 37(7):654–76, 2012.

Dijkman, R.M., Dumas, M., van Dongen, B.F., Käärik, R., and Mendling, J. Similarity of business process models: Metrics and evaluation. *Information Systems* 36(2): 498–516, 2011a.

Dijkman, R., Gfeller, B., Küster, J., and Völzer, H. Identifying refactoring opportunities in process model repositories. *Information and Software Technology,* 53(9): 937–948, 2011b.

Dijkman, R.M., La Rosa, M., and Reijers, H.A. Managing large collections of business process models—Current techniques and challenges. *Computers in Industry* 63(2): 91–7, 2012.

Dumas, M., García-Bañuelos, L., La Rosa, M., and Uba, R. Fast detection of exact clones in business process model repositories. *Information Systems* 38(4): 619–633, 2013a.

Dumas, M., La Rosa, M., Mendling, J. and Reijers, H.A. *Fundamentals of Business Process Management.* Berlin, Germany: Springer, 2013b.

Eid-Sabbagh, R.-H., Dijkman, R.M., and Weske, M. Business process architecture: Use and correctness. *Proceedings of BPM,* LNCS 7481: 65–81, 2012.

Ekanayake, C.C., Dumas, D., García-Bañuelos, L., La Rosa, M. Slice, mine and dice: Complexity-aware automated discovery of business process models. *Proceedings of BPM,* LNCS 8094: 49–64, 2013.

Ekanayake, C.C., Dumas, D., García-Bañuelos, L., La Rosa, M., ter Hofstede, A.H.M. Approximate clone detection in repositories of business process models. *Proceedings of BPM*, LCNS 7481: 302–18, 2012.

Ekanayake, C.C., La Rosa, M., ter Hofstede, A.H.M., and Fauvet, M.C. Fragment-based version management for repositories of business process models, *Proceedings of CoopIS*, LNCS 7044: 20–37, 2011.

Fahland, D. and van der Aalst, W.M.P. Repairing process models to reflect reality, *Proceedings of BPM*, LNCS 7481: 229–45, 2012.

Freedman, D.A. *Statistical Models: Theory and Practice*. United Kingdom: Cambridge University Press, 2005.

Gotts, I. Putting the M back in BPM. *BPTrends*, 2010. Available at www.bptrends.com; accessed on February, 2013).

Gottschalk, F., van der Aalst, W.M.P., and Jansen-Vullers, M.H. Merging event-driven process chains. *Proceedings of CoopIS*, LNCS 5331: 418–26, 2008.

Guo, J. and Zou Y. Detecting Clones in Business Applications. *Reverse Engineering, 2008, WCRE'08*. Proceedings of the 15th Working Conference on Reverse Engineering, October 15–18, Antwerp, Belgium. 91–100, 2008. doi: 10.1109/WCRE.2008.12.

Hilbert, M. and Lopez, P. The world's technological capacity to store, communicate, and compute information. *Science* 332(6025): 60–65, 2011.

IEEE Task Force on Process Mining. Process mining manifesto. In *Business Process Management Workshops*, LNBIP Vol. 99, pp. 169–194, F. Daniel, K. Barkaoui, and S. Dustdar (Eds.). Berlin, Germany: Springer-Verlag, 2012.

Jin, T., Wang, J., La Rosa, M., ter Hofstede, A.H.M., and Wen, L. Efficient querying of large process model repositories. *Computers in Industry* 64(1): 41–9, 2013.

Jin, T., Wang, J., and Wen, L. Querying business process models based on semantics. *DASFAA*, LNCS 6588: 164–78, 2011.

Kunze, M. and Weske, M. Metric trees for efficient similarity search in large process model repositories. *Business Process Management Workshops 2010* LNBIP 66: 535–46, 2011.

Kurniawan, T., Ghose, A., Le, L.-S., and Dam, H.K. On formalizing inter-process relationships, In *Business Management Worshops: BMP 2011 International Workshops*. LNBIP, Vol. 100: 75–86, Berlin, Germany: Springer, 2012.

La Rosa, M., Dumas, M., Uba, R., and Dijkman, R.M. Business process model merging: An approach to business process consolidation. *ACM Transactions on Software Engineering and Methodology* 22(2): 11:1–11:42, 2013.

Mendling, J. and Simon, C. Business process design by view integration. Proceedings of BPM Workshops, LNCS 4103: 55–64, 2006.

Pascalau, E., Awad, A., Sakr, S., Weske, M. On maintaining consistency of process model variants. *Business Process Management Workshops 2010*, LNBIP 66: 289–300, 2011.

Quinlan, J.R., *C4.5: Programs for Machine Learning*. San Francisco, CA: Morgan Kaufmann Publishers, 1993.

Raduescu, C., Tan, H.M., Jayaganesh, M., Bandara, W., zur Muehlen, M., and Lippe, S. A framework of issues in large process modeling projects. *Proceedings of ECIS*, Göteborg, Sweden, Association for Information Systems, 2006.

Reijers, H.A., Mans, R.S., and van der Toorn, R.A. Improved model management with aggregated business process models. *Data and Knowledge Engineering* 68(2): 221–43, 2009.

Reijers, H.A., Mendling, J., and Dijkman, R.M. Human and automatic modularizations of process models to enhance their comprehension. *Information Systems* 36(5): 881–97, 2011.

Rosemann, M. Potential pitfalls of process modeling: Part B. *Business Process Management Journal* 12(3): 377–84, 2006.

Rozinat, A. and van der Aalst, W.M.P. Conformance checking of processes based on monitoring real behavior. *Information Systems*, 33(1), 64–95, 2008.

Rozinat, A., Wynn, M., van der Aalst, W.M.P., ter Hofstede, A.H.M., and Fidge, C. Workflow simulation for operational decision support. *Data and Knowledge Engineering* 68(9): 834–50, 2009.

Suman, B. Study of simulated annealing based algorithms for multi-objective optimization of a constrained problem. *Computers & Chemical Engineering* 28(9): 1849–71, 2004.

Sun, S., Kumar, A., and Yen, J. Merging workflows: A new perspective on connecting business processes. *Decision Support Systems* 42(2): 844–85, 2006.

van der Aalst, W.M.P. Business process management: A comprehensive survey. *ISRN Software Engineering 2013*: 1–37, 2013a. doi:10.1155/2013/507984.

van der Aalst, W.M.P. Business process simulation revisited. In *Enterprise and Organizational Modeling and Simulation*, LNBIP Vol. 63, pp. 1–14, J. Barjis (Ed.). Berlin, Germany: Springer-Verlag, 2010.

van der Aalst, W.M.P. Business Process Simulation Survival Guide. BPM Center Report BPM-13-11, pp. 1–34, 2013b. www.BPMcenter.org.

van der Aalst, W.M.P. Process cubes: Slicing, dicing, rolling up and drilling down event data for process mining, *Proceedings of AP-BPM*, LNBIP 159: 1–22, 2013c.

van der Aalst, W.M.P. *Process Mining: Discovery, Conformance and Enhancement of Business Processes*. Berlin, Germany: Springer, 2011.

van der Aalst W.M.P., Adriansyah, A., and van Dongen, B. Replaying history on process models for conformance checking and performance analysis. *WIREs Data Mining and Knowledge Discovery* 2(2):182–92, 2012.

van Dongen, B.F., Alves de Medeiros, A.K., and Wen, L. Process mining: Overview and outlook of petri net discovery algorithms, *Transactions on Petri Nets and Other Models of Concurrency II*, LNCS 5460: 225–42, 2009.

Weber, B., Reichert, M., Mendling, J., and Reijers, H.A. Refactoring large process model repositories. *Computers in Industry* 62(5): 467–486, 2011.

Weidlich, W., Mendling, J., and Weske, M. A foundational approach for managing process variability. *CAiSE* 2011: 267–82, 2011.

Wolf, C. and Harmon, P. The State of Business Process Management 2012. *BPTrends*, 2012. Available at www.bptrends.com; accessed on February, 2013.

Wu, X., Zhang, C., and Zhang, S. Database classification for multi-database mining. *Information Systems* 30(1): 71–88, 2005.

Yan, Z., Dijkman, R. M., and Grefen, P. Fast business process similarity search. *Distributed and Parallel Databases*, 30(2): 105–144, 2012.

chapter seventeen

Web-based simulation using Cell-DEVS modeling and GIS visualization

Sixuan Wang and Gabriel Wainer

Contents

Introduction

System engineering is an interdisciplinary engineering that consists of
an integrated, life-cycle-balanced set of system solutions that satisfy customer requirements (ANSI/EIA-632 1999). System engineering is the art
and science of creating optimal solutions to complex issues and problems, focusing on designing and managing complex engineering projects
over their life cycles, and handling their work processes, optimization
methods, and risk management tools (Hitchins 2008). On the one hand,
system engineering allows the collaborative development that unifies
all the contributors into one team, following a structured development
process that transforms needs into a set of system product descriptions,
generates information for planners, and provides input for the next level
of development (Leonard 1999). On the other hand, system engineering
facilitates the design phase of projects with a vast amount of data and variables, aiming at the integration of all aspects of the system into a whole.

System engineering encourages the use of modeling and simulation
(M&S) to validate assumptions on systems, handle the interactions within
them, and manage their complexity (Sage and Olson 2001). M&S has been
playing increasingly important roles for analyzing and designing complex systems in system engineering. A model is a physical, mathematical,
or logical abstract representation of a system entity, while a simulation is
the implementation of a model over time that brings the corresponding
model to life (Leonard 2001). It is useful for testing and analyzing the system design before the real project has begun. The other benefits of using

M&S are as follows: it is cheaper and safer than the real case; it is more realistic than traditional experiments; and it is faster than doing it in the real time (Tolk 2010).

Especially in the design phase of the life cycle of a system engineering project, M&S can be used to evaluate the performances of a new product concept, verify design specifications, or suggest improvements for a product. This kind of simulation-based design enables designers in different fields to test whether design specifications are met. It can provide the designer with immediate feedback of different design alternatives and facilitate decision making for optimal performance (Sinha et al. 2001).

For simulation-based design, modeling languages and simulators must take into account the special characteristics of the design process. The modeling language should allow models to be updated and reused easily, and simulators should be well integrated with the design tools for collaborative developments of designers. Among different M&S approaches, discrete event systems specifications (DEVSs) (Zeigler et al. 2000) and Cell-DEVS (Wainer 2009) are two of the most well-defined formal M&S methodologies. DEVS responses to external events based on continuous timing, composed of behavioral (atomic) and structural (coupled) components. Cell-DEVS extends DEVS to the field of Cellular Automata (CA), allowing modeling complex spatial problems.

In recent years, web-based simulation has received increasing attention in the simulation community, which has resulted in the birth of the area of web-based simulation (WBS). WBS is the integration of the Web with the field of simulation, by invocating computer simulation services through the World Wide Web. It has drawn much attention in the simulation community and has been growing for recent years (Huang et al. 2005). It can be defined as the use of resources and technologies provided by the Web with interaction of client and server M&S tools, supported by a browser for graphical interface interaction (Bencomo 2004). We can gain numerous benefits in this way: (1) Users can reproduce the execution of simulation online easily instead of installing complicated configuration of required simulation software. (2) Users can reuse and share simulation resources on site without worrying about the capability of their local machine CPU or memory. (3) With the help of advanced distributed simulation technologies, the simulation can be executed on distributed computers via communication networks, which can further improve interoperability and speed up the execution time. (4) Simulation results can be retrieved or visualized by the emergency crews on site, which can crease the emergency response success.

This chapter presents an effort toward integrating four specific components: geographic information systems (GISs), modeling, simulation, and visualization, to support the simulation-based design process in system engineering. The focus of this chapter is to use WBS and GIS visualization techniques to deal with this kind of integration. The objective is to develop an integration method that enables designers, decision makers, and planners in multidisciplinary fields to easily run simulation models for testing the performance of design and visualize simulated results faster for making fast decision. The basic idea is to provide a general method to extract information from GIS, model with Cell-DEVS theory, run WBS, and visualize results back in Google Earth. To do so, we will discuss the challenges and address the advantages of this integration method, as well as different ways to realize it, and real application case studies.

A GIS is an environmental system that combines hardware, software, and data for managing, analyzing, and displaying all forms of geographically referenced information. Currently, GIS is widely used to manage large spatial databases, to perform statistical analyses, and to produce effective visual data representations. GIS applications have been used to quickly and reliably process spatially referenced data as a decision support tool (Badard and Richard 2001).

The generation of M&S of environmental systems can be traced back to the early days of computer simulation (Botkin et al. 1972). These simulations have been combined with GIS (Band 1986; Desmet and Govers 1995; Van Derknijff et al. 2010; Wang 2005) since the 1980s. M&S in GIS is for characterizing and understanding environmental patterns and processes, and estimating the effects of environmental changes. Many GISs already contain embedded simulation capabilities; however, it usually focuses on specific simulation that is limited by the power of simulator and poor for scaling up. On the other hand, simulating advanced environmental simulation models separated from GIS tools is complex (Zapatero et al. 2011). Therefore, there are growing interests in the potential for integrating GIS technology and environmental simulation models. Several works have been done in the efforts to integrate GIS and DEVS M&S for environmental systems (Gimblett et al. 1995; Hu et al. 2011; Wainer 2006; Zapatero et al. 2011), trying to transform the GIS information into a DEVS/Cell-DEVS model and to visualize simulation results in Google Earth. Although designers use GIS, M&S, and visualization in various projects, the integrative uses between them are still at an elementary stage. Furthermore, most M&S methods and applications run on single-user workstations, which normally cost too much time for installation and configuration of all the software and dependencies needed by the simulation. It is better to have remote access to the simulation resources with Web service (WS) interfaces, improving data accessibility, interoperability, and user experience.

Visualizing large amounts of information interactively is one of the most useful capabilities of GIS. Ware (2000) points out that visualization helps to present mass data, identify patterns or the problems with data, and facilitate understanding of data. Visualizing data using the current computing technology and interactive GIS can create multiple perspectives, enhancing a designer's abilities to better understand the studying phenomenon. On the other hand, visualization of simulation results are usually accompanied by high-fidelity graphics or vivid animation, which can provide a number of benefits: to compare a simulated result to the expected effect, to provide interactive environment to verify models, and to easily refine their solutions with different scenarios. In system engineering, the visualization features can present data views of the present and future. Usually, the visualization provides a graphic user interface to support interactions between the system and the users.

The contributions of this chapter are mainly as follows: (1) It proposes a new integrated framework that combines GIS data collection, Cell-DEVS modeling, WBS, and GIS visualization, which allows users to choose the best available technologies to analyze GIS system behaviors and predict future scenarios. The parts in this framework are loosely coupled and easy to scale up. (2) It introduces details of modeling using Cell-DEVS formalism for analyzing a GIS system in the simulation-based design phase of system engineering. (3) It provides a prototype with different case studies that is implemented based on the RESTful simulation services middleware for WBS and GIS systems with Google Earth visualization. The simulation engines are stored on a server and can be run remotely using our RESTful Interoperability Simulation Environment (RISE) middleware. DEVS is a universal abstract formalism, separating M&S, and RISE middleware separates simulator implementation and underling hardware; therefore, they make it perfect for deployment online remotely.

The rest of the chapter is organized as follows. The Section "Related work" reviews the issues of current approaches for integrating Cell-DEVS modeling, Web-based simulation, and GIS visualization. Then, the Section "Architecture" presents a novel architecture to solve these issues, followed by details in the Sections "Cell-DEVS modeling," "GIS data collection," "Web-based simulation," and "Visualization in Google Earth." Finally, the Section "Applications" shows real cases to demonstrate the advantages of the proposed integration architecture.

Related work

In an integrated system of GIS, M&S, and visualization, each component contributes to the system with distinctive features. GIS provides the functions to manage spatial information between entities. M&S allows representing the dynamic relationships among studying entities and predicting the behaviors in the following period. WBS eases the implementation of simulation in a much easier way using web technologies. Visualization is to represent simulation results in an intuitive and vivid way for fast decision making. In this section, we are going to review the different integration aspects of these three components. First, we will review the M&S method architecture using DEVS in system engineering. We will not only discuss the advantages of using Cell-DEVS for modeling environmental system problems but also review the different categories of WBS for the model execution.

Integration of M&S using DEVS/Cell-DEVS

Many M&S methods can be used to simulate environmental GIS models. DEVS (Zeigler et al. 2000) is a mathematical formalism for specifying discrete events systems using a modular description. DEVS is a perfect option for M&S to system engineering due to many reasons. DEVS is very easy to reuse and integrate with other components. DEVS has been successfully used in this area due to its ease of modeling, the varied ways to combine existing models, and the efficiency of the simulation engines. Other benefits are coupling of components, hierarchical structure, and modularity construction.

Figure 17.1 shows the layered DEVS M&S architecture for system engineering. From bottom to top, it goes from infrastructure hardware, software implementation to conceptual abstraction. DEVS models are closed under coupling, which means a coupled model can be viewed as an atomic one, allowing reusing and integrating to other DEVS models without changes. Each model can be associated with an experimental framework, allowing the individual simulation executing and testing easier. Similarly, WBS, including simulators and supporting middleware, is also independent of the modeling framework, which allows different simulators of particular or customized purposes and a layered view of M&S.

DEVS models are composed of behavioral (atomic) and structural (coupled) components. Cell-DEVS (Wainer 2009) extends DEVS by supporting cellular models in a spatial lattice. Cell-DEVS defines a cell as a DEVS atomic model and a cell space as a coupled model. Each cell holds a state variable and a computing function that updates the cell state based on its present state and its neighborhoods. CD++ (Wainer 2002) is an open-source environment capable of executing DEVS and Cell-DEVS

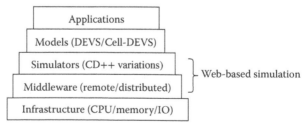

Figure 17.1 A layered DEVS M&S architecture for system engineering.

models, supporting different variations for stand-alone, parallel, or other improved simulators. RISE (Al-Zoubi and Wainer 2011) is a simulation middleware to support RESTful WSs for web-based CD++ simulation. Because DEVS strictly separates models from simulators, and WBS based on RISE Middleware strictly separates simulation from the supporting hardware, each part is loosely coupled and easy to scale up.

The Cell-DEVS formalism combines both CA and DEVS (Zeigler et al. 2000). Cell-DEVS has been widely used in many complex system engineering projects, not to mention GIS system. The advantages of using Cell-DEVS for GIS system are as follows: (1) the inheritance of DEVS, (2) the spatial rule-based features, (3) the event-driven asynchronous execution, and (4) the input or output (I/O) ports for easy integration with each other.

Categories of Web-based simulation

From previous discussion, we have known that WBS gains a lot of benefits for implementing the models: executing repeatedly without complicated installation, resources reuse on site regardless the local hardware constrains, and support for distributed simulation to speed up the execution time. The WBS is one of the focuses of this chapter; now let us see its main categories.

Early WBS efforts began in 1995, as old as the Web itself. However, the area of WBS is still in its infancy with many issues to study (Byrne et al. 2010); besides, the number of real applications and tools for WBS are still very small (Wiedemann 2001). Many authors classified WBS, like Byrne et al. (2010), Myers (2004), and Page (1999). The classification could be developed architecturally and summarized as the following four categories.

Local simulation

Local simulation is where the simulation engine is downloaded directly by the client to the user's local computer. The simulation engine executes the model completely in the client, usually with the capacity to visualize simulation results in a 2D or 3D way. Local simulation makes the server as a central distributor, but it does not do any real work (Bencomo 2004). The common approaches of using this local simulation are through Java applets or executable files. Usually, the user opens a browser and navigates a web page via a uniform resource identifier (URI), which contains an applet or executable file for downloading. After the downloading phase, the user can run the simulation engine within the applet or executable file. Figure 17.2 gives the basic local simulation architecture after this downloading phase. Originally, the simulation engine usually runs on a single local processor (LP). However, as the demand is for larger and more complex models, a single processor becomes very time-consuming. This results in the emergence of parallel simulation, distributing simulation over a set of LPs that are geographically close to each other (e.g., clusters),

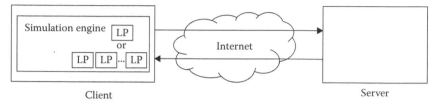

Figure 17.2 Basic local simulation architecture.

to reduce the simulation time (Page 1999). Parallel simulation also handles simulation protocols and synchronization management.

Remote simulation

In remote simulation, the simulation engine is located and executed remotely on the server side (Bencomo 2004). User accesses the simulation engines through a browser on the client side. Users can submit their requests (with specified message/parameters) to the simulation engine through the Web server, then simulation will run remotely, and results are returned to the user once the simulation has finished. The communication and manipulation between the client browser and the simulation engine in the server could be performed by many ways. Following the history of the development of the Web, the means include Common Gateway Interfaces, Java remote method invocation, Common Object Request Broker Architecture (CORBA), remote procedure call (RPC), and WSs (Bencomo 2004). The user opens a browser and navigates to a Web page via a specific URI, follows interface of the Web server operations, communicates to server with message protocol, and then waits for the response from the server on this URI.

In recent years, WSs are increasingly used for remote simulation, improving data accessibility, interoperability, and user experience. WSs are mainly categorized into two classes: RESTful WSs (Richardson and Ruby 2008) (to manipulate Extensible Markup Language [XML] representations of Web resources using a uniform set of *stateless* operations), and arbitrary WSs (Ribault and Wainer 2012) (such as Simple Object Access Protocol [SOAP]-based WS, in which the service exposes an arbitrary set of operations). RESTful WSs imitate the web interoperability style. The major RESTful WSs interoperability principles are universally accepted standards, resource-oriented, uniform channels, message-oriented, and implementation hiding. Figure 17.3 gives the basic RESTful remote simulation architecture. RESTful WSs expose all services as resources with uniform channels where messages are transferred between those resources through those uniform channels. We can access RESTful WS through web resources (URIs) and XML messages using hypertext transfer protocol (HTTP) methods (GET, PUT, POST, and DELETE). RESTful WS is simple, efficient, and scalable. Its strengths of

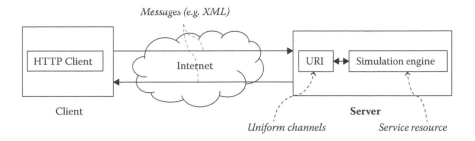

Figure 17.3 Basic remote simulation architecture with RESTful Web service.

simplicity, efficiency, and scalability make RESTful WS an excellent candidate to perform remote simulation. On the basis of these ideas, in Al-Zoubi and Wainer (2011), the authors presented the first existing RISE middleware. The main objective of RISE is to support interoperability and mash-up of distributed simulations regardless of the model formalisms, model languages, or simulation engines. The details of SOAP-based WS and the comparison between SOAP-based WS and RESTful-WS will be discussed later in Section "Distributed simulation."

Remote simulation enables larger simulations to be run on powerful computers in terms of processing, memory, and networks so that users can get the results from any low-end computer with a browser, or any other personal-aided devices (e.g., smartphone, tablet). By this way, thin client is achieved. Another advantage is that the user can reuse the simulation engine that is already available on the server, without worrying about simulation environment setup and other software dependencies issue. Besides, using a familiar interface that separates model and simulator, the user can focus on their model, making maintenance easier and promoting productivity. Some disadvantages of remote simulation are network latency, standardization, and dynamic interaction with simulation (Myers 2004).

Regarding whether the location of visualization/animation engine is on the server side, together with the simulation engine, remote simulation can further be classified into pure remote simulation and hybrid simulation (Byrne 2010).

Note that remote simulation does not specify the number of simulation engines or underling machines. Remote simulation focuses on the user's point of view, providing a way for user to access simulation services easily and efficiently. Conceptually, there is one simulation engine located on a single machine. However, if the simulation runs on various machines located remotely, the simulation is said to be distributed. This topic is discussed in the Section "Distributed simulation."

Distributed simulation

Distributed simulation is created to execute simulations on distributed computer systems (i.e., on multiple processors connected via communication networks) (Fujimoto 2000). Figure 17.4 shows the basic distributed simulation architecture. Its main objective is to interface different simulation resources, allowing synchronization for the same simulation run through different simulation engines across a distributed network, to interoperate heterogeneous simulators or geographically distributed models. The other benefits of using distributed simulation include model reuse, reducing execution time, interoperating different simulation toolkits, and providing fault tolerance and information hiding (Boer et al. 2009; Fujimoto 2000).

By its nature, all of WBS can be described as distributed simulation (Alfonseca et al. 2001; Page et al. 1998). Indeed, Page et al. (1998) classifies distributed simulation as a category of WBS. Web technologies and distributed simulation technologies have grown up largely independently, and influenced each other. The difference of distributed simulation with remote simulation is that distributed simulation aims to speed up simulation time by partitioning of models, focusing on the message transmission between simulation engines that are heterogeneously located and geographically distributed, whereas remote simulation is aiming to provide easy and efficient way for the user to access simulation services. Therefore, we separate them here into two categories for clearance.

The defense sector is one of the main users of distributed simulation technology, providing virtual distributed training environment, relying on the high level architecture (HLA) for simulation interoperability (Khul et al. 1999). Besides, Strassburger et al. (2008) predict the bright future of distributed simulation in the nonmilitary area in the gaming industry, the high-tech industry (e.g., auto, manufacturing, and working training), and emergency and security management. To make distributed simulation more attractive to the industrial community, Boer et al. (2009) suggested that we need a lightweight commercial-off-the-shelf product to interoperate different parts efficiently, effortlessly, and quickly. This demand leads the nonmilitary distributed simulation community reaching out to other

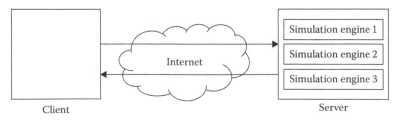

Figure 17.4 Basic distributed simulation architecture.

communication technologies in the Web to overcome HLA shortcomings (e.g., standards complexity, high dependency, and poor scalability). CORBA was used during the 1990s to interoperate heterogeneous simulations, using RPCs style. Those procedures glue different objects operations together, giving the impression that an operation invoked remotely as a local procedure call. The distributed simulation community has also turned to WSs since its birth in year 2000 to provide interoperability between geographically located simulation engines.

As discussed previously, the most widely adopted arbitrary WSs technology is SOAP-based. SOAP-based WSs (Papazoglou 2007) provides a similar way to CORBA RPC-style (see Figure 17.5). It exposes services that encapsulate various procedures on the server side. These services are addressed using URIs and described in XML Web Services Description Language (WSDL) documents. The client side can compile WSDL into procedures stubs. Assume two simulation engines want to interact with each other; the one that starts the request is in the role of the client, while the other that receives the message is the server, and vice versa. Note that these simulation engines will change the roles of client/server during their interactions. At runtime, the client converts the RPC into an SOAP message (XML-based), wraps the SOAP messages in an HTTP message, and POST (HTTP method) to the server. On receiving the HTTP message, the server will extract it reversely into the appropriate procedure call and respond to the client in the same way.

Another popular WS technique for distributed simulation is RESTful WS (Richardson and Ruby 2008), which exposes all services as resources with uniform channels, and messages are transferred between those resources through those uniform channels. Heterogeneous simulation engines can be located on different distributed machines. As shown in Figure 17.6, we can access RESTful WS through the main Web resource (URI) via XML messages

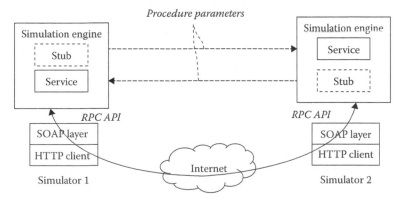

Figure 17.5 Distributed simulation SOAP-based Web service architecture.

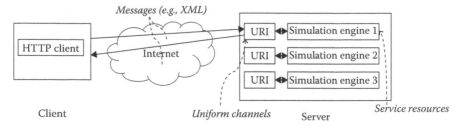

Figure 17.6 Distributed simulation RESTful Web service architecture.

using HTTP methods (GET, PUT, POST, and DELETE), and this main Web resource (URI) acts as manager to coordinate with other simulation engines though uniform channels by their URIs. After simulation finishes, the main Web resource collects all simulation results and forwards them to the client.

In Al-Zoubi and Wainer (2011), the authors identify the shortcomings of using SOAP-based WS comparing with RESTful WS: (1) SOAP-based WS communicates simulation information in the form of procedure parameters that are actually the application programming interface (API) of the simulation component, whereas REST defines them directly as XML message that hides internal implementation. (2) SOAP-based WS transmits all SOAP messages only using HTTP POST channel, while REST uses all HTTP channels for universal interfaces and clear manipulation semantics. (3) SOAP-based WS clients need to have a stub that needs to be written, integrated with existing software and compilers, whereas REST does not require this process. (4) SOAP-based WS groups all services as procedures and exposes them via a port, whereas REST exposes them as resources that are easier to manage and maintain. Therefore, we are using RESTful WS in RISE to interoperate heterogeneous simulators for distributed simulation.

Online model/documentation repository

Another type of Web-based simulation uses online repositories for models or related documentation. This kind of repository aims to use a server-based centralized repository that can be used to store retrieval simulation models (Miller et al. 2000) or online documentation to existing simulations (Narayanan 2000). User can share their work with others, just as a Web document that is hyperlinked to server, improving collaboration and cooperation. Besides, Ribault and Wainer (2012) provide a method using myExperiment (Goble et al. 2010), an online social networking environment, to find, share, and reuse workflows that formulate and automate simulation-based WSs.

Integration of GIS and M&S

Researchers have categorized the integration of GIS and simulation models as *loose* and *deep* coupling (Bell et al. 2000). The loose coupling category consists of most integration efforts through exchanging data files. This approach often needs human interaction, which hinders the automatic operation. The deep coupling approach links GIS and M&S with a friendly user interface, which has drawn much attention in recent years. Some GISs allow getting information from outside using public APIs to support this kind of deep coupling approach. Geographic Resources Analysis Support System (GRASS) (GRASS GIS 2013) is a popular GIS project of the Open Source Geospatial Foundation. GeoTIFF (OSGeo Foundation 2013) is an open standard to establish a TIFF-based interchange format for geo-referenced raster images.

Many of the environmental simulation models use GISs (Gimblett et al. 1995; Hu et al. 2011; Wainer 2006b; Zapatero et al. 2011), which allows manipulating georeferenced information and performing different operations with maps (Longley et al. 2005). However, there is limited data sharing between these GIS system and M&S. Normally, the models for simulation do not use GIS data directly or save model output into a GIS database (Wang 2005). Besides, Ribault and Wainer (2012) state that few of the current solutions can distribute the varied simulation engines based on WBS, and formalize the workflow-like and scalable way to extract data and visualize the results on Google Earth.

GIS is usually organized in multiple data layers, centralizing all the environmental data available and making it accessible in several forms (maps, digital maps, or raw data files). Whatever format is chosen, it is necessary to transform GIS data into a format compatible with the simulation software. In particular, in Wainer (2006), we showed how to simulate environmental systems efficiently using DEVS and Cell-DEVS. Those studies used the CD++ M&S environment (Wainer 2009), and this software stack was recently expanded to allow GIS data to be transformed into CD++ simulation engine (Zapatero et al. 2011). This method relies on the GeoTIFF standard file format, which is supported by most GIS.

Integration of modeling and simulation and geographic information system visualization

We have discussed the benefits of visualization of GIS and simulation results in the Section "Introduction." Here, let us review some literatures of the integration efforts between simulation results and GIS visualization.

Combining the M&S and GIS visualization can significantly increase the representation and understanding of simulation results in an environmental project. This kind of integration goes through system engineering process from separated fields into a whole one. For example, Bishop and Gimblett (2000) present an example of prediction of visitor location and movement patterns in recreational areas. However, as noted by Wang (2005), the integration effort between M&S and GIS visualization is at a preliminary level. An operator often creates visualization and animations separately for simulation results of specific projects.

The limitations of GIS visualization are mainly the quality of presentation and the level of interaction/flexibility of the animation. Many studies have shown the effort to combine the GIS and visualization. Pullar and Tidey (2001) use a 3D GIS for visual impact assessment. Among these efforts, Google Earth (Google 2013) is one of the most popular and massively used for both scientific and generic purposes. Google Earth uses Keyhole Markup Language (KML) (OGC 2013), an XML-based language focused on geographic visualization that includes annotation of maps and images. Its file format can be used to display geographic data in Earth browsers (such as Google Earth), using a tag-based structure with nested elements and attributes. The geographic visualization needs to include not only the presentation of graphical data but also the control of the user's navigation. Once a KML file is created, it can be imported into Google Earth, allowing the visualization of the simulated results on a customized layer impressed over the standard layers (e.g., satellite views and street maps). Google Earth provides mechanisms to make layers evolve forward and backward in time, which is useful to analyze the progress of a simulation interactively (Google 2013). In Zapatero et al. (2011), we adapted Google Earth (Google 2013) as the geospatial visualization system. Here, we will expand our previous work with more advanced WBS techniques and advanced visualization methods of Google Earth.

Architecture

A simulation-based design system in GIS using system engineering enhances the collaboration among stakeholders and interoperability of multidisciplinary experts. This kind of system can facilitate agreement of different kinds of people on the proper alternatives. The integration of GIS, M&S, WBS, and visualization is expected to enhance the analyzing and evaluating process of such a GIS-based decision-making system. Figure 17.7 illustrates the general WBS architecture using Cell-DEVS and GIS. The basic idea is to get data from GIS, model with Cell-DEVS theory, run simulation remotely, and visualize simulation results in Google Earth. The overall approach includes the following four subsystems: Data Collection, Cell-DEVS modeling, WBS, and Visualization. The significance of this integration is to use the best available technologies to analyze GIS system behaviors and predict future scenarios.

1. Data Collection: It is done automatically, generating initial data files from GIS into the Cell-DEVS model. This includes a Dataset Reader for selecting georeferenced raster data from open standard

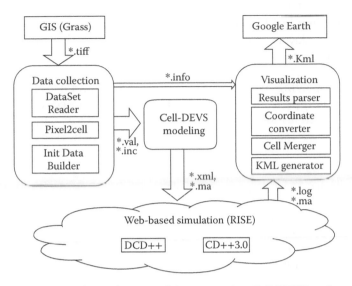

Figure 17.7 Web-based simulation architecture using Cell-DEVS and geographic information system.

GeoTIFF file, and a Pixel2Cell for approximating of collected data into the scale that can be used in Cell-DEVS. Then the Init Data Builder builds initial states of cells and necessary attributes for Cell-DEVS model. The Section "GIS data collection" discusses the details.

2. Cell-DEVS Modeling: It builds a Cell-DEVS environmental model according to the collected data. It defines the cell space size, neighborhood, and rules of the model behaviors using CD++ modeling tool. More details can be seen in the Section "Cell-DEVS modeling."

3. WBS: It submits Cell-DEVS models to RESTful simulation services URI, executes the simulation remotely, and then gets the simulation results. Different simulation engines with CD++ variations are stored on the server and can be run remotely using the RISE middleware. DEVS is a universal abstract formalism separating M&S, and RISE middleware separates the simulator implementation and underling hardware. The Section "Web-based simulation" gives more details.

4. Visualization: It visualizes the simulation results in Google Earth, providing an intuitive and interactive way for analysis. Once simulation results are retrieved from RISE, it parses the simulation log file, optimizes the cells with Cell Merger, and converts the coordinate system into the way used in Google Earth. Then it generates a KML file that can be imported into Google Earth, allowing the visualization

of the simulated results on a customized layer impressed over the standard layers. The Section "Applications" illustrates the details of this subsystem.

In the following sections, we will use a simple land use change example as a prototype to illustrate each step in the proposed WBS architecture. We will first show how to model it using Cell-DEVS, and then show the remaining steps in the process: GIS data collection, WBS, and Google Earth visualization for the model.

Cell-DEVS modeling

Simulation models in GIS are abstract representation for characterizing and understanding environmental patterns and processes. Recent attention has focused on the land surface process models for meteorological study. Changes for land use have drawn much attention in urban planning, engineering, geography, urban economics, and related fields. Some other networks, such as transportation, population, and soil distribution, can play a strong influence on land use pattern changes; on the other hand, new land use patterns can affect the developments of these networks. In this section, we will introduce the Cell-DEVS mechanism and CD++ tool, with a land use changes example to show how to model it.

The Cell-DEVS formalism (Wainer 2006) is defined as an extension to CA combined with DEVS (Zeigler et al. 2000), a formalism for specification of discrete-event models. In DEVS, atomic models describe model behaviors, specified as black boxes, and several DEVS models can integrate together forming coupled models (hierarchical structural models). Cell-DEVS defines a cell as a DEVS atomic model and a cell space as a coupled model. Each cell holds a state variable and a computing function that updates the cell state based on its present state and its neighborhoods. As DEVS models are closed under coupling, the independent and black box-like simulation mechanisms allow these models to communicate each other in single processor, parallel, or distributed simulators without too many changes, which makes it possible for WBS in terms of message transmission and synchronization.

CD++ (Wainer 2002) is a tool for the simulation of DEVS and Cell-DEVS models, and has been widely used to study a variety of models, including architecture, traffic, environmental, emergency, biological, and chemical. The behaviors of a Cell-DEVS atomic model are defined using a set of rules. Each rule indicates the future value of the cell's state if a precondition is satisfied. The precondition is usually checked around the neighborhood of the current cell; Figure 17.8 shows some of the most widely used neighborhoods. Moore's neighborhood contains the origin and its eight adjacent cells; Von Neumann's neighborhood includes the ones to the up, down, left, and right of center. The hexagonal's neighborhood is

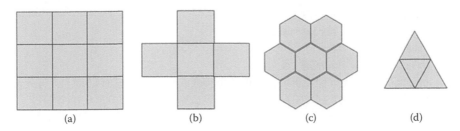

Figure 17.8 Widely used neighborhoods: (a) Moore, (b) Von Neumann, (c) hexagonal topology, and (d) triangular topology.

useful because of the equivalent behavior in every direction, while the triangular neighborhood can cover more varied topology. The local computing function evaluates rules in order, until one of them is satisfied or no more rules. Each rule follows the form: VALUE DELAY {CONDITION}, which means when the CONDITION is satisfied, the state of the cell will change to the designated VALUE, and its output is DELAYed for the specified time.

To forecast land use changes, we use Cell-DEVS theory to model related behaviors. To simulate interaction between land use types and population, we put information into two layers: land use (retrieved from GIS data collection part) and population (predefined data from other system). Figure 17.9 shows the formal specification for this land use model in CD++ (Wang and Chen 2012).

As we can see from Figure 17.9, it defines the size of the cell space (20 × 40), neighbors (extended Moore neighborhood with two layers), and the rules (simple local computing function). We can also notice that there are two zones each with specified rules. The basic idea of the rules is that as the time goes on, the land use pattern will increase/decrease of its intensity according to its neighbors. Take the first rule for example, it means whenever a cell state is 2 and more than one neighbor has the state value of 1, the cell state changes to 1. This state change spreads to the neighbors after 100 milliseconds.

GIS data collection

We have seen how to model using Cell-DEVS theory; now let us deal with how to get information from GIS system, and to link GIS with Cell-DEVS models. The subsystem of GIS Data Collection is responsible for extracting data from GIS and transforming it into inputs used in the Cell-DEVS model.

GIS and GeoTIFF

In Cell-DEVS, the model is always characterized in a specific spatial area that is composed of a set of cells. To integrate GIS and Cell-DEVS modeling, we need to build the initial files (e.g., the layout of studying area) for

```
[land-use]
dim : (20, 20, 2)   border : wrapped ...
initialvalue : 0
initialCellsValue : fromGIS.val
neighbors : (-1,-1,0) (-1,0,0) (-1,1,0) (0,-1,0) (0,0,0) (0,1,0) (1,-1,0) (1,0,0) (1,1,0)
neighbors : (-1,-1,1) (-1,0,1) (-1,1,1) (0,-1,1) (0,0,1) (0,1,1) (1,-1,1) (1,0,1) (1,1,1)
zone : change-rules { (0,0,0)..(19,19,0) }  zone : population { (0,0,1)..(19,19,1) }

[change-rules]
rule : 1 100 {(0,0,0)=2 and stateCount(1)>0 }
rule : 7 100 {(0,0,0)=4 and stateCount(7)>3 } ...
rule : {(0,0,0)} 100 {t}
[population]
rule : {(0,0,0)} 100 {t}
```

Figure 17.9 Land use change model in CD++. (Wang, Y. and Chen, P. 2012. *A Cell-DEVS Geographic Visualization Framework Using Google Earth, Internal Report,* Department of Systems and Computer Engineering, Carleton University, Ottawa, Canada. December 2012.)

Cell-DEVS based on the data extracted from GIS file. The general idea of this GIS Data Collection subsystem is that we are trying to map the geoinformation from real-world geo-ordinates to cell space ordinates, separates the whole region into cells, and stores them as the initial file (*.val file) for CD++. Besides, we also can get a property variable (*.inc) and an information file (*.info) where global geographical references are kept.

In GIS, a geographic dataset always includes a comprehensive collection of vector, or raster, or imagery data covering some parts on Earth. For vector dataset, it often includes hydrographic maps, geological maps, soils, administrative boundaries, and others; for raster dataset, it often includes elevation, slope, aspect, land use, and geology; for imagery dataset, it often includes 1 m resolution orthophoto, land scenes, and daily surface temperature time series (GRASS GIS 2013). GRASS is one of the most popular GIS and can handle with raster, topological vector, image processing, and graphic data. GRASS GIS use Geospatial Data Abstraction Library (GDAL) (GDAL 2013) for raster/vector import and export. Geographic information could be read from raster maps based on the data model of GDAL, such as coordinate system, affine geotransform, and raster band stored information.

Because of the popularity and wide use of GIS, to improve interoperability, a standard file format that is compatible and exchanges well with other different GIS formats is needed. We choose GeoTIFF as the standard file format, as it is supported by most GIS, including GRASS (OSGeo Foundation 2013).

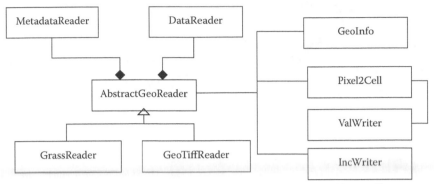

Figure 17.10 Class diagram of geographic information system data collection.

Class diagram

Figure 17.10 shows the implementation class diagram of this part, upgraded from Mariano (2011) and Wang and Chen (2012). AbstractGeoReader implements the general logic of the data input composing of two subclasses: MetadataReader (to obtain geographical references) and DataReader (to get data on each of the pixel). These classes provide an interface through abstract methods, and they can be extended to particular subclasses. Though other systems may have data in various formats, the main logic of this data collection for simulation model does not change. The GeoTiffReader class extends the AbstractGeoReader class, implementing these abstract methods to obtain information from georeferenced files in GeoTIFF format; the GrassReader class retrieves data through the GRASS API. GeoTiffReader implements these operations through the GDAL (GDAL 2013) library for raster geospatial data formats, with ability for efficient handling of large files. GeoInfo is a convenience class for storing geographic contextual data such as coordinates and the resolution of the area (*.info), which is used in visualization part for Google Earth. Pixel2Cell is to approximate the most common value covered in an area that contains multiple pixels, which contains ValWriter class for generating the cell's initial values file (*.val). IncWriter is for generating necessary attributes for Cell-DEVS models from data obtained.

Data collection process

As we have seen the general process and class diagram of this data collection process, now let us go deeper to see the three subprocesses depicted in Figure 17.7.

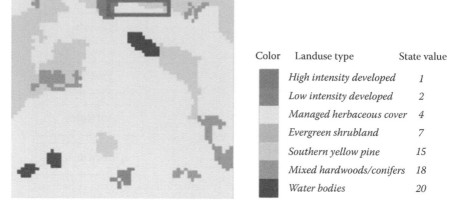

Color	Landuse type	State value
	High intensity developed	*1*
	Low intensity developed	*2*
	Managed herbaceous cover	*4*
	Evergreen shrubland	*7*
	Southern yellow pine	*15*
	Mixed hardwoods/conifers	*18*
	Water bodies	*20*

Figure 17.11 Studied area of *Landuse* map and corresponding states. (Wang, Y. and Chen, P. 2012. *A Cell-DEVS Geographic Visualization Framework Using Google Earth, Internal Report,* Department of Systems and Computer Engineering, Carleton University, Ottawa, Canada. December 2012.)

Step 1: Dataset Reading

Now we use a simple land use changes example to show how this Data Collection subsystem works. The first step is called Dataset Reading. We choose a sample raster dataset named North Carolina (NC, USA) in GeoTIFF format provided by GRASS. This dataset offers raster, vector, LiDAR, and satellite data. We choose the raster format in this GeoTIFF dataset that includes geographic layers of land use, elevation, slope, aspect, watershed basins, and geology. Raster dataset groups different layers into raster bands that contain common information. Each raster band contains the map size, GDAL data types, and a color table for mapping color and land use type values (see Figure 17.11). Each raster band consists of several blocks for efficient access chunk size in GDAL, and each block consists of several pixels. The following pseudo-code shows the algorithm to retrieve data from a raster dataset (Wang and Chen 2012). The general idea of this algorithm is to look through the studying dataset to get each pixel information.

> For the given "landuse.tiff" dataset:
> For each raster band in this dataset:
>   Get this raster band information: Xsize, Ysize, BlockSize.
>   For each block in this band:
>   Get block information: valid block size, stored block data;
>   Output each pixel data in this block.

For the example shown in Figure 17.11, the land use of the studied area has one raster band that contains four blocks, separated in total by 179 × 165 pixels.

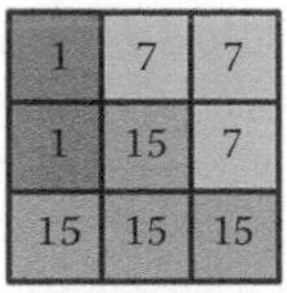

Figure 17.12 Approximate cell state value from multiple pixels.

Step 2: Pixel2Cell

The second step is Pixel2Cell, reading the block data of the raster band of *landuse.tiff*. We need to get the color value of each pixel and generate the initial value of the corresponding cell in Cell-DEVS. Ideally, we can match each pixel to an exact single cell in Cell-DEVS model. However, this is not always the case. In fact, the model behaviors in Cell-DEVS are relatively large (e.g., fire spread in a large forest); therefore, each cell size in Cell-DEVS usually covers multiple pixels that are scaled with minimum unit of a geographical map. Here we add an approximate method to solve this problem, implemented in Pixel2Cell. The idea is to maintain a queue for each cell, sort the pixels shown in this cell according to their colors, and choose the most common color in the queue as the representative value. Figure 17.12 shows an example for this case, in which the cells with state 15 appeared four times in total, more than the other cells (i.e., state 1 and state 7), so state 15 becomes the corresponding cell state value for the Cell-DEVS model.

Step 3: Init Data Building

The third step is Init Data Building. After approximating cell values from the pixels, for this land use change example, we get a Cell-DEVS model with the size of 20 × 20. We store them into an initial information file (*.val) for Cell-DEVS model, along with *.inc (necessary attributes for Cell-DEVS model) from data obtained and *.info (geographic contextual data such as coordinates and the resolution of the area).

Web-based simulation

So far, as we have modeled the rules and extracted the layout from GIS system, the Cell-DEVS land use change model is ready. Now we move to the next step of WBS to execute the model. We use RESTful WS to do so, implemented in RISE. RISE is accessed through web resources (URIs) and XML messages using HTTP methods. Implementations can be hidden in resources, which are represented only via URIs. Users can run multiple instances as needed, which are persistent and repeatable by specific URIs. The HTTP methods are typically four types: GET (to read a resource or get its status), PUT (to create or update a resource), POST (to append data to a resource), and DELETE (to remove a resource). An interface between RISE and different CD++ versions, DCD++ (Al-Zoubi and Wainer 2011) for distributed simulation or CD++ v3.0 (Lopez and Wainer 2004) for improved multiple state variables/ports version, allows running DEVS/Cell-DEVS distributed simulations. RISE API likes a classic website URL http://www.example.com/lopez/sim/workspaces, attached by the following services, and the full RISE design and API described in Al-Zoubi and Wainer (2011):

- ../{userworkspace} contains all simulation services for a given user.
- ../{userworkspace}/{servicetype} contains all frameworks for a given user and simulation engine type (e.g., DCD++ for distributed simulation or CD++ v3.0 for improved CD++ features).
- ../{userworkspace}/{servicetype}/{framework} allows interacting with a framework (including the simulation's initial files, configuration, and source code). The POST channel under this API is used to submit files; PUT is to create a framework or update simulation configuration settings; DELETE is to remove a framework; and GET is to retrieve a framework state.
- ../{userworkspace}/{servicetype}{framework}/simulation interacts with the simulator execution. By calling the Get channel, simulation dedicated to this framework will run. Since RISE supports different servicetypes (simulation engines), in DCD++, this URI is the modeler's single entry to a simulation experiment. It will initialize and communicate with other URIs (e.g., on different machines) to perform distributed simulation, handling synchronization messages.
- ../{userworkspace}/{servicetype}/{framework}/results contains the simulation outputs.
- ../{userworkspace}/{servicetype}/{framework}/debug contains the model-debugging files.

We have two types of servicetype of simulation engines (DCDpp and CD++ v3.0). Figure 17.13 shows an example of having multiple simulation engines at the same time. Requests to .../DCDpp/landuse are sent to *landuse* framework of DCDpp simulator, while requests to .../lopez/landuse are sent to *landuse* framework of CD++ v3.0 simulator.

For a specific simulation, we can realize the remote simulation by using these URIs with HTTP methods. For example, after getting initial model and configuration files for simulation, we use PUT to create a framework with the configuration file and POST to upload these initial model files to this framework. Then, this newly created simulation environment can be executed by using PUT to {framework}/simulation, then we wait for the simulation to finish and GET the simulation results files from {framework}/results.

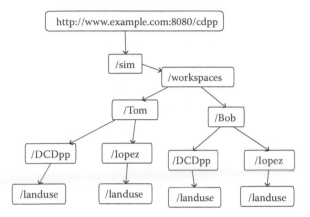

Figure 17.13 RESTful Interoperability Simulation Environment uniform resource identifier example with multiple simulation engines.

In each HTTP response (no matter from which URI), the Response Status is very informative. Normally, it means the request is successful if it returns 200 (OK) and 201 (Created); otherwise, some errors or unexpected exceptions would have happened, such as 400 (bad request), 401 (unauthorized), 403 (forbidden), 404 (not found), 406 (not acceptable), and 501 (not implemented).

In RISE, there are various available CD++ versions, including DCDpp for distributed simulation and CD++ v3.0 for improved CD++ version with different state variables/ports.

DCD++ simulator in RESTful Interoperability Simulation Environment

DCD++ simulator configuration

In RISE, different machines need to coordinate and exchange simulation HTTP messages to perform the distributed simulation. The simulation model is split and assigned into one of these machines. Each physical machine needs to have at least one instance of the RISE middleware (each contains one CD++ engine). DCD++ instances (CD++ Engine on different machines) act as peers to each other. Simulation manager takes responsibility to transmit XML messages and handle synchronization issues between these DCD++ instances, enabling each of them simulating its portion of the model (see Figure 17.14).

Concrete services (e.g., DCD++) are wrapped and accessed through URIs at the middleware level, allowing the middleware to be independent of any specific service. The middleware routes a received request to its appropriate destination resource and apply the required HTTP method on that resource. This middleware design allows additional services

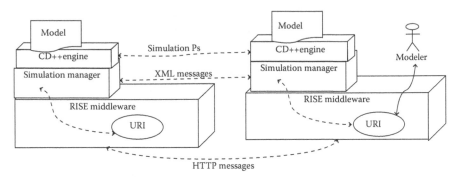

Figure 17.14 RESTful Interoperability Simulation Environment distributed simulation session. (From Al-Zoubi, K. and Wainer, G., *Intelligence-Based Systems Engineering*, 10, 129–157, 2011.)

(e.g., CD++ v3.0 for improved CD++ features) to be plugged into the middleware without affecting other existing services.

The DCD++ is constructed based on the partitioned model under simulation between different machines. The code below shows an example of a part of DCD++ XML configuration information for a Cell-DEVS model. This model-partitioning document describes each cells zone location. Each partition will run the simulation session in dedicated CD++ engine that is located in the belonging machine with the IP and port specified in this document. Note that DCD++ also supports the standard DEVS model, which allows the modeler to customize the partition to an atomic model or coupled model.

```
<ConfigFramework>
    <Doc> This model Simulates Life using Cell-Devs. </Doc>
    <Files>
        <File ftype = "ma">life.ma</File> …
    </Files> …
    <DCDpp>
        <Servers>
            <Server IP = "10.0.40.162" PORT = "8080">
                <Zone>fire (0,0)..(14,29)</Zone>
            </Server>
        </Servers>
        <Servers>
            <Server IP = "10.0.40.175" PORT = "8080">
                <Zone>fire (15,0)..(29,29)</Zone>
            </Server>
        </Servers>
    </DCDpp>
</ConfigFramework>
```

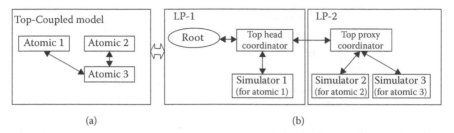

(a)   (b)

Figure 17.15 Head/Proxy Modeling Structure. (a) Coupled model defined. (b) Model hierarchy during simulation. (Al-Zoubi, K. and Wainer, G., Performing distributed simulation with RESTful Web-services, *Proceedings of the 2009 Winter Simulation Conference*, Austin, TX, © 2009 IEEE.)

DCD++ simulation synchronization algorithm

We have discussed that how to configure the model partitions into different DCD++ engines that are located on different machines; it is the time to introduce the synchronization algorithms of DCD++. Figure 17.15a gives a simplified DEVS coupled model that consists of three atomic models. Now suppose that this model is partitioned between two LPs. DCD++ would simulate this model hierarchy as shown in Figure 17.5b (Al-Zoubi and Wainer 2009). The simulator processor simulates an atomic model (e.g., Simulator 1 for Atomic 1 and Simulator 2 for Atomic 2), and the coordinator processor simulates a coupled model. DCD++ uses head/proxy structure to reduce the number of remote messages. In this case, messages exchanged between Simulator 2 and Simulator 3 are handled locally by the proxy coordinator that is responsible for grouping messages, communicating with the Head coordinator if necessary.

Note that this example is a simplified version, but it also shows sufficiently the way DEVS/Cell-DEVS model assigned to this head/proxy structure. Thanks to the coupling closure characteristic of DEVS models, each atomic model can be expended into a complicated coupled model. Also in Cell-DEVS model, each partition zone can be viewed as a coupled model and each cell can be viewed as an atomic model.

Simulation messages (shown in Figure 17.16) for synchronization among processors hierarchy can be grouped into two categories: (1) Content messages represent events generated by a model. There are External messages (X) and Output messages (Y). (2) Synchronization messages for forward simulation phase and time advancing. There are Initialize message (I) to start the initialization phase, Internal message (*) to start the transition phase, Collect message (@) to start collection phase, and Done message (D) to mark the end of a simulation phase.

DCD++ LPs follow the conservative algorithm approach, which guarantees the safe timestamp ordered and always satisfied the local causality constraint. The simulation cycles in phases where LPs are synchronized

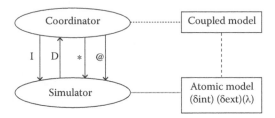

Figure 17.16 Message exchange in a simulation cycle. (From Al-Zoubi, K. and Wainer, G., *Intelligence-Based Systems Engineering*, 10, 129–157, 2011.)

at the beginning of each phase. Each LP (which is a CD++ engine) has its own unprocessed event queue. In this case, after initialization of every atomic model, the Root Coordinator (see Figure 17.17) starts the collection phase by passing a simulation message (@) to the topmost Coordinator (including Top Head/Proxy Coordinator) in the hierarchy, as shown in Figure 17.17. This message is propagated downward in the hierarchy. Next, a DONE message is propagated upward in the hierarchy until it reaches the Root Coordinator. Each model processor uses this DONE message to insert the time of its next change (i.e., an output message to another model, or an internal event message) before passing it to its parent coordinator. A coordinator always passes to its parent the least time change received from its children. Once the Root Coordinator receives a DONE message, it calculates the minimum next change, advances the clock, and starts the Transition phase by passing a simulation message (*). All the collected external messages are executed along with simultaneous internal events. Then, the Root receives a DONE message and starts another cycle again.

CD++ v3.0 simulator in RESTful Interoperability Simulation Environment

CD++ v3.0 simulator configuration

Because of the plug-and-play and scalable resource-oriented design characteristic of RISE middleware, simulation services are wrapped and

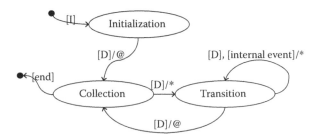

Figure 17.17 Root Coordinator Phases state diagram. (From Al-Zoubi, K. and Wainer, G., *Intelligence-Based Systems Engineering*, 10, 129–157, 2011.)

accessed through their URIs, allowing the middleware independent to any specific simulation engine. Adding a new simulation engine is easy and straightforward in RISE, basically two steps are needed: adding a new CD++ service name in {ServiceName} for RESTful WS URI pattern to recognize it, then adding the path for the new CD++ engine source code directory. In our case, CD++ v3.0 with improved functionality of CD++ is stand-alone version, not geographically distributed, so it can be viewed as single simulation engine. Simulation Manager is enough for manipulation of its execution. From a user's point of view, the modeler manipulates the entire active simulation via HTTP message via framework's URI (including its children's URIs), for example, uploading configuration files, executing simulation, and retrieving results. Figure 17.18 shows the message path when RISE gets other simulation events (e.g., external events), as the following: (1) It passes the simulation event (in XML message) to dedicated simulation manager. (2) Simulation Manager parses the XML message and passes it to Inter Process Communication queue through the local operation system. (3) The CD++ simulation engine (e.g., CD++ v3.0 in our case) executes it properly. Likewise, the RISE gets information from the simulation engine through the reverse way, which is in fact used in DCD++ between two LPs.

The XML configuration file of the CD++ v3.0 is similar as the one in DCD++. Besides multistate variables/ports, it has some other new optional features, like the <OnlyYMsgOP> for allowing to show only Y output messages in log files, and <DrawLog> for viewing 2D visualization results in *.drw file.

CD++ v3.0 simulator features

CD++ is a tool for the simulation of DEVS and Cell-DEVS models, and has been widely used to study a variety of models. CD++ v3.0 (Lopez and Wainer 2004) is an improved CD++ version for Cell-DEVS to allow

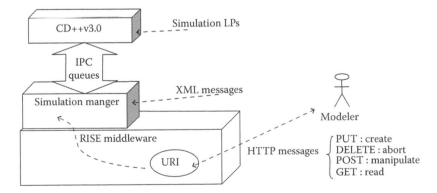

Figure 17.18 CD++ v3.0 execution session.

the cells to use multiple state variables and multiple ports for intercell communication, overcoming some limitations existing in original CD++ implementation, making CD++ more powerful.

One limitation of the original CD++ is that it only supports one state variable in each cell. CD++ v3.0 allows modelers that want to define multiple state variables per cell, avoiding creating extra planes to define as many layers as state variables needed before. Using CD++ v3.0, complex models can be easily integrated and written more clearly, reducing the development time. The format of definition of state variables is as follows:

StateVariables: pend temp vol
StateValues: 3.4 22 -5.2
InitialVariablesValue: init.var

The first line declares the list of state variables for every cell. The second line indicates the default initial values for each state variable. The last line gives the name of a file that stores some initial values for particular cells. State variables can be referred within the rules, by using "$" followed by its name. The identifier ": = " is used to assign values to state variables in a new section in the rules.

Rule: {(0,0) + $pend} {$temp : = $vol/2;} 10 {(0,0) > 4.5 and $vol < 22}

A second limitation is that original CD++ only uses one port for inputs (neighborChange) and one for outputs (out). Different I/O ports can provide a more flexible definition of the cell behavior. CD++ v3.0 supports the use of multiple I/O ports. The format of definition of that is as follows:

NeighborPorts: alarm weight number

The input and output ports share the names, and are generated automatically. An output port from a cell will influence exclusively the input port with that name in each cell of its neighborhood. I/O port can be referred by "~," and similarly, the assignment of a port is using": = ," for example, as this:

Rule: {~weight : = 1;} 100 {(0,1)~alarm ! = 0}

Visualization in Google Earth

After we got the simulation results retrieved from the WBS, we want to visualize them back in the GIS system. Here, we discuss the visualization framework for this purpose, with a particular implementation based on Google Earth using the KML files.

In general, as mentioned in Figure 17.7, this Visualization subsystem has the following four steps:

1. Results parser: it parses the simulation results preserving only output messages for visualization purposes, because the changes information of the cell values in the simulation results are stored in output messages.
2. Coordinate converter: the cells with the same state in an adjacent area can be merged together, to reduce the data size to be visualized.
3. Cell Merger: because coordinates systems used between GIS (information depicted in *.info) and Google Earth are different, to visualize the results, geography references conversion is needed.
4. KML generator: it generates processed simulated results into the desired visualization format (KML).

Once the KML file is created, it can be imported into Google Earth allowing the visualization of the simulated results evolving on a layer impressed over standard layers (e.g., satellite views, and street maps). In Google Earth, we can pause at any point and view layers forward and backward in time, which provides an interactive way to analyze the progress of a simulation.

In the Section "KML introduction," we will briefly introduce KML characteristics. Then we will present the class diagram of developed tool, followed by the details of each step mentioned previously. At last, we will show some results of the studied land use change example.

KML introduction

Google Earth (Google 2013) is a powerful visualization system supporting KML elements (place, marks, images, polygons, 3D models, textual descriptions, etc.) to manage 3D geographic data. Visualization using KML in Google Earth can enhance the communication of results of nontechnical users and provide instantaneous access to layered information. KML (OGC 2013) is based on the XML standard.

We analyzed KML structure in Wang and Chen (2012). KML uses the tags and its attributes to descript the geography information. In our implementation, we use a subset of the elements defined in KML, including <Document>, <Style>, <Folder>, <Placemark>, <Polygon>, and <Timestamp>. The following file segment shows an example. Line 1 is an XML header. Line 2 is a KML namespace declaration. Lines 3–24 state the <Document>, which is our Cell-DEVS simulation unit shown in Google Earth. Lines 5–7 state a Style, which descript a polygon style. We can define different colors for different cell states. Lines 8–23 represent a <Folder> element, which stands for a layer in our Cell-DEVS model. If the model has more than one layer, <Folder> and </Folder> pairs will also appear many times. The <Folder> element contains many <Placemark> elements (lines 9–21). In our design, a <Placemark> element depicts a cell region, which could be a square or irregular polygon. Because the <Polygon> element should have a closed loop, the first coordinate is the same as the last one specification (see lines 15 and 17). In addition, the <Placemark> element contains the <Timestamp> (line 22) element that indicates the <Placemark> showing-up time.

```
1    <?xml version = "1.0" encoding = "UTF-8"?>
2    <Kml xmlns = "http://www.opengis.net/kml/2.2">
3       <Document>
4          <Name>Visualization</Name>
5          <Style id = "CellState">
6             <PolyStyle> <Color>aabbggrr</Color> </PolyStyle>
7          </Style>
8          <Folder id = "#id layer"> <Name>#id layer </Name>
9             <Placemark>
10               <Name>cell name</name>
11               <StyleURL>#CellState</StyleURL>
12               <Polygon>...
13                  <OuterBoundaryIs> <LinearRing>
14                     <Coordinates>
15                        78.7707183652255, 35.809591457038, 0
16                        78.770717636844, 35.8098483367093, 0...
17                        78.7707183652255, 35.809591457038, 0
18                     </Coordinates>
19                  </LinearRing> </OuterBoundaryIs>
20               </Polygon>
21            </Placemark>
22            <TimeStamp> <When> 2013-01-01 ... </When>
     </TimeStamp>
23         </Folder>
24      </Document>
25   </Kml>
```

It is worth to note that the corresponding <Placemark> element will appear on Google Earth until either at the end of the simulation or part of the region covered by another <Placemark> element at a later stamp time, which is explained in Figure 17.19. Therefore, when a cell state changes, we only need to put that <Placemark> element with the corresponding style at that specific time. It will stay in that style until next change occurs.

Class diagram of Visualization tool

Figure 17.20 shows the class diagram of this Visualization part, updated from Wang and Chen (2012) and Zapatero et al. (2011). First, LogParser parses the data generated by the simulator preserving output messages that represent state changes. Parsed results store in the supporting structure class LogInfo. The LogInfo will facilitate and keep changing in further

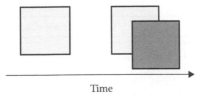

Time

Figure 17.19 A placemark is covered by a later appearance of another one. (Wang, Y. and Chen, P. 2012. *A Cell-DEVS Geographic Visualization Framework Using Google Earth, Internal Report*, Department of Systems and Computer Engineering, Carleton University, Ottawa, Canada. December 2012.)

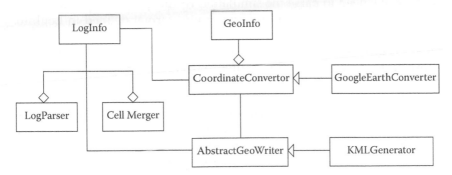

Figure 17.20 Visualization Class diagram.

processing. Then Cell Merger will merge the cells with same state in adjacent area in LogInfo, to reduce the data needed to be visualized, using presented merging algorithm (see details in the Section "Visualization process"). Note that the Cell Merger can be extended for other merging algorithms in the future. Next, CoordinateConverter changes the coordinate in LogInfo into the required way in the output visualization system, an abstract class providing an interface; in our case, its subclass GoogleEarthConverter specifies this step. Finally, AbstractGeoWriter translates LogInfo into the desired output visualization format. Similarly, AbstractGeoWriter is an abstract class providing an interface to translation methods. Here, we focus on the generation of KML files by means of the KMLGenerator class. KMLGenerator takes LogInfo information and the *.info file with the georeferences, processes them, and generates a KML file with georeferenced and timed representation of each simulated cell state change. The process consists of translating each output message into KML tags.

Visualization process

As we have seen the general process and class diagram of this GIS visualization process, now let us go deeper to see its four subprocesses depicted in Figure 17.7.

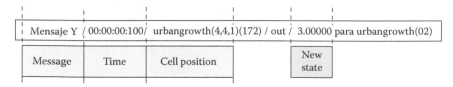

Message	Time	Cell position		New state

Figure 17.21 The output message format in simulation results (CD++ log file).

Step 1: Results parser

As mentioned previously, the changes information of the cell values in the simulation results are stored in output messages. For the visualization purpose, we need to parse the simulation results preserving only output messages, and transfer them into the proper format defined in LogInfo. Figure 17.21 shows the format of the matched message. The message always begins with Mensaje Y followed by the timing and cell position information. A new state value is after the/out/substring.

Step 2: Cell Merger

In each parsed message, each cell uses four positions to descript its geography information, thus we need to record all the four corners for future operation. Besides, each cell requires one <placemark> element in the KML, which makes the KML file size very big and highly redundant, resulting in a long time to render the file when loading to Google Earth. Therefore, it is important to reduce the number of KML elements to be generated. Cell Merger is responsible for solving this issue. The idea is try to merge the cells with same value in adjacent area together as much as possible. Figure 17.22 shows an example, instead of using 16 placemarks (Figure 17.22a) to represent every cell respectively, we could merge the cell with the same state value together and only need to use 4 placemarks (Figure 17.22b), and generate new polygons.

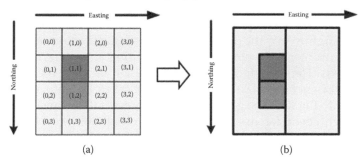

Figure 17.22 Merge cells with the same state into irregular polygons. (a) Before cells merging. (b) After cells merging. (Wang, Y. and Chen, P. 2012. A *Cell-DEVS Geographic Visualization Framework Using Google Earth, Internal Report,* Department of Systems and Computer Engineering, Carleton University, Ottawa, Canada. December 2012.)

Different merging rules can be applied. Ideally, we can find all the adjacent cells for each cell; however, this kind of algorithm is time-consuming, hard to implement, and hard to retrieve the coordinate information. Therefore, we want to merge the cells with the same states as much as possible. In our implementation, we use a heuristic way to design the merging rules. The merging algorithm is shown as follows:

```
If (the current cell has the same state as its left neighbor)
    Merge into its left geometry;
Else if (the current cell has the same state as its upper
neighbor)
    Merge into its upper neighbor;
Else
    Create a new merging geometry;
```

Step 3: Coordinate converter

As we know, the land use change model is sensitive to georeferenced information and its initial data is got from GIS (GRASS), also geo.info stores the geographic information of the studied area. With this information, the geographic visualization could be achieved. The geo.info retrieved from GRASS uses Lambert Conformal Conic geography project system (LCCGPS) (GRASS GIS 2013), and this plane size is fixed by the coordinates of upper left, lower left, upper right, and lower right. Nevertheless, Google Earth uses a different kind of geography reference system called World Geodetic System 1984 (WGS84) (Google 2013). WGS84 uses longitude and latitude pairs to define the unique position on the Earth. So to visualize the simulation results on Google Earth, the geography information along with the initial parameters stored in geo.info has to be converted to longitude and latitude pairs required in Google Earth for each cell.

In the implementation, each cell has four corners with position of (East; North) (see Figure 17.23), and after cells merging step, we only need to know the corners points of all the boundaries in a merged geometry. We use the formulation mentioned in Ghilani (2010) to transform the coordinates from LCCGPS to WGS84.

Step 4: KML generator

After knowing the information needed for each <Placemark> element (all the coordinates of the merged polygon boundaries, Timestamp), we can generate KML file. Basically, this generation step will (1) generate the KML header, (2) write the KML body, and (3) conclude the KML. In the KML header, the state style (state values and associated colors) defined in the Cell-DEVS model is rewritten as polygon style used to paint each

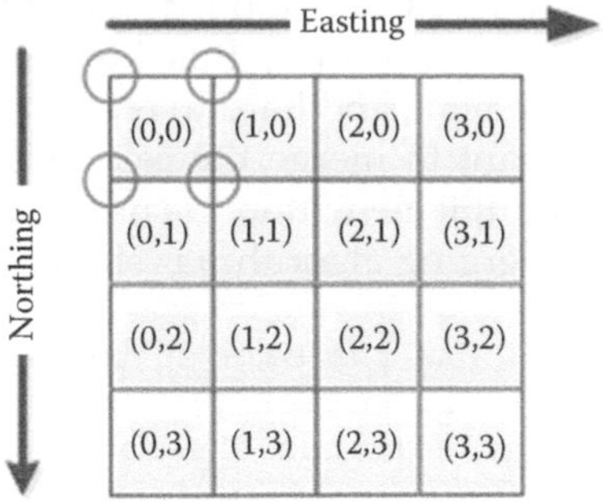

Easting →

(0,0)	(1,0)	(2,0)	(3,0)
(0,1)	(1,1)	(2,1)	(3,1)
(0,2)	(1,2)	(2,2)	(3,2)
(0,3)	(1,3)	(2,3)	(3,3)

Northing ↓

Figure 17.23 The square points of a cell in Cell-DEVS. (Wang, Y. and Chen, P. 2012. *A Cell-DEVS Geographic Visualization Framework Using Google Earth, Internal Report,* Department of Systems and Computer Engineering, Carleton University, Ottawa, Canada. December 2012.)

<Placemark> element. For the body of KML, following simulation time, each <Placemark> element of the merged cells will list all the coordinates of its polygons; and a <Timestamp> element will be inserted to specify the showing-up time of that <Placemark> element. A <Placemark> element with a later <Timestamp> will cover fully or partially of the one in an early <Timestamp>.

Finally, the whole procedure generates geographical areas that emulate simulation in cellular spaces of a Cell-DEVS model. Now we get the KML file for our land use change example. Google Earth version 7.0.1 is used to validate our idea. After using the cell merging, 1157 geometries are needed, instead of 1787 original geometries. Figures 17.24 and 17.25 show some results obtained in Google Earth visualization.

(a)   (b)

Figure 17.24 Google Earth visualization for layer 0 (land use). (a) Initial simulation time. (b) Middle simulation time. (Wang, Y. and Chen, P. 2012. *A Cell-DEVS Geographic Visualization Framework Using Google Earth, Internal Report,* Department of Systems and Computer Engineering, Carleton University, Ottawa, Canada. December 2012.)

(a)   (b)

Figure 17.25 Google Earth visualization for layer 1 (population). (a) Initial simulation time. (b) Final simulation time. (Wang, Y. and Chen, P. 2012. *A Cell-DEVS Geographic Visualization Framework Using Google Earth, Internal Report*, Department of Systems and Computer Engineering, Carleton University, Ottawa, Canada. December 2012.)

Applications

The Section "Architecture" presented the architecture we used to integrate GIS and Cell-DEVS model using WBS, followed by detailed explanation of each steps in the Sections "Cell-DEVS modeling," "Geographic information system data collection," "Web-based simulation," and "Visualization in Google Earth." In this section, we will show two real case studies associated with more complex Cell-DEVS models (Fire Spread and Monkey Pathogen Transmission) in two different engineering fields (environmental engineering and biomedicine engineering) for testing our architecture and show its feasibility and flexibility for the M&S of environmental systems.

Fire Spread

Fire Spread is a 30 × 30 Cell-DEVS model used to study the spreading of fires in forests (Wainer 2006). This model allows foreseeing the propagation and intensity of the fire, which help us to imitate phases of the actual behavior of forest parcels in response to external events (e.g., heat, wind on the area), along with other external parameters. Different parameters affect the fire spread; in this model the following are taken into consideration in terms of the ratio of spread: (1) particles properties (amount of heat, minerals, and density), (2) type of fuel (includes the size of the vegetation), and (3) values involved with the natural environment (wind speed, territory inclination, and humidity). The behavior of each cell depends on its current state whose value is determined by a set of rules after satisfying a precondition of his neighborhood.

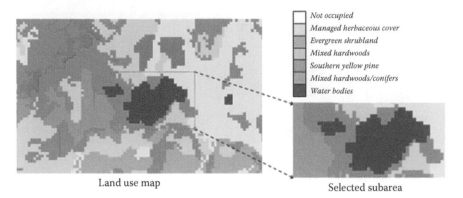

Figure 17.26 Selected subarea of land use map for fire spread simulation. (Zapatero, M. et al., Architecture for integrated modeling, simulation and visualization of environmental systems using GIS and Cell-DEVS, *Proceedings of the 2011 Winter Simulation Conference*, Phoenix, AZ, © 2011 IEEE.)

For GIS data collection, we chose the sample dataset *North Carolina, USA* from the GRASS GIS, which is a dataset readily available with the standard file GeoTIFF. Because of two water surfaces surrounded by flammable land, a section of the *landuse96_28m* map is chosen. This is useful to observe the differences in fire propagation according to the type of surface, giving us how the model sensitive to the land use information provided by GRASS. The propagation of fire is supposed to spread through the land but over water. Figure 17.26 shows the initial map and the selected subarea for simulation (Zapatero et al. 2011).

The simulation of this fire spread model is executed through WBS over RISE middleware. To realize interacting heterogeneous simulators and distributed models, we choose DCD++ for remote/distributed simulation. We first use the PUT method to create a new framework …dcdpp/firespread using distributed simulation engines under an authorized userworkspace in our RISE server, and we use the POST method to upload the initial files (*.xml with partition configuration, *.val with initial values of each cell, *.ma of Cell-DEVS fire spread model, etc.). Then we can run this simulation by using PUT to …/dcdpp/firespread/simulation. We wait for the simulation to finish, then GET simulation results files from …dcdpp/firespread/results. For distributed simulation, DCD++ allows to divide the model into different partitions to be executed on heterogeneous simulations. The model partition configuration can be specified easily in an XML file uploaded at the same time of other initial files POSTed to …dcdpp/firespread.

After retrieving the simulation log from RISE WBS middleware, we can import it into the developed GIS Visualization tool, generating a KML output file. This KML file is then loaded into Google Earth showing the following

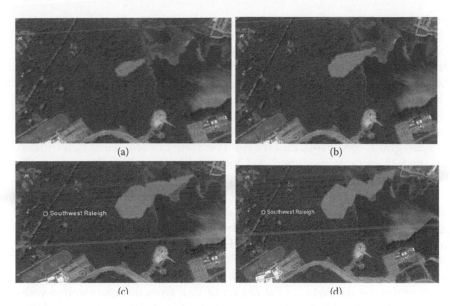

(a)

(b)

(c)

(d)

Figure 17.27 Fire model simulation results shown in Google Earth. Snapshots taken at times (a) 00:15:00, (b) 00:30:00, (c) 00:45:00, and (d) 01:00:00. (Zapatero, M. et al., Architecture for integrated modeling, simulation and visualization of environmental systems using GIS and Cell-DEVS, *Proceedings of the 2011 Winter Simulation Conference*, Phoenix, AZ, © 2011 IEEE.)

results shown in Figure 17.27, following the simulation time during 01:00:00 (Zapatero et al. 2011). It tells us that fire spreads around the lakes as expected, effected by the land use information obtained from GIS. It can also be seen that with the wind direction information extracted from GIS, the large urbanized areas (see the bottom right of Figure 17.27) are not impacted.

Monkey Pathogen Transmission

Besides the environmental system, GIS with advanced M&S techniques has also significantly influenced and improved the field of biomedicine. With the existence of several serious contagious diseases that may cause death, social panic, or even a national crisis, it is becoming essential to study infection transmission to prevent these outbreaks (Kennedy et al. 2009).

Monkey Pathogen Transmission (Wang 2012) is a Cell-DEVS model to study pathogen transmission for Macaques (a kind of monkey) in the region of Bali, Indonesia, by simulating their movement behaviors. Macaques on the island are known to carry a specific pathogen that is transmissible to neighbor macaques. Every monkey is able to host the pathogen that follows the life cycle (susceptible, latent infection, symptomatic infection, and acquired immunity). Their movement, sex, and surrounding environment effect how the pathogen is passed between each other. For example, macaques move at random, but tend to pass through waterways less frequently than forests, and female monkeys are unable to leave their birth *temples*, which have been present in the forests of Bali for centuries (Kennedy et al. 2009). This model is implemented in CD++ v3.0 specification with different state variables per cell (landscape, temple, sex, movement, and pathogen). It uses different phases in each movement cycle (intent, grant, constraint, and move). The state of each cell is determined by the values of its neighborhood followed by a set of rules (where the behavior of each cell is implemented).

This model can monitor the speed of progression and explore the effect of numerous variables on the pattern of disease transmission, allowing scientists understand the behavior of a disease and its movement from one subject to another. This model is sensitive to the landscape geographic information from GIS, requiring the river, coast, and forest information of the island Bali. In the implementation, we get the river and coast datasets for the region of Bali from an open-source website Cloudmade (Cloudmade 2013), and the forest dataset from Carleton University's Library GIS department (Carleton University 2013). Then, we combine the two datasets together using the GRASS GIS raster map calculator, and using our developed tool to get the landscape values form the map for our model. After data collection from GIS, we get the initial file *.val for the following processes. Figure 17.28 shows the initial map of Bali, Indonesia, with initial cell values for simulation (Al-Disi et al. 2013).

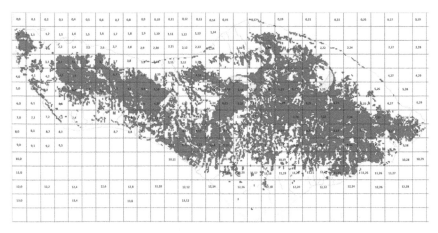

Figure 17.28 Initial cell values of Bali, Indonesia, after geographic information system data collection. (From Al-Disi, E. et al., *Visualizing Models for Biomedical Applications: Disease Transmission, Internal Report*, Department of Systems and Computer Engineering, Carleton University, Ottawa, Canada, April 2013.)

The simulation of this pathogen transmission model is executed through WBS over RISE middleware. To use multistate variables and multiports, we use CD++ v3.0 for our simulation engine. We first use the PUT method to create a new framework ...lopez/pathogen using distributed simulation engines under an authorized userworkspace in our RISE server, and we use the POST method to upload the initial files (*.xml with model configuration, *.val with initial values of each cell, *.ma of Cell-DEVS model, etc.). Then we can run this simulation by using PUT to ... lopez/pathogen/simulation. We wait until the simulation finishes, then GET simulation results files from .../lopez/pathogen/results.

After retrieving the simulation log from RISE WBS middleware, we can see the simulation results in CD++ in a 2D way. Different tests have been performed with various parameters to show the effects of landscape, monkey occupancy, sex, river cross probability, and initial infection ratio on transmission patterns (see Figure 17.29). In each test, the right-most figure shows the changes of the pathogen (i.e., uninfected monkey, latent infection, symptomatic, and acquired immunity) with the movement of the monkeys. Generally, if a cell is uninfected and more than one of its neighbor is not uninfected, this cell will be infected in the next time.

We also can visualize simulation results in Google Earth. This time, we narrow down the scale of our model to a small region to verify our model scalability. Figure 17.30a shows the whole Bali Island, the small white square on the map is the region that was used to test the pathogen transmission. We reuse the previous steps easily, collecting information from GIS and retrieving the log file from RISE. Then we import the log file into our GIS Visualization tool, generating a KML output file. Finally, we load this KML file in Google Earth. The visualization results can be seen in Figure 17.30b.

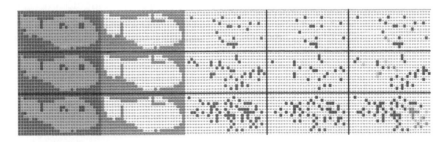

Figure 17.29 Three simulation results of Monkey Pathogen Transmission under different monkey occupancy (top with 10%, middle with 20%, bottom with 30%). (From Al-Disi, E. et al., *Visualizing Models for Biomedical Applications: Disease Transmission, Internal Report*, Department of Systems and Computer Engineering, Carleton University, Ottawa, Canada, April 2013.)

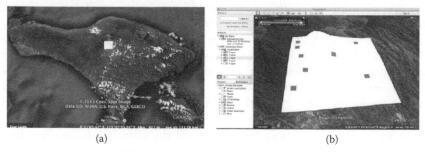

(a)   (b)

Figure 17.30 (a) Bali and (b) Visualization results in Google Earth. (From Al-Disi, E. et al., *Visualizing Models for Biomedical Applications: Disease Transmission, Internal Report*, Department of Systems and Computer Engineering, Carleton University, Ottawa, Canada, April 2013.)

Conclusion

We proposed a general WBS architecture by integrating Cell-DEVS modeling and GIS visualization to study environmental phenomena. We presented a workflow-like process with a land use change prototype to demonstrate the way to extract information from GIS (GRASS), run simulation remotely, and visualize results in Google Earth. The WBS is run on the WSs middleware of RISE, supporting different simulation engines. We also discussed the distributed simulation mechanism (DCD++) using RISE middleware, executing heterogeneous simulations on distributed computer systems. Two applications as case studies applying this proposed architecture were demonstrated: the wildfire spreading in the environmental engineering field and the monkey pathogen transmission in the biomedical engineering field. These applications verified and highlighted the flexibility of the proposed WBS architecture. Some future work may focus on the following aspects: (1) to expand this work with more standard DEVS protocol for distributed simulation synchronization algorithm, (2) to investigate efficient way to interface different components in the proposed architecture and apply related cloud computing technologies, and (3) to develop more applications adapting this architecture in other fields and perform more data and statistical analysis.

Acknowledgments

This chapter is the result of the recent efforts of many students and collaborators, including Mariano Zapatero and Rodrigo Castro (integration of GIS, Cell-DEVS and KLM), Yu Wang, and Peiwen Chen (designing a geographic visualization framework and a land use model), and Eman Al Disi, Joanna Lostracco, and Myriam Younan (implementing monkey pathogen transmission into the proposed architecture).

References

Al-Disi, E., Lostracco, J., and Younan, M. 2013. *Visualizing Models for Biomedical Applications: Disease Transmission. Internal Report.* Department of Systems and Computer Engineering, Carleton University, Ottawa, Canada. April 2013.

Alfonseca, M., De Lara, J., and Vangheluwe, H. 2001. Web II: web-based simulation of systems described by partial differential equations. In *Proceedings of the 2001 Winter Simulation Conference*, Arlington, VA. IEEE.

Al-Zoubi, K. and Wainer, G. 2009. Performing distributed simulation with RESTful Web-services. In *Proceedings of the 2009 Winter Simulation Conference*, Austin, TX. IEEE.

Al-Zoubi, K. and Wainer, G. 2011. Distributed simulation using RESTful interoperability simulation environment (rise) middleware. *Intelligence-Based Systems Engineering* 10, 129–157, Berlin/Heidelberg, Germany, Springer.

ANSI/EIA-632. 1999. *Process for Engineering a System.* Electronic Industry Association, Arlington, VA.

Badard, T. and Richard, D. 2001. Using XML for the exchange of updating information between geographical information systems. *Computers, Environment and Urban Systems*, 25(1), 17–31.

Band, L. 1986. Topographic partition of watersheds with digital elevation models. *Water Resources Research*, 22(1), 15–24.

Bencomo, S. D. 2004. Control learning: present and future. *Annual Reviews in Control*, 28(1), 115–136.

Bell, M., Dean, C., and Blake, M. 2000. Forecasting the pattern of urban growth with PUP: a web-based model interfaced with GIS and 3D animation. *Computers, Environment and Urban Systems*, 24, 559–581.

Bishop, I. D., and Gimblett, H. R. 2000. Management of recreational areas: GIS, autonomous agents, and virtual reality. *Environment and Planning B*, 27(3), 423–436.

Boer C., A. Bruin, and A. Verbraeck. 2009. A survey on distributed simulation in industry. *Journal of Simulation*, 3(1): 3–16.

Botkin, D. B., Janak, J. F., and Wallis, J. R. 1972. Some ecological consequences of a computer model of forest growth. *The Journal of Ecology*, 60(3), 849–872.

Byrne, J., Heavey, C., and Byrne, P. J. 2010. A review of web-based simulation and supporting tools. *Simulation Modelling Practice and Theory*, 18(3), 253–276.

Carleton University. 2013. Carleton University's Library GIS department. Available at http://www.library.carleton.ca/find/gis; accessed April 2013.

CloudMade. 2013. CloudMade. Available at http://cloudmade.com; accessed April 2013.

Desmet, P. J. J. and Govers, G. 1995. GIS-based simulation of erosion and deposition patterns in an agricultural landscape: A comparison of model results with soil map information. *Catena*, 25(1–4), 389–401.

Fujimoto, R. M. 2000. *Parallel and Distribution Simulation Systems.* New York, John Wiley & Sons.

Gimblett, R., Ball, G., Lopes, V., Zeigler, B., Sanders, B., and Marefat, M. 1995. Massively parallel simulations of complex, large scale, high resolution ecosystem models. *Complexity International*, 2, ISSN: 1320–0682.

GDAL. 2013. Geospatial Data Abstraction Library. Available at http://www.gdal.org/; accessed April 2013.

Ghilani, C. D. 2010. *Adjustment Computations: Spatial Data Analysis*. New Jersey, John Wiley & Sons.

Goble, C. A., Bhagat, J., Aleksejevs, S., Cruickshank, D., Michaelides, D., Newman, D., Borkum, et al. 2010. myExperiment: A repository and social network for the sharing of bioinformatics workflows. *Nucleic Acids Research*, 38, Web Server issue (July 2010), W677–82.

Google, Inc. 2013. Google Earth. Available at http://earth.google.com; accessed April 2013.

GRASS GIS. 2013. Geographic Resources Analysis Support System. Available at http://grass.fbk.eu/; accessed April 2013.

Hitchins, D. K. 2008. *Systems Engineering: A 21st Century Systems Methodology*. West Sussex, United Kingdom, John Wiley & Sons.

Hu, X., Sun, Y., and Ntaimo, L. 2011. DEVS-FIRE: Design and application of formal discrete event wildfire spread and suppression models. *Simulation*. 88(3), 259–279, March 2012. doi: 10.1177/0037549711414592.

Huang, Y. and Madey, G. 2005, April. Autonomic web-based simulation. In *Proceedings of the 38th Annual Simulation Symposium* (pp. 160–167). IEEE Computer Society.

Kennedy, R. C., Lane, K. E., Arifin, S. N., Fuentes, A., Hollocher, H., and Madey, G. R. 2009. A GIS aware agent-based model of pathogen transmission. *International Journal of Intelligent Control and Systems*, 14(1), 51–61.

Khul, F., R. Weatherly, J. Dahmann. 1999. *Creating Computer Simulation Systems: An Introduction to High Level Architecture*. Upper Saddle River, NJ, Prentice Hall.

Leonard, J. 1999. *Systems Engineering Fundamentals: Supplementary Text*. Darby, PA, DIANE Publishing.

Longley, P. A, Goodchild, M. F., Maguire, D. J., and Rhind, D. W. 2005. *Geographic Information Systems and Science*. West Sussex, United Kingdom, John Wiley & Sons.

Lopez, A. and Wainer, G. 2004. Improved Cell-DEVS model definition in CD++. In *Cellular Automata*. Berlin/Heidelberg, Germany, Springer, 803–812.

Miller, J. A., Seila, A. F., and Xiang, X. 2000. The JSIM web-based simulation environment. *Future Generation Computer Systems*, 17(2), 119–133.

Myers, D. S. 2004. An extensible component-based architecture for web-based simulation using standards-based web browsers. Master's thesis, Virginia Polytechnic Institute and State University.

Narayanan, S. 2000. Web-based modeling and simulation. In *Proceedings of the 2000 Winter Simulation Conference*, Orlando, FL. IEEE.

OGC. 2013. KML. Available at http://www.opengeospatial.org/standards/kml/; accessed April 2013.

OSGeo Foundation. 2013. Geotiff format specification. Available at http://trac.osgeo.org/geotiff/; accessed April 2013.

Page, E. H. 1999. Beyond speedup: PADS, the HLA and web-based simulation. In *Proceedings of the Thirteenth Workshop on Parallel and Distributed Simulation* (pp. 2–9), Atlanta, GA. IEEE Computer Society.

Page, E. H., Griffin, S. P., and Rother, L. S. (1998). Providing conceptual framework support for distributed web-based simulation within the high level architecture. In *Proceedings of SPIE: Enabling Technologies for Simulation Science II*.

Papazoglou, M. 2007. *Web Services: Principles and Technology*. Upper Saddle River, NJ, Prentice Hall.

Pullar, D. V. and Tidey, M. E. 2001. Coupling 3D visualisation to qualitative assessment of built environment designs. *Landscape and Urban Planning*, 55(1), 29–40.

Ribault, J. and Wainer, G. 2012. Simulation Processes in The Cloud for Emergency Planning. In *Proceedings of the 2012 12th IEEE/ACM International Symposium on Cluster, Cloud and Grid Computing (ccgrid 2012)*, 886–891. IEEE Computer Society.

Richardson, L. and Ruby, S. 2008. *RESTful Web Services.* Sebastopol, CA, O'Reilly Media, Inc.

Sage, A. P., and Olson, S. R. 2001. Modeling and simulation in systems engineering: Whither Simulation based acquisition? *Simulation*, 76(2), 90–91.

Sinha, R., Paredis, C. J., Liang, V. C., and Khosla, P. K. 2001. Modeling and simulation methods for design of engineering systems. *Journal of Computing and Information Science in Engineering*, 1(1), 84–91.

Strassburger, S., Schulze, T., and Fujimoto, R. 2008. Future trends in distributed simulation and distributed virtual environments: results of a peer study. In *Proceedings of the 2008 Winter Simulation Conference*, Austin, TX, eds. S. J. Mason, R. R. Hill, L. Mönch, O. Rose, T. Jefferson, J. W. Fowler, 777–785. Piscataway, NJ, Institute of Electrical and Electronics Engineers, Inc.

Tolk, A. 2010. Engineering management challenges for applying simulation as a green technology. In *Proceedings of the 31st Annual National Conference of the American Society for Engineering Management (ASEM)*, 137–147.

Van Der Knijff, J. M., Younis, J., and De Roo, A. P. J. 2010. LISFLOOD: A GIS-based distributed model for river basin scale water balance and flood simulation. *International Journal of Geographical Information Science*, 24(2), 189–212.

Wainer, G. 2002. CD++: A Toolkit to Develop DEVS Models. *Software: Practice and Experience*, 32(13), 1261–1306.

Wainer, G. 2006. Applying Cell-DEVS Methodology for Modeling the Environment. *Simulation.* 82(10), 635–660.

Wainer, G. 2009. *Discrete-Event Modeling and Simulation: A Practitioner's Approach.* Boca Raton, FL, CRC Press, Taylor & Francis Group.

Wang, S. 2012. *Macaque Pathogen Transmission using Cell-DEVS. Internal Report.* Department of Systems and Computer Engineering, Carleton University, Ottawa, Canada. December 2012.

Wang, X. 2005. Integrating GIS, simulation models, and visualization in traffic impact analysis. *Computers, Environment and Urban Systems*, 29(4), 471–496.

Wang, Y. and Chen, P. 2012. *A Cell-DEVS Geographic Visualization Framework Using Google Earth. Internal Report.* Department of Systems and Computer Engineering, Carleton University, Ottawa, Canada. December 2012.

Ware, C. 2000. *Information visualization: Perception for design.* San Francisco, CA, Morgan Kaufmann Publishers.

Wiedemann, T. 2001. Simulation application service providing (SIM-ASP). In *Proceedings of the 2001 Winter Simulation Conference*, Arlington, VA. IEEE.

Zapatero, M., Castro, R., Wainer, G., and Hussein, M. 2011. Architecture for integrated modeling, simulation and visualization of environmental systems using GIS and Cell-DEVS. In *Proceedings of the 2011 Winter Simulation Conference*, Phoenix, AZ. IEEE.

Zeigler, B. P., Praehofer, H., and Kim, T. G. 2000. *Theory of Modeling and Simulation, 2nd Edition.* Academic Press, San Diego, CA.

Index